Surface Wave Methods for Near-Surface Site Characterization

Surface Wave Methods for Near-Surface Site Characterization

Sebastiano Foti

Carlo G. Lai

Glenn J. Rix

Claudio Strobbia

CRC Press
Taylor & Francis Group
Boca Raton   London   New York

CRC Press is an imprint of the
Taylor & Francis Group, an **Informa** business

A SPON PRESS BOOK

CRC Press
Taylor & Francis Group
6000 Broken Sound Parkway NW, Suite 300
Boca Raton, FL 33487-2742

First issued in paperback 2017

© 2015 by Taylor & Francis Group, LLC
CRC Press is an imprint of Taylor & Francis Group, an Informa business

No claim to original U.S. Government works

Version Date: 20140508

ISBN 13: 978-1-138-07773-7 (pbk)
ISBN 13: 978-0-415-67876-6 (hbk)

Library of Congress Cataloging-in-Publication Data

Foti, Sebastiano.
   Surface wave methods for near-surface site characterization / Sebastiano Foti, Carlo
G. Lai, Glenn J. Rix, and Claudio Strobbia.
      pages cm
   Includes bibliographical references and index.
   ISBN 978-0-415-67876-6 (hardback)
   1. Surface waves (Seismology)--Analysis. 2. Seismic waves--Speed. 3. Inversion
(Geophysics) 4. Earthquake engineering. I. Title.

QE538.5.F68 2015
551.22028'7--dc23   2014015309

Visit the Taylor & Francis Web site at
http://www.taylorandfrancis.com

and the CRC Press Web site at
http://www.crcpress.com

Contents

6 Inversion 273

Preface

More than a century has passed since Lord Rayleigh (1885) first investigated "the behaviour of waves upon the plane free surface of an infinite homogeneous isotropic elastic solid, their character being such that the disturbance is confined to a superficial region, of thickness comparable to the wave-length." Rayleigh immediately recognized their importance stating that "it is not improbable that the surface waves here investigated play an important part in earthquakes, and in the collision of elastic bodies." Since then, surface waves have come to play an important part in many other disciplines as well, including material science, geophysics, nondestructive testing, and engineering site characterization. When used for near-surface site characterization, the objective is usually to determine the shear wave velocity and/or shear damping ratio one-dimensional (1D) profiles for a variety of applications, including earthquake site response, dynamic soil–structure interaction, nondestructive pavement testing, evaluation of ground modification, and so on.

Surface wave testing draws on several bodies of knowledge, including wave propagation theory, signal processing, and inverse modeling. Thus, it is difficult for practicing engineers and geologists with little or no formal training in these subjects to fully understand the test and its interpretation. As a result, for many practitioners, it is not always obvious what can be really attained with surface wave testing and what are the pros and cons of the method. Simultaneously, in past decades, increasing worldwide interest in this geophysical technique, coupled with the lack of standards, has contributed to the development of several test variations—each having a different acronym. The resulting situation is rather confusing to those who want to use surface wave testing for near-surface site characterization at a particular project. For the aforementioned reasons, this book aims to provide a comprehensive, consistent, and clear description of each aspect of the test in addition to guidelines for correctly performing and interpreting the results of this geophysical technique. The book is intended to appeal to two audiences: (1) practitioners who wish to understand the basic concepts and potential of surface wave methods and (2) researchers seeking detailed presentations of the theory and principles underlying each aspect of the test (i.e., wave propagation, signal processing, and inverse theory). Variants of the test are presented within a consistent framework to facilitate comparisons. Several examples and case studies directly drawn from the authors' experience are illustrated to highlight the crucial aspects associated with the standard of practice of surface wave testing for near-surface site characterization. The book also includes a thorough discussion of the uncertainties arising in each phase of surface wave testing—from data measurement to interpretation—a subject that the authors feel has not received appropriate attention thus far.

Chapter 1 explains surface wave testing principles using a minimum of technical jargon and as few equations as possible. This portion of the book is intended for those who would like to have a better understanding of how and why surface wave testing works and to have examples of typical applications of this geophysical technique. This chapter is also intended to serve as a guide for the entire book. From this point of view, it informs the interested reader of the relevant sections of the book where he/she can find greater detail regarding a specific aspect of the test or of the underlying theory.

Chapter 2 details aspects of wave propagation in a vertically heterogeneous half-space, which forms the basis of surface wave testing. The chapter begins by illustrating the theory of wave propagation in elastic continua because the majority of surface wave testing applications for near-surface site characterization require consideration of only the velocity of surface waves. The theory is then extended to the propagation of surface waves in viscoelastic continua, which allows the examination of surface wave velocity and attenuation. The latter can be used to infer the material damping ratio (or quality factor) 1D profile.

In Chapter 3, the principles of acquisition of experimental data are introduced. In particular, the consequences of the physical and technical limitations of surface wave processing are thoroughly analyzed and discussed. The chapter also includes an overview of the equipment commonly used in applications of surface wave testing for near-surface site characterization.

Chapters 4 and 5 discuss a variety of techniques that may be used for processing experimental data to estimate the dispersion and attenuation of surface waves. Increasingly, the availability of affordable multichannel data acquisition systems and sensors has encouraged the use of advanced signal processing algorithms for calculating surface wave velocity and attenuation.

Chapter 6 illustrates the basic theory for the solution of the inverse problem associated with the propagation of surface waves. This is the final step in data interpretation when surface wave testing is used for site characterization. It involves a mathematical operation, called inversion, by which the experimental dispersion and/or attenuation curve is processed to obtain the unknown 1D profile of shear wave velocity and/or material damping ratio at the site. The available procedures include empirical methods, trial and error forward modeling, automated local search algorithms (e.g., least squares uncoupled and coupled inversion), and global search algorithms (e.g., neural networks, simulated annealing, Monte Carlo simulation). The advantages and implications of different algorithms are discussed after a general introduction to the theory of inverse problems.

Chapter 7 illustrates a series of typical surface wave testing applications for near-surface site characterization in engineering. Several examples and case studies at a variety of selected sites are presented to help the interested reader become acquainted with different, sometimes difficult, real-world situations and with experimental data.

Finally, Chapter 8 presents techniques that are on the leading edge of surface wave research and practice; they include offshore and near-shore applications, measurement of Love waves, joint inversion of surface waves with other geophysical data, passive seismic interferometry, and multicomponent surface wave analysis.

Acknowledgments

We thank all those individuals who have contributed either directly or indirectly to build the knowledge upon which the content of this book has been conceived. In particular, we had the privilege to work and share stimulating thoughts with colleagues at our home institutions in addition to many individuals from the scientific community around the world fascinated with surface waves. This stimulating environment contributed significantly to our understanding of surface waves and of their exploitation for site characterization. The number of these individuals is, however, too large for all of them to be explicitly cited without incurring the risk of forgetting someone.

Last but not least, our deepest gratitude goes to our beloveds for their continuous support, without which this book would never have come to an end.

Authors

Dr. Sebastiano Foti is an associate professor in geotechnical engineering at the Politecnico di Torino, Italy. He earned his MSCE and PhD at the same institution in 1996 and 2000, respectively. He has been a research scholar at the Georgia Institute of Technology and a research associate at the University of Western Australia, Perth. His primary interests are in geophysical methods for geotechnical characterization, geotechnical earthquake engineering, and soil–structure interaction.

Dr. Carlo G. Lai is an associate professor in geotechnical engineering at the University of Pavia, Italy. He is also the head of the Geotechnical Earthquake Engineering Section at EUCENTRE and affiliate faculty at the ROSE School, both in Pavia. Dr. Lai earned his MSCE from Politecnico di Torino in 1988 and an MSCE, an MSESM, and a PhD from the Georgia Institute of Technology in 1995, 1997, and 1998, respectively. His primary interests are in modeling of seismic wave propagation in geomaterials, earthquake geotechnics, and engineering seismology.

Dr. Glenn J. Rix is a principal with Geosyntec Consultants Inc. in Atlanta, Georgia. Prior to joining Geosyntec in 2013, Dr. Rix was a professor in the School of Civil and Environmental Engineering at the Georgia Institute of Technology. Dr. Rix earned his BSCE at Purdue University in 1982 and his MSCE and PhD at the University of Texas at Austin in 1984 and 1988, respectively. His primary interests are in soil dynamics, geotechnical earthquake engineering, and seismic hazard and risk analysis.

Dr. Claudio Strobbia is a land processing specialist and seismic processing supervisor with Total, Pau, France. Before joining Total, he was a senior research geophysicist with Schlumberger, stationed in both Cairo, Egypt, and London, U.K. He worked as a researcher for the European Centre for Training and Research in Earthquake Engineering, and he taught exploration seismology at the University of Milan, Bicocca. He earned his PhD at Politecnico di Torino. His primary interests are in wave physics, inverse problems, near-surface geophysics, and seismic processing. Within the exploration seismology, his main contributions are in noise attenuation and near-surface characterization.

Chapter 1

Overview of surface wave methods

Since Lord Rayleigh predicted their existence (Rayleigh 1885), surface waves have attracted the interest of an increasing number of researchers embracing disciplines as diverse as solid-state physics, microwave engineering, geotechnical engineering, nondestructive testing, seismology, geophysics, material science, ultrasonic acoustics, and others. Despite their marked differences, these disciplines share the goal of exploiting surface waves propagating along the boundary of a domain to obtain information about the interior of that domain, usually expressed in terms of one or more scalar fields.

Surface waves are appealing because they are ideal for the development of noninvasive techniques for material characterization from a very small scale, less than a millimeter (e.g., ultrasonic surface waves used to identify material defects), to a very large scale, more than a kilometer (e.g., earthquake-generated surface waves used to investigate the structure of the Earth's crust and upper mantle). At an intermediate scale, geophysicists and geotechnical engineers use surface waves for the characterization of geomaterials. The fundamental idea is the same for all these applications: to use the geometric dispersion of surface waves to infer the relevant medium properties by solving an inverse problem for parameter identification.

The present chapter is devoted to an introduction to surface wave methods. It describes their basic principles and historical developments, with references to geotechnical and geophysical near-surface characterization as well as deep exploration and exploration seismology. The most common applications related to near-surface characterization will be discussed, and some crucial aspects related to the application of the technique will be examined to provide guidance to new users and to identify elements for further research on the topic. Finally, the advantages and limitations of surface waves relative to other seismic methods will be discussed, pointing out the significance and relevance of surface wave tests in the context of engineering site characterization.

Although this chapter mainly focuses on Rayleigh waves, as they are more widely used for characterization purposes, the same concepts apply to the use of Love and Scholte waves for site characterization as discussed in Chapter 8.

1.1 SEISMIC WAVES

Soils and rocks are complex, multiphase, particulate and discontinuous materials; thus, their mechanical behavior cannot in general be described using simple models. Soils exhibit marked nonlinear and irreversible behavior, starting from the very initial stage of loading. Although soils are particulate materials, continuum mechanics approaches are frequently used, accounting for the role of fluid in saturated media through the effective stress concept introduced by von Terzaghi (1936). The dynamic behavior of soil is very complex and depends on a variety of factors. Among these, the dependency on the strain level is of primary interest with respect to the focus of this book. For very small strains, soils exhibit an almost linear stress–strain relationship, so that the assumption of a linear elastic constitutive model is reasonable. Outside the small-strain region, the behavior is far more complex, and its description requires the adoption of advanced constitutive models, which is outside the scope of this book.

The strain level associated with geophysical testing is typically very small, so it is widely accepted that linear elastic theory provides a consistent framework for the interpretation of seismic tests. Linear elastic models can be used directly only for a restricted number of boundary value problems of soil dynamics (e.g., ground vibrations predictions). However, small-strain parameters play a significant and relevant role for a wide range of geotechnical problems (Burland 1989; Atkinson 2000).

Equations of motion for a linear elastic solid will be presented in detail in Chapter 2. The effect of a sharply applied, localized disturbance in a physical medium rapidly spreads in space; this is commonly addressed as wave propagation (Graff 1975). Two different types of body (or bulk) waves propagate in an unbounded, homogeneous, and linear elastic medium: P-waves (primary or compressional or longitudinal waves) and S-waves (secondary or shear or distortional or equivoluminal waves). P-waves propagate with particle motion in the same direction of the propagation and cause volume change without distortion. They propagate at a velocity greater than S-waves, for which the particle motion is perpendicular to the direction of propagation.

The velocities of propagation of seismic waves in a linear elastic solid are associated with the medium's mechanical parameters through simple relationships. Indeed, the mechanical response of an elastic medium is fully characterized by two elastic constants, for example, Lamé's constants λ and μ (the latter being the shear modulus G in engineering notation). Seismic wave velocities can be expressed with simple relationships between such constants and the material mass density ρ (see Section 2.1.3)

$$\begin{cases} V_P = \sqrt{\dfrac{\lambda + 2G}{\rho}} \\ \\ V_S = \sqrt{\dfrac{G}{\rho}} \end{cases} \tag{1.1}$$

The velocity of propagation of body waves is directly linked to the stiffness of the medium and is not frequency dependent in linear elastic materials. In particular, the velocity of propagation of P-waves is associated with the (small-strain) longitudinal modulus, whereas the velocity of propagation of S-waves is associated with the (small-strain) shear modulus. Equation 1.1 forms the basis for the use of seismic waves in material characterization, showing that the elastic constants can be easily determined if the seismic wave velocities are measured experimentally.

P-wave velocities (Equation 1.1) are of limited value in saturated soils because of the role played by the pore fluid in determining the overall response of the soil. This can be explained in detail via Biot's theory of wave propagation in saturated porous media (Biot 1956a, 1956b). It can be shown that the velocity of propagation of P-waves is strongly influenced by the compressibility of the pore fluid rather than the soil skeleton, whereas the influence of the pore fluid on S-wave propagation is negligible and is only linked to the change in mass density because the pore fluid has no shearing resistance. Thus, S-waves (Equation 1.1) are widely used for soil characterization because their velocity is directly related to the shear modulus of the soil skeleton.

Finally, although linear elasticity is adequate for modeling the stiffness of soils at small-strain levels, it is not able to describe the energy dissipation in loading–unloading cycles that is observed in soils even at very small strains. From this point of view, linear viscoelasticity provides a more consistent framework for the interpretation of seismic tests. Wave propagation in linear viscoelastic media can be modeled using the elastic–viscoelastic correspondence theorem as discussed in Section 2.5.

1.1.1 Seismic tests for site characterization

The close link between the velocity of propagation of seismic waves and the elastic constants of the medium makes seismic waves useful for geotechnical site characterization. Several tests have been designed to measure or estimate seismic wave velocities *in situ* with the primary purpose of assessing the variation with depth of the shear wave velocity (i.e., a shear wave velocity profile). In particular, the great advantage of testing geomaterials in their undisturbed state (especially important for hard-to-sample soils) and the large volume of material involved are two main advantages of *in situ* seismic methods. Other applications are discussed by Jamiolkowski (2012). For example, seismic wave velocities can be used to evaluate soil porosity on the basis of Biot's theory (Foti et al. 2002). Although the most common application is the measurement of seismic wave velocities, engineers and seismologists have used *in situ* seismic methods to measure the attenuation of waves and to compute the material damping ratio of soils.

Seismic methods for site characterization are often divided into two broad categories: *invasive* tests and *noninvasive* tests. The former methods require a borehole (cross-hole, down-hole, up-hole, P-S suspension logging) or the insertion of a probe in the soil (seismic cone penetration test or seismic dilatometer), the latter methods are entirely conducted on the ground surface (seismic reflection, seismic refraction, and surface wave tests). In general, the noninvasive tests are affected by a larger degree of uncertainty, but they allow the exploration of larger, more representative volumes, and they are usually more cost-effective than invasive tests.

In cross-hole tests (CHTs), source and receiver(s) are placed at the same depth, and the wave velocity is calculated on the basis of the travel time. Being based on a direct measurement, CHTs are usually considered the most reliable and accurate method. Nevertheless, the need for two or three cased boreholes increases the cost of the test.

In down-hole tests, seismic waves are generated on the ground surface, and the travel times are measured in a borehole at different depths using one or more receivers, which are progressively moved to deeper positions. The interpretation can be based on the trends of travel times with depth or on the travel-time interval between the arrivals at two receivers. Appealing alternatives in geotechnical site characterization are based on receivers mounted within cone penetration test (CPT) or dilatometer test (DMT) probes, which allow the execution of down-hole measurements at different stages during the insertion of the probe in the soil with no need for a dedicated borehole.

In P–S suspension logging, a single, cable-wired instrument is inserted in the borehole. The instrument contains a seismic source and two or more receivers and provides a local measurement of seismic wave propagation at different depths. The method is unique in that the instrument is suspended in a fluid-filled borehole, with the fluid providing the mechanical coupling between the instrument and the surrounding soil. This approach is ideal for measurements in very deep boreholes, and it is widely used in the oil and gas industry, although its use for engineering applications is growing.

Noninvasive seismic techniques include seismic reflection and refraction methods. Both methods have been developed using P-waves but are currently used also with S-waves, although there are some difficulties in generating high-energy, horizontally polarized shear waves.

In seismic refraction methods, soil characterization is based on arrival time on the ground surface of seismic waves that have been critically refracted at interfaces among layers having a different velocity or continuously refracted in an inhomogeneous medium. They suffer from intrinsic limitations related to the presence of velocity inversions or hidden layers (Sheriff and Geldart 1995; Reynolds 1997), which lead to incorrect estimation of interface depth and seismic velocities.

Seismic reflection surveys are primarily aimed at detecting and imaging interfaces among different layers. They require complex processing and interpretation procedures. Although some high-resolution applications for near-surface characterization have been proposed, these methods remain mainly devoted to deep exploration.

Surface wave methods are noninvasive methods based on the solution of the inverse problem of Rayleigh wave propagation, which is aimed at estimating the shear wave velocity profile of the subsurface. Rayleigh waves are easily generated and detected at the ground surface, providing a powerful tool for site characterization.

1.2 SURFACE WAVES

Several of the properties of surface waves make them particularly well suited for geomaterial characterization. They originate from the condition of vanishing stress at a boundary of a domain (e.g., the surface of the Earth), and their radiation pattern (see Figure 1.1) is essentially two-dimensional (2D)

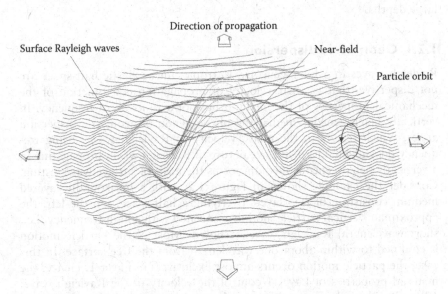

Figure 1.1 2D radiation pattern of Rayleigh surface waves generated by a vertical point source.

and thereby characterized by a much lower rate of geometric attenuation than body (or bulk) waves whose energy spreads in horizontal and vertical directions. As an example, Rayleigh surface waves generated by a line source in a homogeneous elastic half-space do not exhibit any geometric attenuation, whereas for a point load, the rate of spatial decay is proportional to the inverse of the square root of the distance from the source. Conversely, for a point source, the geometric attenuation factor of body waves propagating along the boundary of an elastic half-space is proportional to the inverse of the square of the distance (Ewing et al. 1957). Thus, at distances on the order of one to two wavelengths from the source, the contribution of body waves becomes negligible, and the wave field is dominated by Rayleigh waves. Lamb (1904) was among the first scientists to recognize this fundamental property of surface waves and its implications for the transmission of earthquake energy at large distances.

In the direction orthogonal to that of propagation, the displacement field generated by a surface wave decays exponentially because no energy is propagated in the interior of the half-space. It can be shown that most of the strain energy associated with surface wave motion is confined within a depth of about one wavelength λ from the free boundary (Achenbach 1984). Hence, Rayleigh waves with long wavelengths penetrate deep into the interior of a medium. Because wavelengths are proportional to the inverse of frequency in harmonic waves, this statement can be interpreted as follows: high-frequency waves are confined to shallow depths within the medium, whereas low-frequency components involve motion also at large depths.

1.2.1 Geometric dispersion

Rayleigh waves in a homogeneous, isotropic, linear elastic half-space are not dispersive; that is, their velocity of propagation is a function of the mechanical properties of the medium but not a function of frequency. In vertically heterogeneous media, the phenomenon of geometric dispersion arises, which results in the phase velocity of Rayleigh waves being frequency dependent. The dispersive nature of Rayleigh waves propagating in a vertically heterogeneous medium forms the basis of surface wave testing. Consider the example shown in Figure 1.2, with a horizontally layered medium consisting of two layers overlying a half-space. On the left, the approximate vertical particle motion associated with a high-frequency (i.e., short wavelength) Rayleigh wave is shown. Most of the particle motion is confined to within about one wavelength from the free surface. In this case, the particle motion occurs almost exclusively in Layer 1. Hence, the material properties of Layer 1 control the velocity of the Rayleigh wave. The right side of the figure illustrates the vertical particle motion associated with a low-frequency (i.e., long-wavelength) Rayleigh wave. In this case,

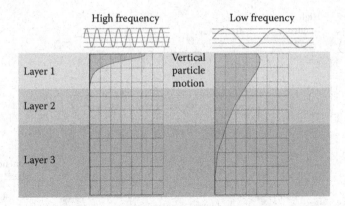

Figure 1.2 Geometric dispersion of Rayleigh waves: trends with depth of the vertical particle motion associated with the propagation of two harmonic waves in a layered medium.

the particle motion extends to a greater depth, and there is significant particle motion in Layers 1 and 2 and less in Layer 3. The velocity of this low-frequency Rayleigh wave is controlled by some combination of the material properties of all three layers, perhaps in rough proportion to the relative amount of particle motion occurring within each layer.

The dispersive nature of Rayleigh waves can be used for identification purposes; by experimentally measuring the dispersion curve (i.e., the variation of Rayleigh phase velocity with frequency) that is associated with a given site, it is possible, via an inversion process, to determine the shear wave velocity profile of the site. More broadly, the objective is to characterize the interior of a medium from measurements of a 2D surface wave field at its free boundary. Mathematical formalization of this objective leads to the theory of *inverse problems*, specifically to a particular branch of this theory known as *system* or *parameter identification problems* (Engl 1993). Solution and properties of such problems depend strongly upon the constitutive model used to describe the response of a medium to the propagation of disturbances, as well as on the properties of its boundaries.

To summarize the concept behind the use of geometric dispersion for site characterization, assume that the stratified medium in Figure 1.3a is characterized by increasing stiffness with depth, so that the shear wave velocity of the top layer is less than the velocity of the second layer, which in turn is less than the velocity of the half-space below. In this situation, a high-frequency Rayleigh wave (Figure 1.3b), traveling in the top layer, will have a velocity of propagation slightly lower than the shear wave velocity of the first layer. On the contrary, a low-frequency wave (Figure 1.3c) will travel at a higher velocity because it is influenced by the underlying

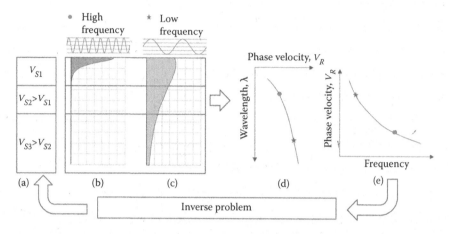

Figure 1.3 Parameter identification on the basis of geometric dispersion.

stiffer materials as well. This concept can be extended to other frequency components. A plot of phase velocity versus wavelength (Figure 1.3d) will hence show an increasing trend for longer wavelengths. Considering the intimate relationship between wavelength and frequency, this informa- tion can also be represented in a plot of phase velocity versus frequency (Figure 1.3e), which is commonly called a dispersion curve. This example shows that, for a vertically heterogeneous medium, the dispersion curve contains information about the variation of medium parameters with depth. This is the so-called *forward problem*, and it will be presented in detail in Chapter 2. If the dispersion curve is obtained experimentally (see Chapter 4), it is then possible to solve the *inverse problem* to obtain the shear wave velocity profile (i.e., the medium parameters are identi- fied on the basis of the experimental data collected on the boundary of the medium; see Chapter 6). This is the essence of surface wave methods; additional details are presented here and in subsequent chapters.

1.3 TEST METHODOLOGY

Most surface wave tests are aimed at measuring the shear wave velocity profile, which is closely linked to the vertical variation of the small-strain shear modulus. Usually, this task is accomplished by first obtaining an experimental dispersion curve from measured field data. These data are subsequently used for the solution of the parameter identification problem. This latter step implies the choice of a reference model for the interpre- tation, which in most cases is a one-dimensional (1D) model comprising a stack of homogeneous, linear elastic layers.

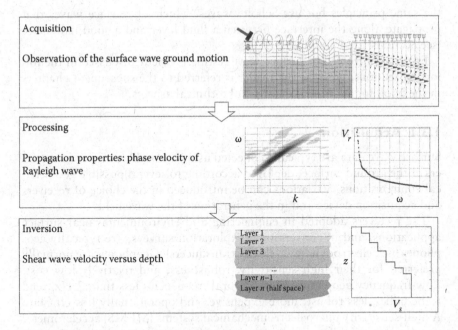

Figure 1.4 Flow chart of surface wave tests.

Figure 1.4 shows the standard procedure for surface wave tests, which can be subdivided into three main steps:

1. Acquisition of experimental data
2. Signal processing to obtain the experimental dispersion curve
3. Inversion process to estimate site properties

It is important to recognize that the individual steps are strongly inter-related and their interaction must be adequately accounted for during the whole interpretation process. Many types of surface wave tests have been developed for near-surface site characterization, often addressed using a wide range of acronyms. They may differ in any one of these steps regarding the type of source and the number of receivers, the signal processing technique, or the inversion strategy and algorithm, but the overall principle is essentially the same.

Surface wave tests are typically performed using Rayleigh waves because of the simplicity of generating and detecting them on the ground surface. Similar applications using Love waves, which originate from horizontally polarized surface waves, are also possible, but they suffer from some limitations related to layer stratigraphy and to the difficulties in generating Love waves. Applications at the seafloor are also becoming popular. They employ the same principles but use Scholte waves, which are surface waves that propagate along the interface between a fluid layer and a solid half-space (see Sections 1.6.4 and 8.2).

Each step involved in surface wave testing is briefly discussed in the following sections, whereas the reader is referred to the subsequent chapters in the book for additional details and technical aspects.

1.3.1 Acquisition

Surface wave data are typically collected using a variable number of receivers at the ground surface, deployed according to several possible geometrical configurations. Variations can be introduced in the choice of receivers and acquisition device and in the generation of the wave fields.

The receivers adopted in engineering and environmental near-surface applications, and in conventional exploration surveys, are typically geophones (i.e., electrodynamic velocity transducers). Geophones are normally preferred for their high sensitivity, robustness, and relatively low cost. Low-frequency geophones (with natural frequencies less than 2 Hz) tend to be bulky, less robust, more expensive, and operationally less efficient. A high-sensitivity microelectromechanical systems (MEMS) accelerometer can be beneficial for low-frequency surface wave acquisition. In very small-scale applications, e.g., the characterization of pavement systems, accelerometers are often used because of the need for high-frequency, high-amplitude signals.

Several types of devices can be used for the acquisition and storage of signals. Basically, any hardware having an A/D converter and the capability to store digital data can be adopted, ranging from seismographs to dynamic signal analyzers to systems comprising data acquisition boards connected to personal computers (PCs) or laptops. Commercial seismographs for geophysical prospecting are typically the first choice because they have high-end specifications and they are designed to be used in the field. Thus, they are robust, waterproof, and resistant to dust. Modern seismographs are composed of scalable acquisition components used in conjunction with field computers, allowing preliminary processing of data on-site.

For the generation of Rayleigh surface waves, several different types of sources can be used, provided they have sufficient energy in the frequency range of interest for the specific application. Transient, impact sources are often preferred because they are inexpensive and rugged. They range from small hammers for generating high-frequency surface waves to large falling weights, which generate more low-frequency components. Seismic guns may be used for shallow applications, although explosives are usually limited to deep exploration because of logistic issues. Appealing alternatives are continuous sources, which are able to generate controlled, harmonic waves. Also in this case, the size of the source varies from relatively small, electromagnetic shakers to large, truck-mounted Vibroseis™. The drawback of such sources is essentially their cost and the need for long acquisition times on site when light sources are used with monochromatic signals.

It is also possible to use passively generated surface waves. In this case, the need for an "active" transient or continuous source is avoided by recording "passive" ambient noise, often called microtremors. Microtremors include cultural noise generated by human activities (traffic on highways, construction activities, and so on) and vibrations arising from natural events. A great advantage is that microtremors are usually rich in low-frequency components. Hence, passive surveys provide useful information for deep characterization (tens or hundreds of meters). However, the level of detail close to the ground surface is typically low because microtremors lack high-frequency energy. This limitation can be overcome by combining active and passive measurements (Figure 1.5). Field acquisition is discussed thoroughly in Chapter 3.

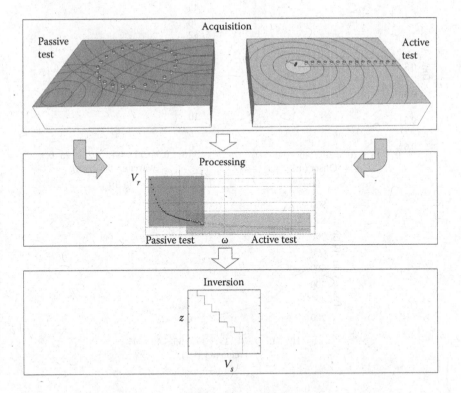

Figure 1.5 Combination of active-source and passive-source surface wave measurements.

1.3.2 Processing

The field data are processed to estimate the surface wave propagation parameters, typically the experimental dispersion curve. Different surface wave test procedures apply a variety of signal processing techniques, but they mainly rely on the Fourier transform to decompose the time history of vertical particle motion into its frequency components. Indeed, using Fourier analysis, it is possible to separate the different frequency components of a signal, which are subsequently processed to estimate phase velocity using different approaches that are a function of the test configuration and the number of receivers. In the example provided in Figure 1.6, the dispersion curve is obtained by selecting the peak amplitudes of the frequency–wavenumber spectrum. Specific processing techniques will be discussed in detail in Chapter 4.

Some types of equipment (e.g., PC-based systems) allow for processing the experimental data directly in the field. Often, the simple visual screening of the recorded seismic traces is not sufficient; surface wave components

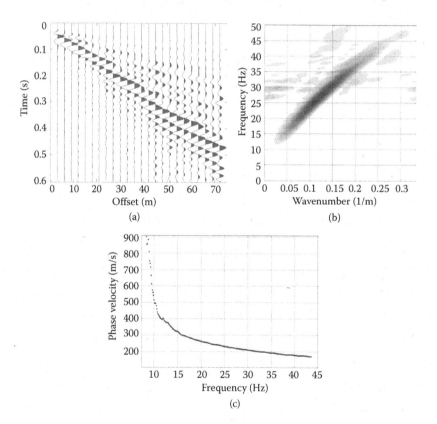

Figure 1.6 Example of processing of experimental data using the frequency–wavenumber analysis: (a) field data; (b) *f–k* spectrum; (c) dispersion curve.

are grouped together, and without signal processing it is not possible to judge the quality of the data over the desired frequency or wavelength range. An assessment of the frequency range with adequate signal quality can be particularly useful to assess the need to change the acquisition setup or to gather additional experimental data.

There are several test variants in which the experimental dispersion curve is replaced by other types of data. These include the inversion of full, time-domain waveforms (e.g., Tran and Hiltunen 2012) and the inversion of the Fourier frequency spectra of observed ground motion (Szelwis and Behle 1987). Such strategies are not commonly used in practice, especially in near-surface site characterization. Moreover, the experimental dispersion curve is informative about trends to be expected in the inverted shear wave velocity profile, so its visual inspection is important for the qualitative validation of the results.

Surface wave data can also be used to characterize the dissipative behavior of soils. The spatial attenuation of surface waves is associated with the internal dissipation of energy. Using a procedure substantially analogous to the one outlined in Figure 1.4, it is possible to extract from field data the experimental *attenuation curve*, that is, the coefficient of attenuation of surface wave as a function of frequency, and then use these data in an inversion process aimed at estimating the material damping ratio (or the quality factor) profile for the site (see Chapters 5 and 6). Furthermore, the procedure for attenuation and dispersion analysis and inversion can be coupled leading to simultaneous determination of shear modulus and material damping ratio profiles (see Chapter 6).

1.3.3 Inversion

The solution of the Rayleigh inverse problem is the final step in test interpretation. The solution of the forward problem (Chapter 2) forms the basis of any inversion strategy. Assuming a model for the soil deposit, model parameters are identified by minimizing an objective function representing the distance between the experimental and the theoretical dispersion curves. The objective function can be expressed in terms of any mathematical norm (usually the root mean square or RMS) of the difference between experimental and theoretical data points. In practice, the set of model parameters that produces a solution of the forward problem (a theoretical dispersion curve) as close as possible to the experimental data (the experimental dispersion curve of the site) is selected as solution of the inverse problem (e.g., Figure 1.7).

This objective can be reached using a variety of strategies. A major distinction arises between local search methods (LSMs), which start from a tentative shear wave velocity profile and search in its vicinity, and global search methods (GSMs), which attempt to explore the entire space of possible solutions. LSMs are undoubtedly faster because they require a limited number

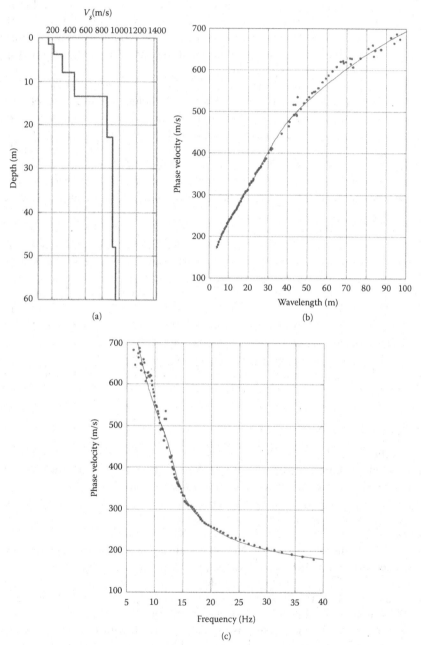

Figure 1.7 Example of inversion process: (a) shear wave velocity profile; (b) comparison between the theoretical and experimental dispersion curves in wavelength–phase velocity domain and (c) in frequency–phase velocity domain.

of runs of the forward problem of Rayleigh wave propagation. However, because a solution is sought in the vicinity of a tentative profile, there is the risk of being trapped in local minima. Conversely, GSMs require substantial computational effort because a large number of runs of the forward problem is required, so that the approach may be quite time consuming.

In general, inverse problems are inherently ill-posed, and a unique solution does not exist. A major consequence is the so-called equivalence problem (i.e., several shear wave velocity profiles may be equivalent in the sense that the theoretical dispersion curves associated with these profiles are at the same distance from the experimental dispersion curve). A meaningful evaluation of equivalent profiles must take into account the uncertainties in the experimental data. Additional constraints and a priori information from borehole logs or other geophysical tests are useful elements to mitigate the equivalence problem.

Infact, engineering judgment plays an important role in the entire test procedure, and the results of fully automated interpretation methods should be carefully examined, with special attention to intermediate results during each step of the test procedure. Knowledge of the theoretical aspects of wave propagation, signal processing, and inversion, and experience are hence essential ingredients for the successful application of surface wave tests.

1.4 HISTORICAL PERSPECTIVE

The interpretation of surface wave data requires the availability of digital records and many computationally intensive tasks. It is therefore not surprising that most advancements in surface wave testing and their widespread application are closely linked to the historical progress in electronics, for data acquisition devices and computers. In particular, the advent of reasonably priced data collection devices and PCs has led to their use in near-surface geophysical and engineering site characterization. Most of the tools for the analysis of seismic records and for the solution of the forward and inverse Rayleigh problems come from the field of seismology and have been successively adapted for engineering applications. Wave field transforms and multichannel data processing inherited contributions from seismic data processing for oil and gas exploration. Nevertheless, the spread of surface wave methods in near-surface applications has led to a number of subsequent developments, which take into account the different scale of the applications.

1.4.1 Global seismology

Energy released by earthquakes travels to a teleseismic distance mainly in the form of long-period surface waves, which represent by far the largest component of seismic records at great distance from the epicenter. Surface waves have been studied in seismology for the characterization of Earth's interior since the 1920s, but their widespread use started during the 1950s and 1960s thanks to the increased possibilities of numerical analysis and improvements in instrumentation for recording seismic events (Ben-Menhaem 1995; Aki and Richards 2002). In particular, the growth of long-period and broadband seismic networks, started in the 1970s, has led to recent large-scale and global studies on upper mantle structure (Romanowicz 2002).

Most seismological applications have been traditionally based on the analysis of Rayleigh wave components because they are easily identified in the vertical component of seismograms, whereas only with the advent of digital records that allow for a clearer identification of the transverse motion Love wave components have begun to be widely used.

The propagation characteristics of surface waves contain relevant information about the structure of the Earth along the source-station path and the characteristic of the earthquake source. The separation of these two effects is a major challenge for seismologists (Romanowicz 2002). Most crustal and mantle studies use a two-step procedure consisting of the evaluation of surface wave velocity dispersion followed by inversion. The availability of an increasing number of measurement stations has led to different approaches in the evaluation of surface wave velocity.

Originally, the evaluation of group velocity was based on the time interval between peaks and troughs in a single dispersed wave train, while the phase velocity determination was based on the analysis of its Fourier spectrum (Romanowicz 2002). This raised some issues related to the necessity of separating source and path effects. In order to address this issue, many studies began using a "two-station method," in which the dispersion was measured between two recording stations approximately aligned with the epicenter, eliminating the common source phase. The next step was the use of multiple stations spanning a geologically homogenous region (Dziewonski and Hales 1972).

An important problem to be solved was related to the influence of higher modes of propagation in the recorded signals. This motivated the development of several sophisticated filtering techniques based on group velocity features, aimed at separating modal components in teleseismic signals (Dziewonski et al. 1969; Levshin et al. 1994). The inversion of higher modes, whenever they can be extracted from the recording, is advantageous because, as shown in Chapter 2, the particle motion associated with higher modes extends to greater depths than that of the fundamental mode.

The attenuation of surface waves has also been extensively studied for the determination of the anelastic structure of the upper mantle. Working in the frequency domain and exploiting the real and imaginary part of the spectrum, Dziewonski and Steim (1982) obtained information related not only to the velocity model but also to the quality factor (dissipation) model with a formulation based on transfer functions (see also Section 5.3).

Approaches based on a time-domain waveform inversion started to be developed at the end of the 1970s with the formulation of a first-order perturbation theory allowing the computation of synthetic seismograms for a reference model and the evaluation of derivatives with respect to the model parameters. The waveform approach permits the evaluation of model parameters directly from seismograms via a single-step procedure, but the necessity of correcting for crustal structure poses some challenges (Romanowicz 2002).

1.4.2 Exploration geophysics

Traditionally, deep geophysical surveying for hydrocarbon exploration and for regional structure identification has employed seismic reflection and seismic refraction techniques. Sources and receivers are typically placed on the ground surface. Although these techniques are based on body wave propagation, seismograms collected for deep exploration surveys exhibit significant surface wave components. In the gather of Figure 1.8, the first 1500 m

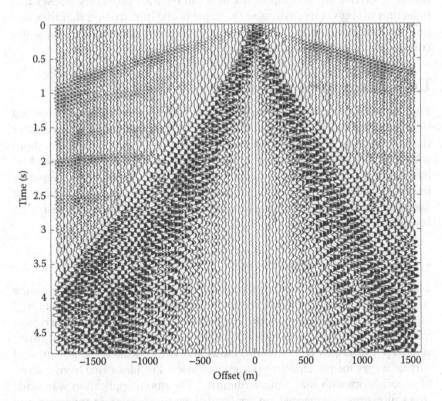

Figure 1.8 Example of common shot gather for seismic reflection surveying. Despite the presence of other events (refracted and reflected P-waves), the gather is dominated by the surface waves.

of offset (i.e., source-to-receiver distance) is dominated by surface waves with multiple modes of propagation.

Surface waves carry a large portion of the energy imparted by shallow sources and they have less geometric attenuation than body waves. Thus, they dominate at small offsets. Moreover, the geometric dispersion in heterogeneous media creates long wave trains with multiple modes that mask reflection signals in the seismograms. For the aforementioned reasons, surface waves are a tedious source of coherent noise in seismic reflection surveys—referred to as "ground-roll." Because of its coherent nature, it is quite difficult to eliminate ground-roll from the shot gathers, and several techniques have been developed to achieve this result (Doyle 1995). As it turns out, the tools used to eliminate ground-roll can be also profitably used to extract information for characterization purposes (Nolet and Pansa 1976; McMechan and Yedlin 1981). An example is shown in Chapter 4 for analyses in the frequency–wavenumber domain.

The interest in the geophysical exploration community toward the exploitation of surface wave components in seismic gathers collected for seismic reflection surveys is increasing as the value of the information that they can provide is recognized (e.g., for the near-surface characterization, for static corrections, or velocity modeling in reflection surveys).

1.4.3 Near-surface applications

Engineering applications of surface wave testing were initially proposed in the 1950s, but they started to become popular and widely used after the introduction of the spectral analysis of surface waves (SASW) method, which for the first time enabled a rapid, theoretically sound test procedure by exploiting the capabilities of new electronic equipment and computers in the late 1970s. Today, the technique is widely used, and the adoption of advanced acquisition and interpretation schemes has made the test even faster and more reliable.

1.4.3.1 Pioneering applications

The first method developed for engineering site characterization was the steady-state Rayleigh method (Section 4.2) proposed by Jones (1958, 1962) and then adopted at the U.S. Army Corps of Engineers Waterways Experiment Station (Ballard 1964). This method was very simple, particularly in the inversion step. Nevertheless, it was ingenious and demonstrated the potential of surface waves for site characterization purposes. The idea came from a series of experiments with mechanical vibrators. The initial application was made using ultrasonic frequencies on concrete slabs to assess their thickness and elastic properties. The success of this technique led to the extension to soil deposits using lower frequencies.

In his field experiments, Jones attempted to use both Rayleigh and Love waves to characterize the soil. The field equipment was composed of a mechanical vibrator and a single receiver. To investigate Rayleigh wave propagation, the vibrator and receiver were placed vertically, while for Love waves the vibrator and receiver were oriented to produce and detect vibrations in a horizontal direction transverse to the testing line.

The experimental dispersion curve was obtained by moving the receiver away from the harmonic source along a straight line, looking for positions where the vibrator and the receiver were in phase. The average distance between two in-phase positions is the wavelength associated with the frequency of the harmonic source. The Rayleigh phase velocity was calculated by multiplying the frequency and observed wavelength. Repeating the process for different frequencies, the dispersion curve was obtained. Once the experimental dispersion curves were obtained, approximate procedures (see Section 6.3) based on theoretical analysis of surface wave propagation were used to infer the shear modulus profile.

1.4.3.2 Spectral analysis of surface waves

The SASW was introduced at the University of Texas at Austin during the late 1970s and early 1980s (Nazarian and Stokoe 1984; Stokoe et al. 1994). Significant advances were introduced with respect to data acquisition/processing and inversion procedures. The acronym itself could in general be used for any surface wave method, but it is associated mainly to this specific technique, because of the large popularity it has achieved.

The SASW test is based on a two-receiver configuration and straight-forward signal processing tools (see Section 4.3). The dispersion curve is evaluated by estimating for each frequency component the time delay between the arrivals at the two receivers (expressed by the phase shift) of the wave generated by an active source. Because the two-receiver approach imposes limitations on the frequency range over which the test is effective, the experimental dispersion curve at a site is estimated using several receiver spacings. The individual experimental dispersion curves from different spacings are then combined to obtain a single, composite curve to be used in the inversion process. This procedure may be time consuming and the production rate in the field is adversely affected by the required changes in receiver positions.

Moreover, the interpretation of two-station measurements requires a certain amount of engineering judgment and cannot be easily automated because the periodicity of the phase shift between receivers leads to the necessity of "unwrapping" the phase of the cross-power spectrum used to calculate the Rayleigh phase velocity. The most important limitations are related to the effect of incoherent and coherent noise. The distortions of the experimental local phase induced by the presence of body waves, near-field effects, lateral variations, and, above all, higher modes can be particularly critical. Despite these limitations, the SASW method occupies a significant place in the historical development of surface wave methods for engineering site characterization and is still in widespread use today.

I.4.3.3 Multistation approaches

The use of multiple receivers enhances the production rate in the field and makes the processing of the data faster, less subjective, and more robust. Early applications of multistation surface wave tests for near-surface characterization were developed in the 1980s (McMechan and Yedlin 1981; Gabriels et al. 1987), but their widespread adoption started in late 1990s. Today, most near-surface applications make use of multistation approaches. They are often identified with the acronym MASW (multistation analysis of surface waves), which was initially introduced by researchers at the Kansas Geological Survey (Park et al. 1999).

Multistation approaches employ a linear array of geophones in line with an active source. Several techniques can be used to process the data, the most common being transform-based approaches. Field data collected in the time–space domain are transformed into a domain (e.g., the frequency–wavenumber domain) where the phase velocities associated with different frequencies are easily chosen by picking the spectral maxima (see Section 4.6).

I.4.3.4 Microtremor surveys

The SASW and MASW methods are active-source tests in which the wave field is generated by a seismic source acting on the ground surface. As mentioned previously, microtremors and ambient vibrations generated by natural events or human activities contain low-frequency energy to characterize the subsurface at depth without resorting to large, heavy active sources. Microtremor surveys, also called passive surface wave tests, record ambient noise using an array of geophones and then analyze these data to extract the experimental dispersion curve. The data are typically collected using a 2D array of receivers because the position of the sources is not known a-priori. Several processing techniques are available, with the most prevalent being the frequency-domain beamforming (Lacoss et al. 1969) and the spatial autocorrelation (Aki 1957) techniques.

The refraction microtremors (ReMi) approach (Louie 2001) is a variant of a microtremor survey using a simple linear array of receivers, although the method has nothing in common with refraction surveys apart from the receiver configuration. In this case, the dispersion curve is estimated using the much simpler approach adopted for active surface wave tests, relying on the assumption that sources of ambient vibrations are uniformly distributed in space. Although this approach is very simple and fast, it can lead to significant overestimation of the shear wave velocity profile if the background noise is traveling along a preferential direction not in line with the receiver array (Strobbia and Cassiani 2011).

For a given source, the amplitude spectrum of surface waves is a function of the site velocity structure and the source spectrum. However, inverting the amplitude spectrum is not practical because deconvolving the effect of the source is not straightforward due to the fact that the effective source radiation pattern in the far-field is not known even with controlled sources. The use of the ratio between the vertical and horizontal-radial component of the surface waves solves the issues related to the estimation of the source. The polarization, or ellipticity, of the surface wave as a function of the frequency can be inverted to estimate a velocity profile. This approach, used in multichannel, multicomponent data (Muyzert 2007a, 2007b), is also the basis of single-station, multicomponent passive surveys. The average horizontal-to-vertical spectral ratio (HVSR or simply H/V) of passive data is measured and interpreted to estimate the site natural frequency or is inverted (Fäh et al. 2003).

1.5 CHALLENGES OF SURFACE WAVE METHODS

As described earlier, surface wave tests are indirect methods in which the soil properties are estimated on the basis of observed field data by assuming a soil model and inferring its parameters via an inverse problem. The interpretation of experimental data is not straightforward because it is based on a series of complex processing steps. Some assumptions are implicit in the interpretation procedure and their relevance and implications should be evaluated with care. In the following, some specific aspects that play a major role in the interpretation are discussed.

1.5.1 Sampling in space: Apparent phase velocity (mode superposition)

Surface wave tests are based on the geometric dispersion of Rayleigh waves in vertically heterogeneous media. From a mathematical point of view, the propagation of surface waves generated by a point source in vertically heterogeneous media can be represented as the superposition of free Rayleigh modes (see Section 2.4). A dispersion curve is associated with each free Rayleigh mode; thus, the representation shown in Figure 1.3 was simplified in the sense that only a single mode of propagation was displayed.

An ideal experimental survey would be able to extract and identify the dispersion curve for each mode. These data could be used for a robust and effective inversion process (e.g., Gabriels et al. 1987). In reality, the number of receivers is limited and other factors, such as wave attenuation caused by dissipative behavior of soils, make this ideal survey unachievable (see Chapter 3). Using a limited number of receivers, a single, *apparent* dispersion curve is typically extracted from the field data.

In some situations, the fundamental mode assumes a predominant role in the propagation of Rayleigh waves, meaning that higher modes are not excited by the applied source or, if they are, the energy associated with them is negligible. In such situations, the fundamental mode will be easily extracted via signal processing and the inversion process can be simplified.

Whenever higher modes play an important role in the propagation, the processing of experimental data recovers an apparent or effective phase velocity that is influenced by several modes (see Figure 1.9). In such situations, the inversion process cannot be based on the free Rayleigh modes and the modal superposition must be accounted for in the solution of the forward problem (see Section 2.4). In this case, the inversion becomes more onerous and the stability of automated search algorithms is adversely affected. An alternative strategy can be based on more sophisticated processing techniques allowing for filtering different mode components before the extraction of the experimental dispersion curve.

A fundamental mode approach is more easily implemented and less computationally intensive so that, whenever appropriate, it is the preferred choice. In general, it is not possible to define a rigorous rule to assess the relevance of higher modes. The fundamental mode is typically predominant when the shear wave velocity profile increases gradually with depth. Higher modes

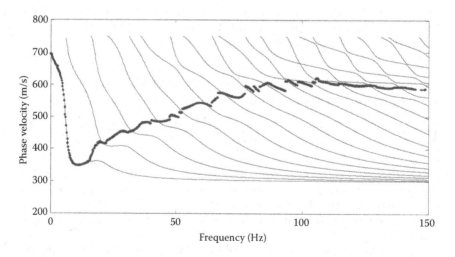

Figure 1.9 Example of modal superposition: an apparent curve results from the superposition of multiple modes at an inversely dispersive site.

become prevalent in the high-frequency range in the presence of a stiff, top layer (e.g., a stiff soil crust or a pavement system as in Figure 1.9), whereas they can be important in the low-frequency band if an abrupt increase in stiffness is present (e.g., shallow bedrock overlain by soft soils).

1.5.2 Near-field effects

A common assumption in surface wave tests is that the wave field comprises only plane surface wave components. In reality, for most applications, the waves are generated by a point source on the ground surface and the receivers are placed in the vicinity of the source so that so-called near-field effects come to play a role. Near-field effects are mainly due to the following:

1. Body wave interference
2. Cylindrical wave front of Rayleigh waves

The wave field in the vicinity of the source is typically quite complex because body wave components and surface wave components are not well separated at this stage. Body wave components attenuate at a much faster rate than surface waves. This is because the wave front is spherical for body waves and cylindrical for surface waves, inducing a different spreading of energy in space. As a consequence, the amplitude of particle motion associated with body waves attenuates proportionally to the inverse of the distance from the source, whereas the particle motion of surface waves attenuates proportionally to the inverse of the square root of the distance from the source. As the wave train travels away from the source, the relative contribution of body wave components decreases so that in the far-field it is acceptable to neglect the influence of body waves.

The second aspect is also related to the spreading of surface waves. In most techniques adopted for determining the experimental dispersion curve, it is assumed that the wave front is plane. Again this assumption is valid only in the far-field, whereas in the near-field the cylindrical shape of the wave front cannot be neglected and accurate analysis requires the use of transforms employing cylindrical coordinates (Zywicki and Rix 2005).

Near-field effects have been extensively studied for the two-station SASW method. The body wave components exert a larger influence on the closest receiver, and as a consequence the overall estimate of the phase velocity is biased. This is one of the major reasons for using several testing configurations in the SASW test as discussed in Section 4.3. Numerical simulations of the complete wave field generated by a point source have shown that near-field effects can affect the dispersion curve estimate if the closest receiver is placed within a certain critical distance from the source. This critical distance can range from one-half to two wavelengths depending on the soil profile (Sànchez-Salinero 1987; Tokimatsu 1995).

Near-field effects can be taken into account by using forward models able to reproduce the complete wave field, but this inevitably leads to computationally intensive inversion methods (e.g., Ganji et al. 1998).

The adverse influence of near-field effects in multistation approaches is alleviated because longer receiver arrays are used so that a larger number of receivers are in the far-field. Moreover, the processing techniques can separate body and surface wave components and identify the portion of data that can be used. Low-frequency components exhibit more near-field effects, requiring the adoption of adequate countermeasures (Strobbia and Foti 2006). A typical strategy is to increase the offset of the first receiver of the array with respect to the source position (Yoon and Rix 2009). Although effective in mitigating near-field effects, this approach also causes the loss of high-frequency components because of internal energy dissipation in soils and hence a loss of resolution in the shallow part of the profile.

1.5.3 Model errors

As discussed previously, surface wave testing is based on the solution of a model parameter identification process. The choice of the reference model is a critical aspect because the experimental data will be fit to the model regardless of whether it is appropriate for the specific site. The most common choice for the model is a stack of homogenous linear elastic layers (Figure 1.10). This is a 1D model that neglects lateral variations in soil properties. Clearly, this is only an approximation of reality, and the more the real conditions differ from this model, the less accurate will be the shear wave velocity profile. The reasons for choosing such a simplified model are: (1) the solution of the forward problem can be obtained very efficiently using a variety of algorithms such as propagator matrix (Thomson 1950; Haskell 1953; Gilbert and Backus 1966)

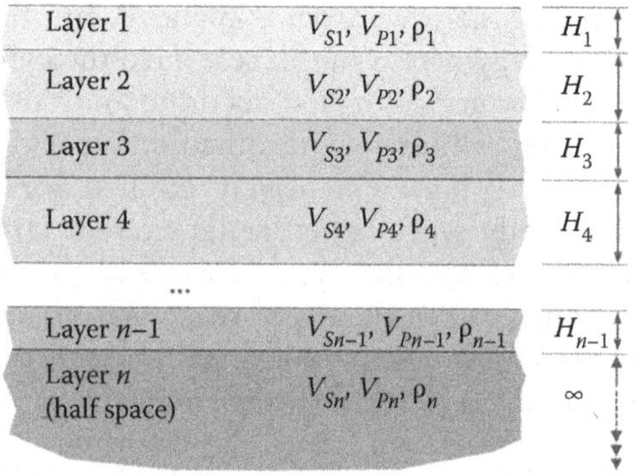

Layer 1	V_{S1}, V_{P1}, ρ_1	H_1
Layer 2	V_{S2}, V_{P2}, ρ_2	H_2
Layer 3	V_{S3}, V_{P3}, ρ_3	H_3
Layer 4	V_{S4}, V_{P4}, ρ_4	H_4
Layer $n-1$	$V_{Sn-1}, V_{Pn-1}, \rho_{n-1}$	H_{n-1}
Layer n (half space)	V_{Sn}, V_{Pn}, ρ_n	∞

Figure 1.10 Subsurface model commonly adopted for inversion: an 1D stack of linear elastic, homogeneous layers.

or stiffness matrix (Kausel and Roesset 1981) algorithms and (2) the number of unknowns is limited. Both features are helpful when formulating and solving the inverse problem because they reduce the computational effort and reduce the ill-posedness of the inverse problem, respectively.

The number of unknowns may be further reduced by selecting a priori the parameters to be estimated in the inversion process. For the layered model, four parameters are needed to fully characterize each layer (Figure 1.10): density ρ, thickness h, and two elastic constants (e.g., shear modulus G and Poisson's ratio v). Often, the elastic parameters are defined in terms of the P-wave and S-wave velocities for each layer. On the basis of sensitivity analyses (Nazarian 1984), two of these parameters are typically fixed a priori for each layer, that is, the density and Poisson's ratio (or a given ratio between P-wave and S-wave velocities) of each layer. This strategy reduces the number of unknowns from $4n-1$ (because the thickness of the half-space is not defined) to $2n-1$, where n is the number of layers including the half-space. Sometimes the layer thicknesses are also fixed on the basis of a priori information (e.g., borehole logs) or when a large number of layers are introduced to adequately reproduce the variation of S-wave velocity with depth.

Special care must be used for the selection of the Poisson's ratio. The presence of pore water in a saturated, porous medium is reflected in a continuum mechanics model by a variation of the Poisson's ratio because soil behavior is undrained under dynamic loads. Hence the appropriate value of Poisson's ratio, or equivalently the P-wave velocity, should be assumed taking into account the presence and position of the water table (see Section 7.1.2). Above the water table, the P-wave velocity is governed by the stiffness of the soil skeleton and the Poisson's ratio is typically between 0.2 and 0.3, while in a saturated, porous medium, the velocity of P-wave is governed by the bulk compressibility of the water (for soils) and the Poisson's ratio is close to 0.5.

Model errors arise whenever the model is not appropriate for the real subsurface conditions. Problems may be caused by lateral inhomogeneities or inclined bedding in the strata. Surface wave testing based on a 1D model is not appropriate if these features are marked. It can be useful to perform measurements along perpendicular lines to evaluate the differences in the measured dispersion curves. This strategy may allow an assessment of the hypothesis of plane and parallel layers. An approach for the evaluation of lateral variation is based on tests with moving receiver arrays or moving spatial windows in datasets with a large number of receivers. It is then possible to estimate a series of shear wave velocity profiles for adjacent locations, which provides a pseudo-2D or pseudo-3D model (Socco et al. 2009).

Another situation in which the usual layered model is not appropriate is when the stiffness profile is not characterized by sharp contrasts in properties between adjacent layers but rather by smooth variations with depth. This is the case for homogenous soils in which the variation of stiffness with depth is associated with the increase of confining stresses.

In such conditions, the choice of a different model with smooth variations of properties can be more appropriate (Rix and Lai 2013). An alternative strategy is to use a large number of layers with a constraint in the inversion requiring smoothness of the solution (see Chapter 6). The final solution will in this case resemble the continuous variation with a steplike gradual increase of stiffness in the layered model.

1.5.4 Resolution and depth of investigation

It is intuitive that the resolution of surface wave tests decreases with depth. Thin layers are well resolved when they are close to the ground surface, whereas at great depth the resolution is limited and only large changes can be detected. Indeed, small perturbations of the stiffness profile at great depth (e.g., thin layers of different material) have a modest effect on the dispersion curve for the site. This is illustrated in Figure 1.11, where the percent change of the fundamental-mode phase velocity dispersion curve due to a 10% variation in the velocity of different layers is shown. The magnitude of the change in the phase velocity and the affected frequency range decreases as the depth of the layer increases. The reduction of the sensitivity with depth results in a loss of resolution or in the ability to identify the properties of thin layers. Thus, these features cannot be accurately resolved, especially in the presence of uncertainty in the experimental data.

The distribution of frequencies contained in the experimental dispersion curve also plays a role. It is instructive to plot the experimental dispersion curve in terms of phase velocity versus wavelength rather than in terms of phase velocity versus frequency (Figure 1.12). If the dispersion curve is sampled at evenly spaced frequencies over a particular interval, this will result in a concentration of data points at shorter wavelengths within that interval. Conversely, a wide range of wavelengths (zone "a" in Figure 1.12) is represented by a narrow range of low frequencies. Because the depth of penetration is closely related to wavelength, the consequence is that a relatively large amount of information is available for the portion of the soil profile close to the ground surface, whereas few data points are available to assess the medium properties at depth. The distribution of the information in the experimental dispersion curve is reflected in the solution of the inversion problem. Most inversion algorithms search for the best fitting profile by minimizing the distance between theoretical and experimental dispersion curves. The uneven distribution of information means that the agreement in the short wavelength range, where most data points are, will be most influential in the choice of the final profile.

Several strategies can be devised to overcome this problem. One possibility is to sample or resample the dispersion curve during post-processing with a nonuniform distribution of frequencies (e.g., with an exponential distribution).

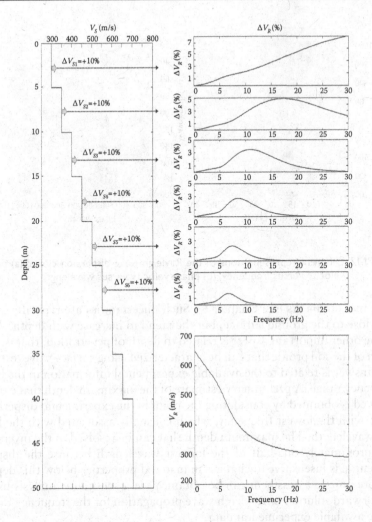

Figure 1.11 Sensitivity of the dispersion curve to changes in the shear wave velocity of layers at different depths.

Another possibility is to assess the fit between the theoretical and experimental dispersion curves in the wavelength domain, but this poses practical difficulties because the theoretical and experimental data are not sampled at the same wavelength.

A third factor that causes loss of resolution with increase depth is random error in the experimental dispersion curve. These random errors increase with decreasing frequency (Lai et al. 2005).

The loss of resolution with depth is ultimately reflected in the uncertainties associated with the shear wave velocity profile estimated from surface

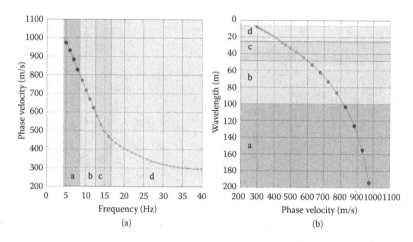

Figure 1.12 Different representations of the Rayleigh wave dispersion curve: (a) phase velocity versus frequency; (b) phase velocity versus wavelength.

wave measurements (see Chapter 6). Such uncertainties are typically quite low close to the ground surface, but they tend to increase with depth.

The other important aspect is related to depth of penetration, that is, the depth of the soil profile that can be characterized using surface wave testing. The answer is related to the available experimental information in the low-frequency range. A preliminary estimate of the maximum depth that can be resolved is obtained by considering the point in the experimental dispersion curve with the lowest frequency, which is usually associated with the longest wavelength. The maximum depth that can be resolved in the inversion is approximately one-half of the longest wavelength because the dispersion curve is insensitive to changes in material properties below this depth. A more precise assessment can be obtained via a sensitivity analysis using the forward solution of Rayleigh wave propagation for the frequency range of the available experimental data.

As a rule of thumb, it is often said that active surface wave methods can resolve the soil profile up to about one-half of the total aperture of the receiver array. The possibility of obtaining information on deep layers is thus limited by logistic aspects and by the necessity of using high-energy sources able to generate surface waves at long offsets. As mentioned previously, passive surface wave tests based on microtremors can be of significant help to extend the depth of penetration.

Although the focus is often on increasing the depth of penetration, the importance of high-frequency information in the solution of the inversion problem should not be overlooked, even if a high level of detail close to the ground surface is not a primary objective of a specific characterization survey. Shallow layers exert an influence on the entire experimental dispersion curve, and a lack of information for shallow layers is propagated throughout the entire inversion process. In practice, the inability to adequately characterize the shallow portion of the soil deposit will result in a biased model for deeper layers. For this reason, it is important to employ a broad frequency range in the experimental test, and combining active and passive surface wave data is a good strategy in this respect.

1.6 TYPICAL APPLICATIONS

The growing use of surface wave methods in practice and research is demonstrated by the increasing number in scientific papers in the geophysical, engineering, and seismological literature (Socco et al. 2010). In the following sections, a brief review of typical applications is presented with special attention to engineering problems.

1.6.1 Site characterization

The primary use of surface wave tests today is to determine the shear wave velocity profile for site characterization purposes. As mentioned in Section 1.1, shear wave velocity is directly linked to the small-strain shear modulus of soils. The shear wave velocity profile is of primary interest for seismic site response studies and for studies of vibration of foundations and vibration transmission in soils. Other applications are related to the prediction of settlements and to soil–structure interaction.

With respect to the evaluation of seismic site response, it is worth noting the affinity between the model used for the interpretation of surface wave tests and the model adopted for most ground response analyses. The application of equivalent linear elastic methods is often associated with 1D, layered models (e.g., the code SHAKE by Schnabel et al. 1972 and its successors). This affinity is also particularly important in the light of equivalence problems, which arise because of the nonuniqueness of the solution in inverse problems. Indeed, profiles that are equivalent in terms of Rayleigh wave propagation are typically also equivalent in terms of seismic amplification (Foti et al. 2009).

Many seismic building codes use the weighted average of the shear wave velocity profile in the upper 30 m of the soil profile ($V_{S,30}$) to discriminate classes of soils that have similar site amplification characteristics. Because it is an average of the properties in the upper 30 m, $V_{S,30}$ can be evaluated very efficiently with surface wave methods because it does not require the high degree of accuracy that can be obtained with seismic borehole methods. Indeed, surface wave tests are more economical to perform because they are noninvasive, and they are often the only possible choice when geological or geotechnical considerations do not allow invasive tests to be used.

Figure 1.13 shows a comparison between a surface wave test and invasive methods for a large number of sites in the United States and Italy that exhibit good agreement in the estimate of $V_{S,30}$ for a wide range of soil deposits with different stiffness.

Measurements of surface wave attenuation provide a means to determine the *in situ* material damping ratio profile of near-surface soils. Frequency-dependent surface wave attenuation coefficients can be determined from measurements of Rayleigh wave amplitudes at various offsets (see Chapter 5). The attenuation curve can be subsequently used in an inversion process with an approach similar to the one used for evaluating the shear wave velocity profile. The two inversions can also be performed simultaneously in a coupled analysis of surface wave dispersion and attenuation (see Chapter 6).

In situ measurements provide the opportunity to assess low-strain material damping ratio free from the adverse effects of specimen disturbance. In particular, surface wave techniques offer several advantages compared to borehole methods for damping measurements. First, the presence of the borehole together with poor coupling between borehole and receiver may adversely affect measurements of particle motion amplitudes. Surface wave tests eliminate these problems because the receivers are on the ground surface where proper soil–receiver coupling can be verified. The frequencies used in surface wave tests are also closer to those of interest in earthquake site response analyses than frequencies used in cross-hole or down-hole tests.

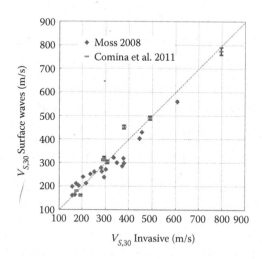

Figure 1.13 Comparison of $V_{S,30}$ determined with invasive tests and surface wave tests at sites in the United States (Moss 2008) and Italy (Comina et al. 2011). For the Italian sites, an estimate of the uncertainty associated with surface wave analysis is provided.

The use of surface wave tests for characterization of municipal solid waste is particularly appealing (Kavazanjian et al. 1996). In such cases, the use of noninvasive methods to obtain the small-strain stiffness and damping is advantageous due to the difficulties and danger associated with the collection of samples and to the realization of boreholes and intrusive testing. Moreover, the advantage of measuring average properties of the waste material is in this case emphasized because of the large variability in particle dimensions (Haegeman and Van Impe 1998; Rix et al. 1998).

1.6.2 Soil improvement

Surface wave tests can be used as a monitoring tool to check the effectiveness of soil improvement techniques at a site by repeated realizations of the test. The comparison of the experimental dispersion curve at different times gives a preliminary and rapid evaluation of the evolution of ground properties, whereas detailed information concerning the effectiveness at different depths is provided by the shear wave velocity profiles obtained via inversion. Successful applications are reported in the literature for monitoring compaction for liquefaction mitigation in sands (Andrus et al. 1998) and for effects of preloading in clays (Raptakis 2012). The relationship between shear wave velocity and soil density can also be used for quality assurance of soil compaction in embankments and fills (Kim et al. 2001).

Similar to the case of soil improvement, applications of surface wave testing to study permafrost evolution in cold regions are reported in the literature. The repetition of the test at different times of the year can yield information about the seasonal variations of soil stiffness caused by freezing cycles (Alkire 1992). The investigation of the permafrost zone is particularly significant for applications related to slope stability and foundation systems.

1.6.3 Nondestructive testing of pavements

The characterization of pavements is a straightforward application of surface wave methods because they are typically horizontally layered media that conform well to the model used in interpretation. Indeed, the early applications of the two-station SASW method were mainly related to pavement systems (Heisey et al. 1982; Nazarian 1984). The primary objectives are quality assurance during construction and evaluating the degradation of the pavement system during its life cycle to optimize maintenance activities.

High-frequency information is necessary because of the scale of the application, and this is usually accomplished from an experimental point of view using accelerometers rather than geophones as receivers. The inversely dispersive nature of pavements (i.e., stiffness decreasing with depth) complicates the inversion procedure (Al-Hunaidi 1998; Ryden et al. 2004; Ryden and Park 2006). Apparent or effective experimental dispersion curves are typically obtained in such situations, and the contribution of higher modes must be appropriately accounted for (see Section 1.5.1). On the contrary, the geometry of the system is often known a priori, which reduces the number of unknown parameters and allows the inversion to focus on obtaining the elastic moduli of the pavement layers.

1.6.4 Offshore and near-shore site characterization

Applications of surface wave testing underwater are based on the generation and detection of Scholte waves, which are the equivalent of Rayleigh waves on the seafloor (i.e., the interface between a fluid layer and elastic half-space). The geometric dispersion of Scholte waves is influenced by the thickness of the fluid layer. The dispersion curve for Scholte waves approaches the Rayleigh wave dispersion curve as the thickness of the fluid layer diminishes. Proper interpretation of marine tests requires the implementation of the forward problem for Scholte wave propagation by modeling a fluid layer above linear elastic layers.

Theoretical and experimental studies have been carried out by researchers at the University of Texas at Austin to assess the possibility of applying the two-station SASW method on the seafloor (Manesh 1991; Luke 1994). In their application, the source and receiver were manually placed on the seafloor (Luke and Stokoe 1998). Obviously this strategy is only viable for shallow water applications.

Many applications are reported in the literature using multistation data acquired from seismic vessels typically used for deep exploration for the oil and gas industry. Sources are typically towed by vessels, and receivers can be embedded in a towed streamer or deployed at the sea bottom via cables or independent stations.

1.6.5 Near-surface characterization in seismic exploration

In seismic reflection surveying, shallow subsurface deposits are responsible for two of the main challenges associated with land applications: perturbations and distortions induced by the near-surface and coherent noise consisting of high-energy, long-duration, near-surface modes.

In large-scale seismic exploration, the shallow subsurface often has large lateral variability. Heterogeneities in the near-surface can be responsible for large distortions of the body-wave wave field, and can impact the final images of the exploration targets and the extracted seismic attributes. In seismic reflection surveying, P-waves must travel twice through this heterogeneous, usually low-velocity portion of the subsurface, downward from sources at the surface to the deep reflectors and then upward to the receivers. The two-way travel time contains the contribution of the near-surface on the source and on the receiver side. Traditionally, the correction of the near-surface perturbations is performed with static corrections derived from near-surface models obtained via refraction data.

The near-surface is also the portion of the subsurface in which the surface waves propagate. The complexity of the surface wave waveform, with dispersion and multiple modes, in fact depends directly on the velocity variations within the subsoil. Surface waves have been traditionally considered as coherent noise to be removed, and they are conventionally referred to as "ground-roll." Mitigation of these effects is achieved with a combination of field acquisition practices, such as the use of source and receiver arrays, and with digital filters during data processing.

Recently, progress in seismic data acquisition and data processing has changed this conventional perspective on surface waves (Strobbia et al. 2011). Broadband, point-receiver data with three-dimensional (3D) geometries offer an optimal sampling of the surface waves. Low-frequency sources and receivers allow the extension of the frequency band down to 1 Hz and, therefore, the acquisition of long wavelengths needed for a meaningful investigation depth. At the same time, finely spaced point receivers provide a dense spatial sampling, enabling the observation of short wavelengths needed for the resolution in the shallow subsurface and for stable surface wave inversion.

Without substantial differences in the general principles, surface wave methods in reflection surveying benefit from the optimal sampling and extreme redundancy of modern 3D geometries. The sampling is not 1D and evenly spaced, and the processing requires advanced approaches, but the number of traces that can be used is very large. With a local array of receivers, thousands of shots can be used. Moreover, powerful land seismic sources generate high-energy surface waves and can create a broad spectrum, spanning two orders of magnitude from 1 to 100 Hz. For these reasons, the analysis can produce accurate, high-resolution data describing the surface wave propagation properties at the scale of interest. The results can yield velocity models of the near-surface over large areas, revealing geological complexity such as structural features, lineaments, and lithological boundaries. Figure 1.14 shows a slice of a dispersion volume, which is the equivalent of a dispersion curve for a 3D survey. It represents the phase velocity as a function of the wavelength (or the frequency) for a specific surface location.

The velocity models obtained from surface wave inversion can be used to compute perturbation corrections or merged with deeper velocity models for depth imaging. The detailed identification of the propagation properties also allows the generation of synthetic model traces of the surface waves, which can be subtracted or adaptively subtracted from the raw data. This model-based approach to the surface wave removal has the advantage of avoiding large, multichannel filters, which can distort useful signal features. Moreover, the model-based approach allows dealing with aliased surface waves.

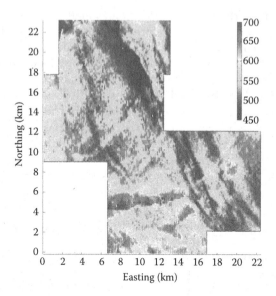

Figure 1.14 Slice of a 3D dispersion volume representing the phase velocity at constant wavelength over a large area.

1.6.6 Anomaly detection

Although estimating the soil stiffness and/or damping profile is the most common objective of surface wave testing, other applications have been proposed that are related to the detection of voids and inclusions in soil deposits. Numerical simulations (typically using the finite element method) and lab-scale tests have been conducted to identify perturbations of the wave field caused by anomalies (Gucunski et al. 1996, 1998; Ganji et al. 1997; Nasseri-Moghaddam et al. 2005). The identification of the anomaly is typically based on the recognition of particular features in the signals (using several signal analysis techniques) rather than on the solutions of a formal inverse problem. Some applications based on pseudo-2D (two-dimensional) modeling are also reported in the literature (Xu et al. 2008).

In large-scale surveying, the spatial distribution of near-surface velocity can be used to identify faults and structural elements (in the geologic meaning). This is useful for the geological modeling and also for drilling safety in hydrocarbon exploration.

1.7 ADVANTAGES AND LIMITATIONS

This last section is devoted to briefly outlining advantages and limitations of surface wave tests. As with any geophysical method, it is important to understand what can and what cannot be achieved.

The first limitation of standard surface wave methods is related to the underlying geophysical model: the assumption of a laterally homogeneous model affects the processing and the inversion. Typically, a stack of linear elastic layers is used as a forward model. A single surface wave test cannot identify lateral variation, and the final result is biased if the soil deposit does not resemble reasonably a 1D, layered medium at the scale of the test. Most of the approaches proposed for the construction of 2D and 3D models from surface wave data are still based on a series of 1D analyses, and as such they should be used with particular care and with a clear understanding of the actual procedure (see Chapter 6). True 2D or 3D testing is possible, but it requires more advanced processing and inversion strategies.

Because inverse problems are mathematically ill-posed, the nonuniqueness of the solution is a limitation. Several profiles that yield theoretical dispersion curves that have a similar distance from the experimental dispersion curve can be identified. This problem is well known as equivalence in geophysical tests based on inverse problems. The implication is a certain degree of uncertainty in the final shear wave velocity profile. For example, surface wave tests are likely not the first choice when the objective is the exact location of an interface between different layers.

The assessment of the uncertainty of the final result is important to evaluate the reliability of the solution. The resolution of the shear wave velocity profile that can be obtained with the surface wave method decreases with depth; thin layers can be resolved if they are close enough to the ground surface, but they are not detected if they are at depth.

Despite their limitations, surface wave methods provide an excellent tool for soil characterization if the overall behavior of the medium is the objective. One important advantage derives from the noninvasive nature of the test: source and receivers are located on the ground surface, which eliminates the need for boreholes. For this reason, they are cost and time effective and can be performed where it is not advisable to invade the medium (e.g., waste materials).

Compared to other seismic techniques, surface wave methods are robust with respect to data acquisition. Compared to shear wave refraction and reflection methods, the acquisition of high-quality surface wave data is easier. Due to the high-energy nature of the surface waves, acquisition is possible even in noisy environments (e.g., urban areas or industrial sites). Other seismic tests based on the evaluation of first arrivals and travel times are more difficult to interpret in the presence of background noise. In surface wave tests, background noise can even be used as a source of information using microtremors surveys.

Surface wave methods are also robust with respect to the model complexity and do not have limitations related to the stratigraphy of a site; they are able to characterize the medium irrespective of the sequence of stiffer (faster) and softer (slower) layers. They work in the case of large contrasts of velocity and in the case of smooth profiles. Refraction-based techniques, also popular in near-surface applications, have intrinsic limitations (velocity inversions, hidden layers) that can result in ambiguities. The joint inversion or interpretation of refracted and surface waves offers many synergies (Foti et al. 2003).

From a geotechnical engineering perspective, large volumes of soils are tested and the test results reflect the overall dynamic behavior of the soil deposit. The degree of accuracy obtained with a surface wave test is typically in line with the assumptions and the simplifications adopted in geotechnical design. Moreover, the 1D model used for the interpretation is also common in many engineering analysis and design procedures (e.g., the code SHAKE for seismic site response analysis). From an exploration geophysics viewpoint, surface waves can be inverted for the near-surface characterization and then removed for the reflection processing. Thus, surface wave tests can be used for evaluating subsurface profiles at many different engineering scales, ranging from pavements to deep soil profiles. They are a flexible tool to explore the subsurface conditions.

Chapter 2

Linear wave propagation in vertically inhomogeneous continua

This chapter illustrates the theory of surface wave propagation in linear elastic and in linear viscoelastic forward modeling. Although the basic theory will be developed for both Love and Rayleigh waves, the main focus will be given to Rayleigh waves because of their greater importance in practical applications of surface waves in exploration geophysics and geotechnical engineering. The chapter is subdivided into five main sections:

Section 2.1 briefly describes some features of two main classes of wave motion, hyperbolic and dispersive waves. Next, the equations of motion in elastic solids and the propagation of body waves in unbounded homogeneous, linear elastic (isotropic) continua are briefly reviewed. This will allow for the introduction of harmonic waves and wave-related parameters.

Section 2.2 shows the existence of Rayleigh waves in linear elastic homogeneous continua are briefly reveiwed. The dispersion relation of Rayleigh waves is derived using the method of potentials of classical elasticity.

Section 2.3 illustrates the conditions for the existence of Love waves propagating in a layer overlying an elastic, homogeneous half-space. The nondispersive features of Rayleigh waves are compared and contrasted with the dispersive properties of Love waves.

Section 2.4 reviews the theory of surface wave propagation in linear elastic, vertically inhomogeneous continua. Well-known results are rederived for Love and Rayleigh waves. An explicit formula for the calculation of the apparent (effective) Rayleigh phase velocity resulting from mode superposition is illustrated together with the geometric spreading function. The classical Lamb's problem is briefly revisited when discussing the solution of the inhomogeneous boundary value problem associated with the propagation of surface waves (i.e., the so-called source problem).

Section 2.5 introduces the theme of wave propagation in dissipative media. After a brief discussion on constitutive modeling of dissipative materials, the main properties of linear viscoelastic body waves are obtained by using the elastic–viscoelastic correspondence principle. Next, the theory of surface wave propagation in linear, vertically heterogeneous dissipative continua is illustrated. Attention is focused on the solution of the Rayleigh eigenproblem in weakly dissipative media. This result forms the basis of the most common procedures used by seismologists and geophysicists to solve surface wave propagation problems in inelastic continua.

2.1 BASIC NOTIONS OF WAVE PROPAGATION

2.1.1 Two categories of wave motion

A wave may be defined as any recognizable disturbance that is transferred from one part of a medium to another with a recognizable velocity of propagation. The disturbance may distort, attenuate, and change its velocity provided it is still recognizable (Whitham 1999). Although there are varieties of mathematical models describing different types of waves, an important classification criterion is the distinction between *hyperbolic* and *dispersive* waves. This discrimination is valid even if restraining the attention to linear waves. Hyperbolic waves are described by hyperbolic partial differential equations and represent the simplest type of wave motion. Their definition is precise, and it depends only on the coefficients of the corresponding equation. It is independent of whether explicit solutions can be found.

The classical, one-dimensional (1D) wave equation exemplifies the paradigm of linear hyperbolic waves

$$\frac{\partial^2 \phi}{\partial x^2} = \frac{1}{c_0^2} \frac{\partial^2 \phi}{\partial t^2} \tag{2.1}$$

where $\phi(x, t)$ is the unknown function; x and t are the spatial and the temporal variables, respectively; and c_0 is the speed of propagation of the signal $\phi(x, t)$. The general integral of Equation 2.1 can be easily obtained after introducing a change of variables, namely $\zeta_1 = x - c_0 t$ and $\zeta_2 = x + c_0 t$, which allows the standard *d'Alembert solution* of the wave equation to be obtained

$$\phi(x, t) = f(x - c_0 t) + g(x + c_0 t) \tag{2.2}$$

where $g(\cdot)$ and $f(\cdot)$ are two, twice differentiable arbitrary functions. Equation 2.2 represents a superposition of two waves moving in opposite directions: $f(x - c_0 t)$ moves to the right, whereas $g(x + c_0 t)$ moves to the left,

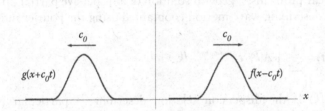

Figure 2.1 Solution of the classical 1D wave equation.

with speed c_0 (Figure 2.1). In fact, $\phi(x + \Delta x, t + \Delta t) = \phi(x, t)$ with $\Delta x / \Delta t = c_0$ and the motion at $x+\Delta x$ is a delayed replica of the motion at x.

A fundamental characteristic of the solution represented by Equation 2.2 is that the two waves propagate at a velocity c_0 without distortion; that is, the wavelets preserve their shape as they propagate through the medium. This is happening because the speed of propagation of linear hyperbolic waves does not depend on frequency; thus, all the waves of any wavenumber or frequency travel at the same phase velocity c_0.

Dispersive waves, on the contrary, have different characteristics. To begin with, they are not classifiable as easily as hyperbolic waves. In fact, it is simpler to describe them by first analyzing the features of the solutions of the corresponding partial differential equation. A linear dispersive wave equation is any equation admitting solutions of the type

$$\phi(x,t) = A \cdot e^{i[kx-\omega(k)t]} \tag{2.3}$$

where A is the amplitude of the wave and the circular frequency $\omega(k)$ is not a constant but is a function of the particular wavenumber k characterizing the wave. The phase speed is the velocity of the wave front, which is the locus of points having constant phase, thus

$$[kx - \omega(k)t] = \text{constant} \tag{2.4}$$

and the phase velocity of the wave is given by

$$c_0 = \frac{dx}{dt} = \frac{\omega(k)}{k} \tag{2.5}$$

which shows that, unless $\omega'(k) \neq 0$, the phase velocity is not a constant but will actually depend on k.

Waves of this type are said to be *dispersive*. An immediate implication of Equation 2.5 is that because waves with different wavenumbers will travel at different speeds, nonmonochromatic signals change shape as they propagate, or they *disperse*.

For linear problems, a general solution of a dispersive partial differential equation describing wave motion is obtained using the Fourier integral

$$\phi(x, t) = \frac{1}{2\pi} \int\limits_{-\infty}^{+\infty} A(k) \cdot e^{i[kx - \omega(k) \cdot t]} \, dk \tag{2.6}$$

where $\omega(k)$ is the dispersion relation describing a particular problem. Equation 2.6 is a superposition of wave trains of different wavenumbers, each traveling with its own phase velocity given by Equation 2.5. As time evolves, the different monochromatic components of Equation 2.6 spread out and, for example, a localized, narrow-band signal disperses as it propagates into a dispersive medium (Figure 2.2).

Table 2.1 reports the dispersion relation $\omega = \omega(k)$ for linear water waves for three cases of deep water approximation, shallow water, and intermediate depth. In the formulas, g is the acceleration of gravity, h is the undisturbed depth, and λ is the wavelength.

Figure 2.3 shows the plot of the dispersion curves for $h = 10$ m. As expected for small wavenumbers (i.e., large wavelengths), the shallow water approximation fits well with the rigorous solution (intermediate depth). The same situation occurs for large wavenumber values (i.e., short wavelengths) for the deep water approximation.

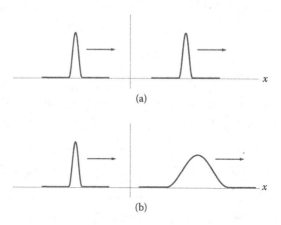

Figure 2.2 Propagation of a signal in a (a) nondispersive and (b) dispersive medium.

Table 2.1 Dispersion relations of linear water waves

Deep water (h > 0.5 λ)	Shallow water (h < 0.05 λ)	Intermediate depth (all λ and h)
$\omega = \pm\sqrt{gk}$	$\omega = \pm k\sqrt{gh}$	$\omega = \pm\sqrt{gk \cdot \tanh kh}$

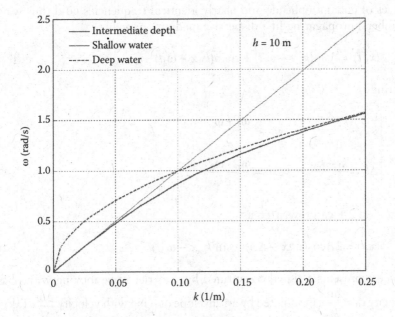

Figure 2.3 Dispersion curves associated with linear water waves.

The categories of hyperbolic and dispersive waves are not mutually exclusive. There is some degree of intersection in that there are dispersive waves that are also solutions of hyperbolic partial differential equations.

As will be discussed thoroughly in this chapter, body waves propagating in linear elastic, isotropic media are nondispersive. However, the same type of waves propagating in dissipative or multicomponent (e.g., porous) media is dispersive. *Material dispersion* should be distinguished from *geometric dispersion*—a phenomenon by which, in inhomogeneous continua, the phase velocity of surface waves is a multivalued function of the frequency of excitation and is responsible for the existence of several modes of propagation. In Section 2.2.2, it will be demonstrated that surface Rayleigh waves propagating along the free surface of a homogeneous half-space are nondispersive.

2.1.2 Group velocity

A key concept that comes out quite naturally from the study of dispersive waves is that of *group velocity*, which is formally defined as

$$c_g = \frac{d\omega(k)}{dk} = c_0 + k\frac{dc_0}{dk} \tag{2.7}$$

This definition was first introduced by Stokes (1880), who considered the following argument: let $\phi(x, t)$ denote the superposition of two monochromatic waves of equal amplitude and nearly identical frequencies (and thus wave-numbers) propagating in a dispersive medium, namely

$$\phi(x, t) = A\sin(k_1 x - \omega_1 t) + A\sin(k_2 x - \omega_2 t) \tag{2.8}$$

Setting

$$\begin{cases} k_m = \dfrac{k_1 + k_2}{2} & \omega_m = \dfrac{\omega_1 + \omega_2}{2} \\[2mm] \Delta k = \dfrac{k_1 - k_2}{2} & \Delta\omega = \dfrac{\omega_1 - \omega_2}{2} \end{cases} \tag{2.9}$$

Equation 2.8 can rewritten as

$$\phi(x,t) = 2A\cos(\Delta k x - \Delta\omega t) \cdot \sin(k_m x - \omega_m t) \tag{2.10}$$

which represents a signal constituted by a carrier wave moving with phase velocity $c_0 = \dfrac{\omega_m}{k_m}$, modulated by an envelope moving with velocity $\dfrac{\Delta\omega}{\Delta k}$. Taking the limit of this ratio for $\Delta k \to 0$ yields Equation 2.7 for the group velocity c_g. Figure 2.4 shows a plot of the waveform represented by Equation 2.10 for $A = 1$, $\Delta k = 10^{-1}$, $\Delta\omega t = 10^{-1}$, $k_m = 2$, and $\omega_m t = 1$.

This reasoning can be generalized by considering the superposition of more than two monochromatic waves having almost identical frequencies. This would generate a localized waveform rather than a succession of waveforms (Figure 2.4). However, the conclusion would still be the same,

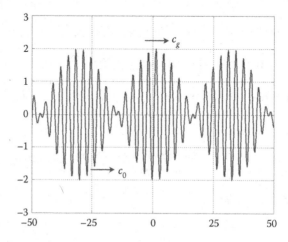

Figure 2.4 Distinction between phase (c_0) and group (c_g) velocity.

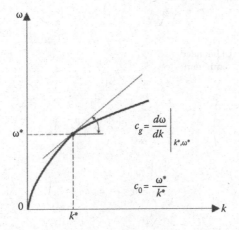

Figure 2.5 Frequency spectrum in the ω–k plane and interpretation of c_0 and c_g.

with a carrier wave moving with phase velocity c_0 and the modulation with group velocity c_g. It can also be shown that c_g also represents the velocity at which the energy associated with an arbitrary wave motion propagates in a medium (Achenbach 1984).

Depending on the properties of a particular dispersive system, phase velocity c_0 may be greater or less than the group velocity c_g.[01] If $c_0 > c_g$ the carrier wave will appear to originate at the rear of the group, travel to the front, and disappear. This is a common situation, and a good example is represented by linear water waves in deep ocean. In the case of $c_0 < c_g$, the carrier wave will appear to originate at the front of the group, travel to the rear, and then disappear (Graff 1975).

Phase and group velocities admit interesting geometrical interpretations as is illustrated in Figure 2.5. The plot of the dispersion relation $\omega = \omega(k)$ is called the *frequency spectrum*. The group velocity c_g is obtained by differentiation of the dispersion curve.

2.1.3 Body waves in unbounded, homogeneous, linear elastic, isotropic continua

The essential features of wave propagation in unbounded, homogeneous, linear elastic, isotropic continua are briefly reviewed in this section. This will allow the introduction of the existence of body P- and S-waves, which are associated with volumetric and distortional modes of deformation, respectively.

To derive the Navier's equations of motion of linear elastodynamics, consider the equilibrium, in the sense of d'Alembert, of an infinitesimal

---

[01] In a nondispersive system, $c_0 = c_g$ and the signal propagates with no distortion.

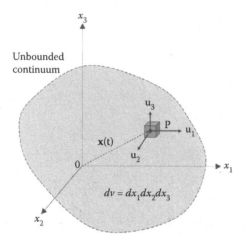

Figure 2.6 Displacement components of an infinitesimal volume element *dv* in an unbounded, homogeneous, linear elastic, isotropic continuum.

volume element *dv* surrounding an arbitrary point P (Figure 2.6) of an unbounded continuum subjected to a dynamic excitation. Let u_j ($j = 1,3$) denote the components of the displacement vector $\mathbf{u}(\mathbf{x}(t))$, where $\mathbf{x}(t)$ is the position vector in Cartesian coordinates and t is time.

Imposing the dynamic equilibrium in the absence of body forces of the volume element *dv* yields, in direct (or Gibbs) notation

$$div\, \sigma = \rho \frac{\partial^2 \mathbf{u}}{\partial t^2} \tag{2.11}$$

where σ is the Cauchy stress tensor, $div(\cdot)$ is the *divergence* differential operator, and ρ is the mass density, which is assumed constant with time.

Assuming the validity of the small-strain theory, *Hooke's law* of linear elasticity applied to an isotropic body can be written as follows

$$\sigma = \lambda tr(\varepsilon)\mathbf{1} + 2\mu\varepsilon \tag{2.12}$$

where $tr(\cdot)$ is the *trace* matrix operator, $\mathbf{1}$ is the identity tensor, λ and μ are the *Lamé's* elastic constants (μ is the shear modulus), and ε is the infinitesimal strain tensor, which is defined as

$$\varepsilon = \frac{1}{2}(grad\, \mathbf{u} + (grad\, \mathbf{u})^T) \tag{2.13}$$

where $grad(\cdot)$ is the *gradient* differential operator and $(\cdot)^T$ is the transpose matrix operator. In general, Lamé's parameters and mass density are functions of the coordinates, namely $\lambda = \lambda(\mathbf{x})$, $\mu = \mu(\mathbf{x})$, and $\rho = \rho(\mathbf{x})$. However, for a homogeneous medium, the elastic moduli and material density are invariant in space. Substituting Equation 2.13 into Equation 2.12 yields

$$\sigma = 1 \cdot \lambda div \ \mathbf{u} + \mu \ (grad \ \mathbf{u} + (grad \ \mathbf{u})^T) \tag{2.14}$$

Differentiation of Equation 2.14 with respect to x and substitution into Equation 2.11 finally allow obtaining

$$\mu \nabla^2 \mathbf{u} + (\lambda + \mu) grad \ div \ \mathbf{u} = \rho \frac{\partial^2 \mathbf{u}}{\partial t^2} \tag{2.15}$$

which are the *Navier's displacement equations of motion* for homogeneous, isotropic, linear elastic continua in absence of body forces written in direct notation where $\nabla^2(\cdot)$ denotes the *Laplacian* differential operator in Cartesian coordinates.

Applying the *divergence* and *curl* operators to Equation 2.15 yields the following two wave equations

$$\begin{cases} (\lambda + 2\mu)\nabla^2(div \ \mathbf{u}) = \rho \cdot \dfrac{\partial^2}{\partial t^2} div \ \mathbf{u} \\[2mm] \mu \nabla^2(curl \ \mathbf{u}) = \dfrac{\partial^2}{\partial t^2} curl \ \mathbf{u} \end{cases} \tag{2.16}$$

which shows that in linear elastic isotropic media, *volumetric* (associated with divergence of displacement) and *distortional* deformations (associated with curl of displacement) are uncoupled. Alternatively, Equation 2.16 could have been obtained from Equation 2.15 by using the Helmoltz's decomposition theorem (Achenbach 1984).

Relationships (Equation 2.16) are two linear, second-order, hyperbolic, constant coefficients, partial differential equations. On inspection, it can be immediately recognized that they represent two standard wave equations for *div* u and *curl* u with speeds of propagation given by the following relations

$$\begin{cases} V_P = \sqrt{\dfrac{\lambda + 2\mu}{\rho}} \\[4mm] V_S = \sqrt{\dfrac{\mu}{\rho}} \end{cases} \tag{2.17}$$

where V_P represents the velocity of propagation of *longitudinal* (or irrotational or dilatational or compressional) waves (also called *primary* or *P-waves* in seismology because they are the fastest of body waves). The value is proportional to the square root of the ratio of the oedometric

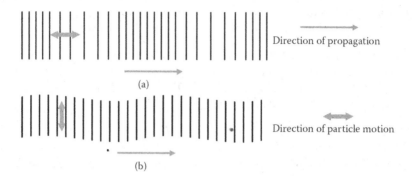

Figure 2.7 Deformation pattern generated by the passage of (a) longitudinal and (b) transversal waves in a linear elastic, homogeneous, isotropic, unbounded medium.

or constrained modulus ($\lambda + 2\mu$) and mass density ρ. Longitudinal waves propagate in fluids and in solids, and at their passage, the material undergoes a deformation constituted by alternate patterns of contractions and dilations along the direction of propagation, which is parallel to the direction of polarization or particle motion (Figure 2.7).

Transversal (or distortional, equivoluminal, solenoidal, or shear) waves, (also called *secondary* or *S-waves* in seismology), propagate at a speed that is proportional to the square root of the ratio of the shear modulus μ and mass density ρ. Transversal waves cannot be transmitted in (perfect) fluids because they induce shear stresses in the material where they propagate. For transversal waves, the direction of propagation is orthogonal to the direction of polarization (Figure 2.7). In homogeneous, linear elastic, isotropic continua, P- and S-waves are nondispersive because their speed of propagation is frequency independent.

The ratio of V_P and V_S can be expressed solely as a function of the Poisson ratio ν

$$\left(\frac{V_P}{V_S}\right)^2 = \frac{\lambda + 2\mu}{\mu} = \frac{2(1-\nu)}{(1-2\nu)} > 1 \tag{2.18}$$

which shows that it is always $V_S < V_P$. For $\nu = 1/4$ (a typical value for several materials), $V_P = \sqrt{3} \cdot V_S$. Equation 2.18 allows the computation of V_S if the values of V_P and ν are known. However, this approach to estimate V_S should be avoided in geomaterials—particularly in water-saturated soils—because it may lead to gross errors. This happens because under undrained conditions (occurring when the water cannot move relative to the soil skeleton), the speed of propagation of P-waves in water-saturated porous materials is mainly controlled by the compressibility of water, which is very low. Under these circumstances, the Poisson ratio approaches 0.5 (a value ideally reached by incompressible materials) and the velocity of propagation of P-waves would tend to be infinite (Figure 2.8).

The values of V_P and V_S in soils and rocks vary considerably. At depths of only a few kilometers from the Earth's surface, the values of V_P are typically in the range of 6000 to 7000 m/s. Table 2.2 shows typical values of V_P and V_S for near-surface, dry geomaterials.

As mentioned earlier, water-saturated soil deposits exhibit values of V_P that are much larger if compared with the corresponding V_P associated with dry (i.e., above the water table) materials. Typical values of V_P in fully

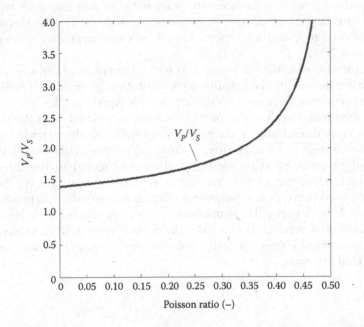

Figure 2.8 Functional dependence of the ratio between the speed of propagation of longitudinal (V_P) and transversal (V_S) waves and Poisson ratio ν.

Table 2.2 Typical values of V_P and V_S for near-surface dry geomaterials

Geomaterial	V_P (m/s)	V_S (m/s)	Poisson ratio
Crystalline rocks	4000÷6500	2500÷3500	0.20÷0.30
Calcareous, fractured rocks	1600÷3000	1000÷1500	0.20÷0.30
Soft rocks, very dense gravels	800÷2000	500÷1000	0.20÷0.30
Medium to dense gravels	650÷1500	400÷800	0.20÷0.30
Medium to dense sands	350÷750	200÷400	0.20÷0.30
NC clays and silts	250÷500	150÷300	0.15÷0.25
Very soft clays	80÷200	50÷100	0.15÷0.25

saturated alluvial soils are in the order of 1500 m/s, which is close to the speed of propagation of sound in fresh water.

A harmonic solution of Equation 2.16 may be written as follows

$$\mathbf{u}(\mathbf{x}, t) = \mathbf{A}_1 \exp[i(\omega t - \mathbf{k}_\chi \cdot \mathbf{x})] + \mathbf{A}_2 \exp[i(\omega t + \mathbf{k}_\chi \cdot \mathbf{x})] \tag{2.19}$$

where $i = \sqrt{-1}$, $\mathbf{A}_1$ and $\mathbf{A}_2$ are two arbitrary constant vectors to be determined from boundary conditions, $\chi = P, S$ is a subscript denoting either longitudinal or transversal wave motion. Finally, $\mathbf{k}_\chi$ is the vector wavenumber, which defines the direction of propagation for the χ-wave. Equation 2.19 represents a general monochromatic wave with circular frequency ω. This solution is important because in combination with the Fourier theorem it can be used to construct a more general (i.e., nonharmonic) solution for Equation 2.16.

It can be shown that the vector $\mathbf{k}_\chi$ is normal to planes of constant phase, which are defined by the equation $\mathbf{k}_\chi \cdot \mathbf{x} =$ constant (Achenbach 1984). The *phase velocity* of the monochromatic χ-wave is equal to $\omega/|\mathbf{k}_\chi|$.

As mentioned earlier, the particle motion associated with P-waves is in the same direction of wave propagation, whereas the particle motion associated with S-waves is along a direction perpendicular to the direction of propagation. Considering a plane orthogonal to the direction of propagation, the particle motion associated with an S-wave can be decomposed into two components that are mutually perpendicular (Figure 2.9). A particular situation is the one in which the direction of propagation is vertical. In this case, the particle motion of an S-wave can be decomposed into a vertically polarized SV-wave and a horizontally polarized SH-wave.

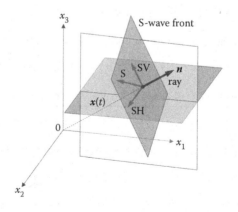

Figure 2.9 Polarization of an S-wave into the SV and SH transversal components of particle motion.

In general, due to Snell's law, an incident P- or SV-wave hitting an interface generates both reflected (and refracted or transmitted) P- and SV-waves whereas an incident SH-wave generates only reflected (and refracted or transmitted) SH-waves (Figure 2.10). The *mode conversion* of P- and SV-waves has important implications at the free surface of a half-space (a stress-free boundary condition interface) for the formation of surface Rayleigh waves (with particle motion in the vertical plane) as will be shown later in this chapter. On the contrary, Love waves, which are another type of surface wave, are horizontally polarized SH-waves.

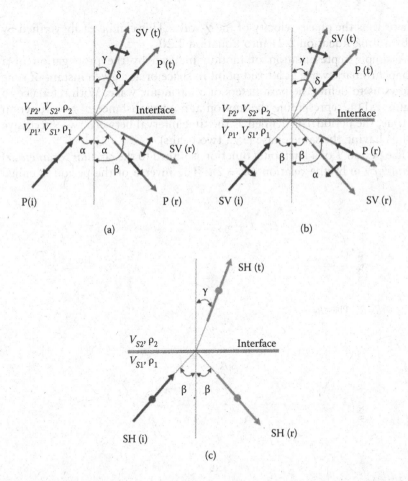

Figure 2.10 Incidence (*i*), reflection (*r*), transmission (*t*), and mode conversion of P-, SV-, and SH-waves at an interface between two elastic media according to Snell law $\dfrac{\sin\alpha}{V_{PI}} = \dfrac{\sin\beta}{V_{SI}} = \dfrac{\sin\delta}{V_{P2}} = \dfrac{\sin\gamma}{V_{S2}}$: (a) incident P-wave, (b) incident SV-wave, (c) incident SH-wave.

If Equation 2.19 is specialized for 1D wave propagation, all vectors degenerate into corresponding scalar quantities yielding the following relation

$$u(x,t) = A_1 \exp[i(\omega t - k_\chi \cdot x)] + A_2 \exp[i(\omega t + k_\chi \cdot x)] \tag{2.20}$$

where the scalar wavenumber k_χ associated with the propagation of the χ-wave is defined by

$$k_\chi = \frac{\omega}{V_\chi} \tag{2.21}$$

where V_χ is the phase velocity of the χ-wave. This can be easily verified by substituting Equation 2.21 into Equation 2.20.

A simple representation of motion induced by the propagation of a monochromatic wave at a fixed point in space or at a given instant of time allows us to define the parameters of a harmonic wave. With reference to Figure 2.11a, representing the motion at a given distance x^* as a function of time, the period T is defined as the time interval between two successive points having the same phase (e.g., two peaks).

The period T of a harmonic function is linked to the *circular* (or *angular*) *frequency* ω by the relation $\omega T = 2\pi$. The inverse of the period is called

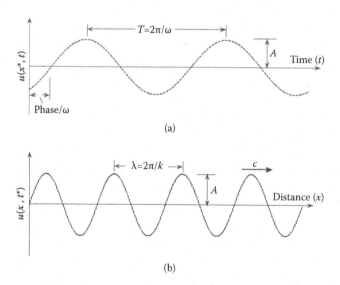

(a)

(b)

Figure 2.11 Displacement profile associated with the propagation of a harmonic wave (a) as a function of time at a fixed point in space and (b) as a function of distance at a given instant of time.

Table 2.3 Definitions and units for wave parameters of monochromatic waves

Meaning	Symbol	Quantity	Dimensions	SI Units
Temporal parameters	T	Period	[time]	[s]
	f	Cyclic frequency	[1/time]	[Hz = 1/s]
	ω	Circular frequency	[1/time]	[rad/s]
Spatial parameters	λ	Wavelength	[length]	[m]
	v	Cyclic wavenumber	[1/length]	[1/m]
	k	Circular wavenumber	[1/length]	[rad/m]
Other parameters	c	Phase velocity	[length/time]	[m/s]
	A	Amplitude	Any unit	Any unit
Relations		$c = \omega/k = \lambda/T = \lambda f = f/v = (1/T)\cdot(1/v)$		
		$\omega = 2\pi f = 2\pi/T \qquad k = 2\pi v = 2\pi/\lambda$		

cyclic frequency or simply *frequency*, and it is related to the circular frequency by the relationship $\omega = 2\pi f$. With reference to Figure 2.11b, showing a snapshot of motion at a given instant of time t^*, the distance in space between any two points having the same phase is referred to as the *wavelength* and is linked to the *wavenumber* k by the expression $k\lambda = 2\pi$. Combining this relation with Equation 2.21 yields

$$\lambda_\chi = \frac{2\pi}{k_\chi} = \frac{V_\chi}{f} \tag{2.22}$$

which is an important relation linking wave parameters. Table 2.3 summarizes a few definitions, symbols, and relationships for (harmonic) wave parameters that will be used systematically throughout this book.

2.2 RAYLEIGH WAVES IN HOMOGENEOUS ELASTIC HALF-SPACES

2.2.1 Overview

Unbounded, elastic, isotropic continua can only support the propagation of longitudinal and transversal (body) waves. However, the introduction of a boundary in a continuum gives rise to the existence of other types of waves named *surface waves* because they propagate along the boundary of a deformable body rather than in its interior.

The first scientist who predicted the existence of surface waves in elastic solids was Lord Rayleigh in 1885 (Whitham 1999), in whose honor they were named. Surface waves originate from the condition of vanishing

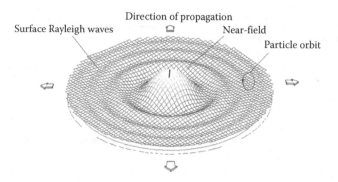

Figure 2.12 Two-dimensional radiation pattern of surface Rayleigh waves generated by a vertical point source.

stress at a boundary of a domain. Their radiation pattern (Figure 2.12) is essentially two-dimensional (2D) and is characterized by a much lesser rate of geometric attenuation than body waves the energy of which spreads in horizontal and vertical directions.

In a homogeneous, elastic half-space, Rayleigh surface waves generated by a line vertical source do not suffer any geometric attenuation, whereas for a vertical point load, the rate of spatial decay is proportional to the inverse of the square root of the distance from the source. In contrast, for the same geometry of the source, the geometric attenuation factor of body waves propagating along the boundary of an elastic half-space is proportional to the inverse of the square of the distance (Ewing et al. 1957). Thus, at distances in the order of one to two wavelengths from the source, the contribution of body waves becomes negligible and the wave field is dominated by Rayleigh waves. Lamb (1904) was among the first to recognize this fundamental property of surface waves and its implications in the transmission of earthquake energy at large distances.

In the direction orthogonal to the direction of propagation, the displacement field generated by a surface wave decays exponentially because no energy is propagated in the interior of the half-space. As a matter of fact, this property is often used as a definition of surface wave. It can be shown that most of the strain energy associated with surface wave motion is confined within a depth of about a wavelength from the free boundary (Achenbach 1984). One means of describing this property is through the concept of *skin depth*, the depth at which the amplitude decreases by a factor of 1/e. For Rayleigh waves in a homogeneous medium, the skin depth is approximately 0.94λ. Hence, Rayleigh waves of larger wavelengths penetrate deeper into the interior of a medium. The opposite is true for shorter wavelengths as shown in Figure 2.13.

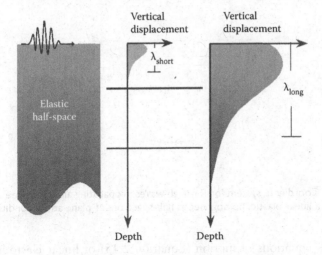

Figure 2.13 Dependence of skin depth from wavelength in a Rayleigh surface wave.

2.2.2 Dispersion relation of Rayleigh waves

To investigate the existence of Rayleigh waves, it is instructive to consider the propagation of 2D plane waves along the free surface (x_1 direction) of a linear elastic, homogeneous half-space, characterized by the following displacement field (Figure 2.14)

$$
\begin{cases}
u_1 = C_1 \cdot e^{-\alpha x_2} \cdot e^{ik(x_1 - ct)} \\
u_2 = C_2 \cdot e^{-\alpha x_2} \cdot e^{ik(x_1 - ct)} \\
u_3 = 0
\end{cases}
\tag{2.23}
$$

where k and c are the wavenumber and the speed of propagation of the plane wave, respectively; α is a constant, which may assume complex values. However, the real part of α is assumed to be greater than zero so that the displacement components, u_1 and u_2, decrease exponentially with the increase of x_2, so as to represent a surface wave. C_1 and C_2 are arbitrary constants to be determined from boundary conditions.

Equation 2.23 corresponds to a plane strain field with motion confined in x_1–x_2 plane (Figure 2.14). This assumption does not imply any loss of generality because it can be shown that cylindrical Rayleigh waves (which may be, for instance, those generated by a vertically oscillating concentrated point load) share the same x_2-dependence indicated by Equation 2.23 (Aki and Richards 2002).

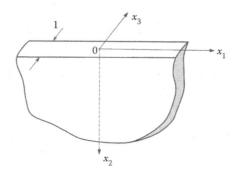

Figure 2.14 Coordinate system for Rayleigh waves propagating along the free surface of a linear elastic, homogeneous half-space under plane strain conditions.

Navier's equations of motion (Equation 2.15) of linear elastodynamics (Section 2.1.3) applied to a plane strain field can be successfully uncoupled using the Helmoltz's decomposition theorem (Achenbach 1984) for the unknown vector displacement field $\mathbf{u}(\mathbf{x}, t)$

$$u = grad\,\phi + curl\,\mathbf{\Psi} \tag{2.24}$$

where ϕ and $\mathbf{\Psi}$ are a scalar and a vector displacement potential function, respectively. For plane strain conditions, $u_3 = 0$ and the Helmoltz's decomposition simplifies as follows

$$\begin{cases} u_1 = \dfrac{\partial \phi}{\partial x_1} - \dfrac{\partial \psi}{\partial x_2} \\[2mm] u_2 = \dfrac{\partial \phi}{\partial x_2} + \dfrac{\partial \psi}{\partial x_1} \end{cases} \tag{2.25}$$

where $\mathbf{\Psi} = [0\ 0\ \psi_3 = \psi]$. If Equation 2.25 is substituted into Equation 2.15, the result is the following pair of linear, uncoupled, hyperbolic partial differential equations

$$\begin{cases} \dfrac{\partial^2 \phi}{\partial x_1^2} + \dfrac{\partial^2 \phi}{\partial x_2^2} = \dfrac{1}{V_P^2} \cdot \dfrac{\partial^2 \phi}{\partial t^2} \\[3mm] \dfrac{\partial^2 \psi}{\partial x_1^2} + \dfrac{\partial^2 \psi}{\partial x_2^2} = \dfrac{1}{V_S^2} \cdot \dfrac{\partial^2 \psi}{\partial t^2} \end{cases} \tag{2.26}$$

Now, the scalar displacement potentials ϕ and ψ are chosen in such a way that they are compatible with the wave field represented by Equation 2.23, namely

$$\begin{cases} \phi = \Phi(x_2).e^{ik(x_1-ct)} \\ \psi = \Psi(x_2).e^{ik(x_1-ct)} \end{cases} \tag{2.27}$$

Substituting Equation 2.27 into Equation 2.26 yields

$$\begin{cases} \dfrac{d^2\Phi}{dx_2^2} + \left(\dfrac{\omega^2}{V_P^2} - k^2 \right) \cdot \Phi(x_2) = 0 \\ \dfrac{d^2\Psi}{dx_2^2} + \left(\dfrac{\omega^2}{V_S^2} - k^2 \right) \cdot \Psi(x_2) = 0 \end{cases} \tag{2.28}$$

which is a pair of linear, second-order, constant coefficients, ordinary differential equations and where $\omega = k \cdot c$. If $p^2 = \dfrac{\omega^2}{V_P^2} - k^2 > 0$ and $q^2 = \dfrac{\omega^2}{V_S^2} - k^2 > 0$[02] are set in Equation 2.28, the solution of these equations would not correspond to a depth dependence of a surface wave because it would be given by a linear combination of harmonic functions. However, if $p < 0$ and $q < 0$, then the solution of Equation 2.28 becomes

$$\begin{cases} \Phi(x_2) = A_3 \cdot e^{-px_2} + A_4 \cdot e^{px_2} \\ \Psi(x_2) = B_3 \cdot e^{-qx_2} + B_4 \cdot e^{qx_2} \end{cases} \tag{2.29}$$

which shows the exponential decaying modal shape typical of a surface wave.[03]

Substituting Equation 2.29 into Equation 2.27 yields

$$\begin{cases} \phi = \Phi(x_2) \cdot e^{ik(x_1-ct)} = \left(A_3 \cdot e^{-px_2} + A_4 \cdot e^{px_2} \right) \cdot e^{i(kx_1-\omega t)} \\ \psi = \Psi(x_2) \cdot e^{ik(x_1-ct)} = \left(B_3 \cdot e^{-qx_2} + B_4 \cdot e^{qx_2} \right) \cdot e^{i(kx_1-\omega t)} \end{cases} \tag{2.30}$$

---

[02] Note that because $V_P > V_S$, the assumption $p > 0$ implies necessarily $q > 0$.
[03] In reality, this would require the disappearance of the growing terms $A_3 \cdot e^{-px_2}$ and $B_3 \cdot e^{-qx_2}$; however, this issue will be thoroughly discussed shortly when introducing the boundary conditions the solution must satisfy.

The displacement field is finally obtained by substituting Equation 2.30 into Equation 2.24. The result is

$$
\begin{cases}
u_1 = \dfrac{\partial \phi}{\partial x_1} - \dfrac{\partial \psi}{\partial x_2} = \left[\left(A_3 \cdot e^{-px_2} + A_4 \cdot e^{px_2} \right) \cdot ik + \left(B_3 \cdot e^{-qx_2} - B_4 \cdot e^{qx_2} \right) \cdot q \right] \cdot e^{i(kx_1 - \omega t)} \\[3mm]
u_2 = \dfrac{\partial \phi}{\partial x_2} + \dfrac{\partial \psi}{\partial x_1} = \left[\left(A_4 \cdot e^{px_2} - A_3 \cdot e^{-px_2} \right) \cdot p + \left(B_3 \cdot e^{-qx_2} + B_4 \cdot e^{qx_2} \right) \cdot ik \right] \cdot e^{i(kx_1 - \omega t)}
\end{cases}
$$

$$(2.31)$$

The constants A_3, A_4 and B_3, B_4 appearing in Equation 2.31 are determined from the boundary conditions, which are the stress-free conditions at the ground surface of the half-space and the radiation (or *Sommerfeld* or source free) condition as $x_2 \to \infty$. The latter may be expressed by the following equations

$$
\begin{cases}
u_1 \to 0 \\
u_2 \to 0
\end{cases}
\quad \text{as } x_2 \to \infty
\tag{2.32}
$$

which implies that in Equation 2.31, the constants A_3 and B_3 must vanish, yielding

$$
\begin{cases}
u_1 = \left(ik \cdot A_4 \cdot e^{-rx_2} + s \cdot B_4 \cdot e^{-sx_2} \right) \cdot e^{i(kx_1 - \omega t)} \\[2mm]
u_2 = \left(ik \cdot B_4 \cdot e^{-sx_2} - r \cdot A_4 \cdot e^{-rx_2} \right) \cdot e^{i(kx_1 - \omega t)}
\end{cases}
\tag{2.33}
$$

where $r^2 = k^2 - \dfrac{\omega^2}{V_P^2}$ and $s^2 = k^2 - \dfrac{\omega^2}{V_S^2}$.

The stress-free boundary condition at the ground surface of the half-space may be expressed by the following equations

$$
\begin{cases}
\sigma_{21} = 0 \\
\sigma_{22} = 0
\end{cases}
\quad \text{at } x_2 = 0
\tag{2.34}
$$

When Hooke's law $\sigma_{ij} = \lambda \varepsilon_{kk} \delta_{ij} + 2\mu\varepsilon_{ij}$ ($\delta_{ij} = 0$ if $i \ne j$ and $\delta_{ij} = 1$ if $i = j$) is explicitly written for the stress components, σ_{21} and σ_{22} yields

$$
\begin{cases}
\sigma_{22} = \lambda \dfrac{\partial u_1}{\partial x_1} + (\lambda + 2\mu)\dfrac{\partial u_2}{\partial x_2} \\[3mm]
\sigma_{12} = \sigma_{21} = \mu \left(\dfrac{\partial u_1}{\partial x_2} + \dfrac{\partial u_2}{\partial x_1} \right)
\end{cases}
\tag{2.35}
$$

which after substituting Equation 2.33 becomes

$$\begin{cases} \sigma_{22} = \lambda\left(ikA_4 e^{-rx_2} + sB_4 e^{-sx_2}\right) \cdot ike^{i(kx_1 - \omega t)} + (\lambda + 2\mu) \cdot \left[r^2 A_4 e^{-rx_2} - iksB_4 e^{-sx_2}\right] \cdot e^{i(kx_1 - \omega t)} \\ \sigma_{12} = \mu\left[\left(-ikrA_4 e^{-rx_2} - s^2 B_4 e^{-sx_2}\right) \cdot e^{i(kx_1 - \omega t)}\right] + \left(ikB_4 e^{-sx_2} - rA_4 e^{-rx_2}\right) \cdot e^{i(kx_1 - \omega t)} \end{cases}$$

$$(2.36)$$

The stress-free boundary condition requires that at $x_2 = 0$

$$\begin{cases} \sigma_{22} = \left[\lambda\left(ikA_4 + sB_4\right) \cdot ik + (\lambda + 2\mu) \cdot \left(r^2 A_4 - iksB_4\right)\right] \cdot e^{i(kx_1 - \omega t)} = 0 \\ \sigma_{12} = \mu\left[\left(-ikrA_4 - s^2 B_4\right) + \left(ikB_4 - rA_4\right) \cdot ik\right] \cdot e^{i(kx_1 - \omega t)} = 0 \end{cases}$$

$$(2.37)$$

Equation 2.37 can now be rearranged as a homogeneous system of two linear algebraic equations in two unknowns as follows

$$\begin{cases} \left[r^2\left(\lambda + 2\mu\right) - \lambda k^2\right] \cdot A_4 + \left[\lambda sik - (\lambda + 2\mu)iks\right] \cdot B_4 = 0 \\ \left[-2rik\right] \cdot A_4 - \left[k^2 + s^2\right] \cdot B_4 = 0 \end{cases}$$

$$(2.38)$$

which can be written in matrix form:

$$\begin{bmatrix} \left[r^2\left(\lambda + 2\mu\right) - \lambda k^2\right] & -2iks \cdot \mu \\ -2rik & -\left(k^2 + s^2\right) \end{bmatrix} \begin{bmatrix} A_4 \\ B_4 \end{bmatrix} = \begin{bmatrix} 0 \\ 0 \end{bmatrix}$$

$$(2.39)$$

A nontrivial solution to this system of equations exists if the determinant of the coefficients vanishes, leading to the well-known *Rayleigh dispersion equation*

$$-\left(k^2 + s^2\right) \cdot \left[r^2\left(\lambda + 2\mu\right) - \lambda k^2\right] + 4k^2 \mu rs = 0 \qquad (2.40)$$

Recalling that $r^2 = k^2 - \dfrac{\omega^2}{V_P^2}$ and $s^2 = k^2 - \dfrac{\omega^2}{V_S^2}$, Equation 2.40 becomes

$$4k^2 \mu \left[\left(k^2 - \frac{\omega^2}{V_P^2}\right) \cdot \left(k^2 - \frac{\omega^2}{V_S^2}\right)\right]^{1/2} - \left[2k^2 - \frac{\omega^2}{V_S^2}\right] \cdot \left[\left(k^2 - \frac{\omega^2}{V_P^2}\right) \cdot (\lambda + 2\mu) - \lambda k^2\right] = 0$$

$$(2.41)$$

which, after the substitution $k = \omega/c = \omega/V_R$ and recalling Equation 2.17, can finally be rearranged as follows

$$\left(\frac{V_R}{V_S}\right)^6 - 8\left(\frac{V_R}{V_S}\right)^4 + 8\left(\frac{V_R}{V_S}\right)^2 \cdot \left[1 + 2\cdot\left(1 - \frac{V_S^2}{V_P^2}\right)\right] - 16\left(1 - \frac{V_S^2}{V_P^2}\right) = 0 \qquad (2.42)$$

In Equation 2.42, V_R represents the speed of propagation of a wave moving along the free surface of the half-space (Figure 2.14); therefore, it is a *surface wave*, which has been named *Rayleigh wave* in honor of its discoverer. From Equation 2.42, V_R depends solely on V_P and V_S, which are intrinsic, frequency-independent properties of the medium. Thus, V_R does not exhibit any dependence upon the wavenumber k, and Rayleigh waves in a linear elastic, homogeneous, isotropic half-space are *nondispersive* (that is, their speed of propagation V_R is independent of frequency). This happens because a homogeneous half-space does not possess an intrinsic length scale (Aki and Richards 2002).

Using Equation 2.18, it is easy to show that Equation 2.42 can be rewritten with the ratio V_R/V_S expressed solely as a function of the *Poisson ratio* ν of the medium. Figure 2.15 shows a plot of the variation of V_R/V_S with ν.

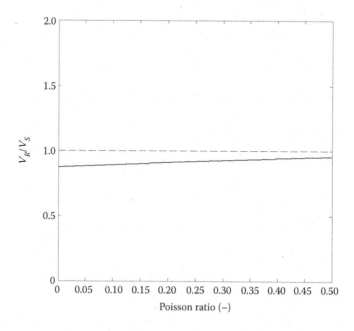

Figure 2.15 Variation of the ratios V_R/V_S with Poisson ratio. For ν varying from 0 to 0.5, the ratio V_R/V_S increases from 0.862 to 0.955.

The dependence upon v exhibited by the ratio V_R/V_S is rather weak. As v varies from 0 to 0.5, V_R/V_S increases from 0.862 to 0.955. For the special case of $v = 1/4$, $V_P/V_S = \sqrt{3}$ and Equation 2.42 yields $V_R/V_S = \sqrt{2 - 2/\sqrt{3}} \approx 0.92$.

Equation 2.42 is an algebraic equation of sixth degree in V_R/V_S and a question arises about the number of its roots along the real axis and their physical meaning. The problem is nontrivial, and a careful analysis is beyond the scope of this monograph. The matter has been thoroughly investigated by several authors (see, for instance, Achenbach 1984; Hudson 1980) using the elegant method based on the *principle of the argument* of complex variable theory (Remmert 1997). The conclusion is that Equation 2.42 may only have two *real* roots of which only the positive one is physically meaningful because it corresponds to the speed of propagation of Rayleigh waves in linear elastic, homogeneous media.

After computing V_R from Equation 2.42, the last step to completely resolve the Rayleigh eigenproblem is the computation of displacement and stress eigenfunctions, a task that is accomplished by means of Equations 2.33, 2.36, and 2.39. For the displacement field, the result is

$$
\begin{cases}
u_1 = \dfrac{ics \cdot B_4}{\omega\left(1 - c^2/2V_S^2\right)}\left[\left(1 - \dfrac{c^2}{2V_S^2}\right)\cdot e^{-s\cdot x_2} - e^{-r\cdot x_2}\right]\cdot e^{i(kx_1 - \omega t)} \\[4mm]
u_2 = \dfrac{c^2 rs \cdot B_4}{\omega^2\left(1 - c^2/2V_S^2\right)}\left[\left(1 - \dfrac{c^2}{2V_S^2}\right)^{-1}\cdot e^{-s\cdot x_2} - e^{-r\cdot x_2}\right]\cdot e^{i(kx_1 - \omega t)}
\end{cases}
\tag{2.43}
$$

where $c = V_R$. Equation 2.43 can be combined to yield the equation of an ellipse

$$
\frac{u_1^2}{r_1^2(x_2)} + \frac{u_2^2}{r_2^2(x_2)} = 1
\tag{2.44}
$$

where $r_1(x_2)$ and $r_2(x_2)$ are the displacement eigenfunctions defined as follows

$$
\begin{cases}
r_1(x_2) = \dfrac{cs \cdot B_4}{\omega\left(1 - c^2/2V_S^2\right)}\left[e^{-r\cdot x_2} - \left(1 - \dfrac{c^2}{2V_S^2}\right)\cdot e^{-s\cdot x_2}\right] \\[4mm]
r_2(x_2) = \dfrac{c^2 rs \cdot B_4}{\omega^2\left(1 - c^2/2V_S^2\right)}\left[\left(1 - \dfrac{c^2}{2V_S^2}\right)^{-1}\cdot e^{-s\cdot x_2} - e^{-r\cdot x_2}\right]
\end{cases}
\tag{2.45}
$$

Therefore, the particle orbit described by the passage of a Rayleigh wave is an ellipse on the $\{x_1, x_2\}$ plane (Figure 2.16) because the horizontal and the vertical displacement are out of phase by $\pi/2$.

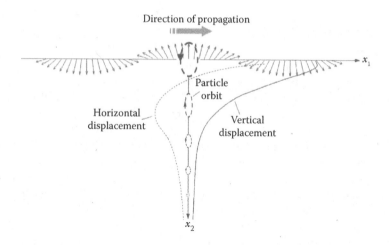

Figure 2.16 Elliptical polarization of particle motion in a Rayleigh wave in a homogeneous, linear elastic half-space. At the free boundary, the particle orbit is *retrograde*; at a depth of ~0.2·λ, it becomes *prograde*.

From Equation 2.43, it can be shown that the particle trajectory, which is described by $\vartheta = \tan^{-1} u_1/u_2$, is *retrograde*, that is counterclockwise, for depths $0 \leq x_2 \leq x_2^c$, and is *prograde*, that is clockwise, for depths greater than x_2^c. The *critical* depth $x_2^c \approx 0.2\lambda$ corresponds to the depth at which the horizontal displacement vanishes and the motion becomes purely vertical. Its precise definition is given by the following relation

$$x_2^c = \frac{\ln\left(1 - \dfrac{V_R^2}{2V_S^2}\right)}{\omega\left(\sqrt{\dfrac{1}{V_R^2} - \dfrac{1}{V_S^2}} - \sqrt{\dfrac{1}{V_R^2} - \dfrac{1}{V_P^2}}\right)} \tag{2.46}$$

where $\ln(\cdot)$ is the natural algorithm of the argument. Figure 2.16 shows qualitatively the variability of the horizontal and vertical displacement with depth induced by the passage of a Rayleigh wave. At the free surface, the ratio of vertical to horizontal displacement is approximately 1.5.

2.3 EXISTENCE OF LOVE WAVES

Rayleigh waves are the only type of surface waves that can propagate along the free surface of a linear elastic, homogeneous half-space. If, however, the homogeneity constraint is relaxed and some degree of heterogeneity in the medium is allowed, another type of surface wave exists, named *Love waves* after A. E. H. Love who predicted their existence mathematically (Love 1911).

Love waves are horizontally polarized, transversal waves (SH-waves) that arise from the phenomena of constructive interference occurring among upgoing and downgoing waves in inhomogeneous media. In stratified media, the ray paths are rectilinear, yet interference phenomena occur among waves undergoing multiple reflections at the layer interfaces. A shallow stratum traversed by a Love wave behaves essentially as a *waveguide* with horizontal particle motion confined in the outcropping layer and decaying exponentially with depth in the half-space.

The simplest geometric configuration that may support the existence of Love waves is that of an elastic homogeneous layer of finite thickness "welded" at the interface with an elastic homogeneous half-space. For such a system, the Love dispersion equation will now be derived following a procedure similar to that used for Rayleigh waves.

Let Figure 2.17 represent the setting for studying the conditions of the existence of 2D surface waves propagating along the free surface (x_1 direction) of a linear elastic, layered half-space with displacement orthogonal to the plane of propagation.

The following displacement field is assumed to satisfy Navier's equations of motion (Equation 2.15)

$$\begin{cases} u_3 = C_1 \cdot e^{\beta x_2} e^{ik(x_1 - ct)} & 0 \le x_2 \le h \\ u_3 = C_2 \cdot e^{-\alpha x_2} e^{ik(x_1 - ct)} & x_2 > h \\ u_1 = u_2 = 0 \end{cases} \qquad (2.47)$$

where α and β are complex-valued constants; however, the real part of α is assumed greater than zero so that the displacement component u_3 decreases exponentially with depth so as to represent a surface wave. C_1 and C_2 are arbitrary constants to be determined from boundary conditions.

If in Equation 2.15, u_1 and u_2 are set equal to zero, then a wave equation similar to the second of Equation 2.26 is obtained for the displacement

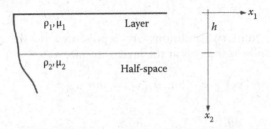

Figure 2.17 Coordinate system for Love waves propagating along the free surface of an elastic, homogeneous layer overlying a homogeneous elastic half-space.

component u_3. A solution of this equation compatible with the displacement field represented by Equation 2.47 is

$$u_3 = U_3(x_2) \cdot e^{ik(x_1 - ct)} \tag{2.48}$$

Substitution of Equation 2.48 into the wave equation for u_3 yields

$$\frac{d^2 U_3}{dx_2^2} + \left(\frac{\omega^2}{V_S^2} - k^2 \right) \cdot U_3(x_2) = 0 \tag{2.49}$$

which is an ordinary differential equation of the same type as Equation 2.28. A solution of Equation 2.49 satisfying the constraints of Equation 2.47 is the following

$$\begin{cases} U_3 = \left(D_1 \cdot e^{-i \cdot q_1 x_2} + D_2 \cdot e^{i \cdot q_1 x_2} \right) & 0 \leq x_2 \leq h \\ U_3 = \left(D_3 \cdot e^{-s_2 x_2} + D_4 \cdot e^{s_2 x_2} \right) & x_2 > h \end{cases} \tag{2.50}$$

where $q_1^2 = \dfrac{\omega^2}{V_{S1}^2} - k^2 > 0, V_{S1}^2 = \dfrac{\mu_1}{\rho_1}$ (Figure 2.17), $s_2^2 = k^2 - \dfrac{\omega^2}{V_{S2}^2} > 0$, and $V_{S2}^2 = \dfrac{\mu_2}{\rho_2}$. The constants D_1, D_2, D_3, and D_4 are determined from the boundary conditions. Substitution of Equation 2.50 into Equation 2.48 yields

$$\begin{cases} u_3 = \left(D_1 \cdot e^{-i \cdot q_1 x_2} + D_2 \cdot e^{i \cdot q_1 x_2} \right) \cdot e^{ik(x_1 - ct)} & 0 \leq x_2 \leq h \\ u_3 = \left(D_3 \cdot e^{-s_2 x_2} + D_4 \cdot e^{s_2 x_2} \right) \cdot e^{ik(x_1 - ct)} & x_2 > h \end{cases} \tag{2.51}$$

The condition of the shear stress vanishing at the ground surface as well as the radiation condition may be formally expressed as follows

$$\begin{cases} \mu_1 \dfrac{\partial u_3}{\partial x_2} = 0 & \text{for } x_2 = 0 \\ u_3 \rightarrow 0 & \text{as } x_2 \rightarrow \infty \end{cases} \tag{2.52}$$

The other boundary conditions are represented by the continuity of displacement and shear stress at the layer interface $x_2 = h$

$$\begin{cases} u_3^-(x_2) = u_3^+(x_2) & \Rightarrow \lim_{x_2 \to h^-} u_3(x_2) = \lim_{x_2 \to h^+} u_3(x_2) \\ \mu_1 \dfrac{\partial u_3}{\partial x_2}\bigg|^- = \mu_1 \dfrac{\partial u_3}{\partial x_2}\bigg|^+ & \Rightarrow \lim_{x_2 \to h^-} \mu_1 \dfrac{\partial u_3}{\partial x_2} = \lim_{x_2 \to h^+} \mu_2 \dfrac{\partial u_3}{\partial x_2} \end{cases} \tag{2.53}$$

Combination of Equations 2.52 and 2.53 allows us to construct a homogeneous system of linear algebraic equations similar to Equation 2.39. A nontrivial solution for this system is found by setting the determinant of the coefficients equal to zero, which leads to the *Love dispersion equation*

$$\tan\left(\frac{\omega h}{V_{S1}}\sqrt{1-\left(\frac{V_{S1}}{c}\right)^2}\right) - \frac{V_{S2}}{V_{S1}}\cdot\frac{\rho_2}{\rho_1}\sqrt{\frac{\left(\frac{V_{S2}}{V_{S1}}\right)^2 - \left(\frac{c}{V_{S1}}\right)^2}{\left(\frac{c}{V_{S1}}\right)^2 - 1}} = 0 \qquad (2.54)$$

Equation 2.54 is a nonlinear, transcendental relationship in the unknown parameter $c = \omega/k = V_L$, which represents the phase velocity of a wave moving along the free surface of the layered half-space in the x_2 direction and inducing particle motion in the x_3 direction (Figure 2.17). Thus, it is a surface wave, which has been named a *Love wave*. The dispersion equation of Love waves is the mathematical statement of constructive interference occurring among the elastic waves trapped in the outcropping layer. Some authors have used this argument as an alternative procedure for its derivation (Achenbach 1984).

The general solutions of Equation 2.54 cannot be found in closed form; however, a close inspection will still allow the inference of important properties. Perhaps, the most important one is that $c = V_L$ is frequency dependent; thus, Love waves are *dispersive* because their speed of propagation depends upon the frequency (or the wavenumber) of excitation. This result contrasts with that obtained for Rayleigh waves in homogenous media, which were shown to be nondispersive (Section 2.2.2).

Second, the unknown parameter c appearing as argument of the trigonometric function tan ($\cdot$) in Equation 2.54 will determine the existence of multiple roots corresponding to the periodicity and multiple branches of the tangent function. Therefore, the solution for Equation 2.54 will give rise, in general, to multiple dispersion curves (in fact, a countable infinity) and to frequency spectra corresponding to different modes of propagation.

Third, depending on the values of the ratio ρ_2/ρ_1 and more significantly on V_{S2}/V_{S1}, solutions to Equation 2.54 may not exist, at least those with physical meaning. Thus, even in layered systems, the existence of Love waves may not always be guaranteed. A closer look at Equation 2.54 would, in fact, reveal that no real roots can be found if $V_{S2} \leq V_{S1}$ (Hudson 1980); that is, if, loosely speaking, the outcropping layer (Figure 2.17) is stiffer than the underlying half-space.

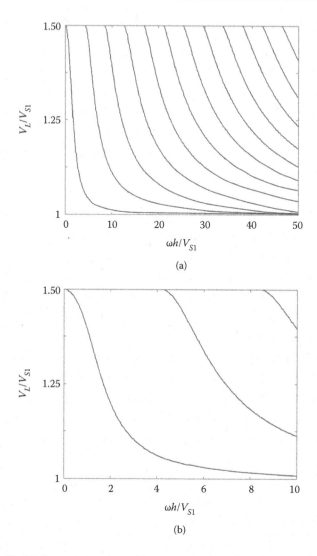

Figure 2.18 Modal dispersion curves of Love waves plotted as function of dimension-less frequency $\omega h/V_{S1}$ for a ratio $V_{S2}/V_{S1} = 1.5$: (a) frequency range 0–50 Hz; (b) close-up on the range 0–10 Hz.

Equation 2.54 has been solved numerically for a ratio $V_{S2}/V_{S1} = 1.5$. Figure 2.18 illustrates a plot of the normalized phase velocity V_L/V_{S1} of Love waves versus the dimensionless frequency $\omega h/V_{S1}$ for various modes of propagation. In this calculation, the ratio ρ_2/ρ_1 has been assumed equal to 1. The figure on the right shows a close-up view of the dispersion curves for the dimensionless frequency ranging from 0 to 10.

Whereas the fundamental mode exists for the whole range of frequencies, higher modes exist only above a *cutoff frequency*. For the nth mode, the cutoff frequency is given by the following relation

$$\frac{\omega_c^n \cdot h}{V_{S1}} = \frac{\pi(n-1)}{\sqrt{1 - \left(\dfrac{V_{S1}}{V_{S2}}\right)^2}} \quad n = 1, 2 \ldots \tag{2.55}$$

Figure 2.18 shows that, as the frequency decreases, the phase velocity of Love waves approaches V_{S2} (in all modes)—namely, the shear wave velocity of the half-space. This is consistent with the fact that a decrease in frequency is accompanied by an increase in the wavelength; thus, the speed of propagation of Love waves is influenced more by the mechanical properties of the half-space than those of the outcropping layer. The reverse occurs at large frequencies (or short wavelengths) where the value V_L is mostly affected by the shear wave velocity of the outcropping layer V_{S1}. Thus, in Figure 2.18, the dispersion curves of all modes approach a normalized Love phase velocity of 1 as the dimensionless frequency becomes arbitrary large.

2.4 SURFACE WAVES IN VERTICALLY INHOMOGENEOUS ELASTIC CONTINUA

This section describes the most relevant features of Love and Rayleigh wave propagation in vertically inhomogeneous, linear elastic continua delimited by a plane boundary. Particular focus will be devoted to Rayleigh waves because of their greater importance in the applications.

The first subject to be discussed is the eigenvalue problem associated with the propagation of free surface waves. Solution of the eigenvalue problem leads to the important concept of geometric dispersion, a phenomenon by which, in inhomogeneous continua, the phase velocity of surface waves is a multivalued function of the frequency of excitation.

The phenomenon of geometric dispersion is caused by the effects of constructive interference occurring in media that are either bounded (e.g., rods, plates, and other types of waveguides) or inhomogeneous. It is responsible for the existence of several modes of propagation each traveling at a different phase and group modal velocity. Another effect produced by geometric dispersion is to alter the geometric spreading law governing the spatial attenuation of surface waves in elastic inhomogeneous media.

Some peculiar properties of surface wave propagation in inhomogeneous media can be derived from the application of certain variational principles the applicability conditions of which are rather general. Among the results obtained from this principle, there is a closed-form expression for the Jacobian of Love and Rayleigh phase velocity with respect to medium parameters, which is of fundamental importance in the solution of the inverse problem (Chapter 6) and an approximate solution of Love and Rayleigh eigenproblem in weakly dissipative media (Section 2.5.3).

For surface waves generated by harmonic sources applied at the boundary or in the interior of a vertically inhomogeneous half-space, the various modes of propagation of Love and Rayleigh waves are superimposed such as in a spatial Fourier series. The phase velocity of the resulting waveform is called *apparent* (or effective) phase velocity, and it can be obtained from an appropriate superposition of modal quantities (i.e., phase and group velocities, displacement, and stress eigenfunctions). The notion of apparent phase velocity is particularly relevant in applications of surface waves for near-surface site characterization. Treatment of the source problem will also include a discussion of the classical Lamb's problem, which will be briefly revisited for its implications in the surface wave radiation pattern induced by a vertical point load in the far-field.

The solutions presented in this section will be given in the frequency domain. The mathematics of wave propagation problems are often complex, and explicit nonintegral solutions can rarely be obtained. Among the few exceptions are boundary value problems where the boundary conditions and body forces are specified as harmonic functions of time. However, working in the frequency domain does not necessarily imply a loss of generality because the availability of harmonic solutions is often a sufficient condition to obtain more general results using the Fourier integral theorem.

In the following section, the properties of the medium are assumed to be arbitrary (hence, not necessarily continuous) functions of depth. Explicit solutions, however, are presented only for the case of a finite number of homogeneous strata overlaying a homogeneous half-space (i.e., a horizontally layered system).

2.4.1 Eigenvalue problem associated with free surface waves

In Section 2.1.3, the Navier's displacement equations of motion (Equation 2.15) have been obtained under the assumptions of homogeneous, isotropic, linear elastic media in the absence of body forces. In general, Lamé's parameters and mass density are functions of the coordinates. If the elastic medium is assumed to be vertically heterogeneous (Figure 2.19)—namely $\lambda = \lambda\,(x_2)$, $\mu = \mu\,(x_2)$, and $\rho = \rho\,(x_2)$—Navier's equations of motion in vector notation become (Ben-Menahem and Singh 2000)

$$\mu \nabla^2 \mathbf{u} + (\lambda + \mu)\, grad\,(div\,\mathbf{u}) + \mathbf{e}_2 \frac{d\lambda}{dx_2} div\,\mathbf{u} + \frac{d\mu}{dx_2}\left(\mathbf{e}_2 \times curl\,\mathbf{u} + 2\frac{\partial \mathbf{u}}{\partial x_2}\right) = \rho \frac{\partial^2 \mathbf{u}}{\partial t^2}$$

$$(2.56)$$

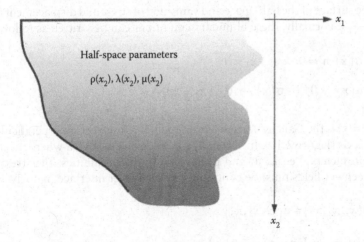

Figure 2.19 Coordinate system for surface waves propagating along the free surface of an elastic, vertically inhomogeneous half-space.

In Equation 2.56, the term e_2 denotes the base vector acting along the x_2 axis, and the symbol $(\cdot) \times (\cdot)$ is used to indicate the vector product.

To seek solutions for Equation 2.56, representing surface waves, let the following form of the displacement field be assumed

For *Love waves*:
$$\begin{cases} u_1 = 0 \\ u_2 = 0 \\ u_3 = l_1\left(x_2, k, \omega\right) \cdot e^{i\left(kx_1 - \omega t\right)} \end{cases} \tag{2.57}$$

For *Rayleigh waves*:
$$\begin{cases} u_1 = r_1\left(x_2, k, \omega\right) \cdot e^{i\left(kx_1 - \omega t\right)} \\ u_2 = i \cdot r_2\left(x_2, k, \omega\right) \cdot e^{i\left(kx_1 - \omega t\right)} \\ u_3 = 0 \end{cases} \tag{2.58}$$

Equations 2.57 and 2.58 correspond to the analogous displacement fields assumed in Sections 2.2 and 2.3, respectively, for studying the propagation of Rayleigh waves in a homogeneous half-space and Love waves in a homogeneous layer overlying a homogeneous half-space.

In Equation 2.58, the ellipticity of Rayleigh particle motion has been explicitly recognized by setting the horizontal and vertical components of the displacement field $\pi/2$ radians out of phase.

To represent surface waves, Equations 2.57 and 2.58 must be supplemented with appropriate boundary conditions that are vanishing of the stress field at the free surface of the half-space and vanishing of stress and displacement fields as $x_2 \to \infty$. Formally, these boundary conditions can be written as follows

$$\begin{cases} \sigma(\mathbf{x}) \cdot \mathbf{n} = 0 & \text{at} \quad x_2 = 0 \\ \mathbf{u}(\mathbf{x}) \to 0 \qquad \sigma(\mathbf{x}) \to 0 & \text{as} \quad x_2 \to \infty \end{cases} \tag{2.59}$$

where $\sigma(\mathbf{x})$ is the Cauchy's stress tensor and $\mathbf{n}$ is a unit vector perpendicular to the x_1 axis (Figure 2.19). In vertically inhomogeneous media where the material parameters (i.e., λ, μ, and ρ) have jump discontinuities, the stress and displacement fields must be continuous at each layer interface, namely

$$\mathbf{u}\left(x_1, x_2^-, x_3\right) = \mathbf{u}\left(x_1, x_2^+, x_3\right)$$
$$\sigma\left(x_1, x_2^-, x_3\right) \cdot \mathbf{n} = \sigma\left(x_1, x_2^+, x_3\right) \cdot \mathbf{n} \tag{2.60}$$

otherwise the stress and displacement discontinuities would correspond to the existence of seismic sources.[04]

From the displacement field defined by Equations 2.57 and 2.58, the stress field can be easily computed using Hooke's law (Equation 2.12) to obtain

For *Love waves*:
$$\begin{cases} \sigma_{11} = \sigma_{33} = \sigma_{22} = \sigma_{21} = 0 \\ \sigma_{23} = \mu \dfrac{dl_1}{dx_2} \cdot e^{i(kx_1 - \omega t)} \\ \sigma_{13} = -ik\mu l_1 \cdot e^{i(kx_1 - \omega t)} \end{cases} \tag{2.61}$$

For *Rayleigh waves*:
$$\begin{cases} \sigma_{23} = \sigma_{13} = 0 \\ \sigma_{11} = i\left[\lambda \dfrac{dr_2}{dx_2} - k(\lambda + 2\mu) \cdot r_1 \right] \cdot e^{i(kx_1 - \omega t)} \\ \sigma_{33} = i\left(\lambda \dfrac{dr_2}{dx_2} - k\lambda r_1 \right) \cdot e^{i(kx_1 - \omega t)} \\ \sigma_{22} = i\left[(\lambda + 2\mu) \cdot \dfrac{dr_2}{dx_2} - k\lambda r_1 \right] \cdot e^{i(kx_1 - \omega t)} \\ \sigma_{21} = \mu \left(\dfrac{dr_1}{dx_2} + k r_2 \right) \cdot e^{i(kx_1 - \omega t)} \end{cases} \tag{2.62}$$

---

[04] A discontinuity in the displacement field across a layer interface would be equivalent to a mechanical dislocation.

In Equation 2.60, the stress components σ_{23} for Love waves and σ_{12}, σ_{22} for Rayleigh waves are continuous in x_2. It is convenient to rewrite these stress components as follows

For *Love waves*: $\sigma_{23} = l_2\left(x_2, k, \omega\right) \cdot e^{i(kx_1 - \omega t)}$ (2.63)

For *Rayleigh waves*:
$$\begin{cases} \sigma_{12} = r_3\left(x_2, k, \omega\right) \cdot e^{i(kx_1 - \omega t)} \\ \sigma_{22} = i \cdot r_4\left(x_2, k, \omega\right) \cdot e^{i(kx_1 - \omega t)} \end{cases}$$ (2.64)

Substituting Equations 2.57 and 2.58 into Navier's equations of motion (Equation 2.56) and rearranging Equations 2.61 and 2.62 in light of Equations 2.63 and 2.64 yields (Aki and Richards 2002)

For *Love waves*: $\dfrac{d}{dx_2}\begin{bmatrix} l_1 \\ l_2 \end{bmatrix} = \begin{bmatrix} 0 & \mu(x_2)^{-1} \\ k^2 \cdot \mu(x_2) - \omega^2 \cdot \rho(x_2) & 0 \end{bmatrix} \cdot \begin{bmatrix} l_1 \\ l_2 \end{bmatrix}$ (2.65)

and

$$\frac{d}{dx_2}\begin{bmatrix} r_1 \\ r_2 \\ r_3 \\ r_4 \end{bmatrix} =$$

$$\begin{bmatrix} 0 & k(\omega) & \mu^{-1}(x_2) & 0 \\ -k(\omega)\lambda(x_2)\cdot\left[\lambda(x_2)+2\mu(x_2)\right]^{-1} & 0 & 0 & \left[\lambda(x_2)+2\mu(x_2)\right]^{-1} \\ \left[k(\omega)\right]^2 \zeta(x_2) - \omega^2\rho(x_2) & 0 & 0 & k(\omega)\lambda(x_2)\left[\lambda(x_2)+2\mu(x_2)\right]^{-1} \\ 0 & -\omega^2\rho(x_2) & -k(\omega) & 0 \end{bmatrix}$$

$$\cdot \begin{bmatrix} r_1 \\ r_2 \\ r_3 \\ r_4 \end{bmatrix}$$

(2.66)

for *Rayleigh waves*. The functions l_2 for Love waves and r_3, r_4 for Rayleigh waves can be expressed in terms of the eigenfunctions l_1 and r_1, r_2, which are defined by Equations 2.57 and 2.58, respectively

For *Love waves*: $l_2\left(x_2, k, \omega\right) = \mu(x_2) \cdot \dfrac{dl_1}{dx_2}$ (2.67)

For *Rayleigh waves*:
$$\begin{cases} r_3\left(x_2,k,\omega\right)=\mu(x_2)\cdot\left(\dfrac{dr_1}{dx_2}-kr_2\right) \\[4mm] r_4\left(x_2,k,\omega\right)=\left[\left(\lambda(x_2)+2\mu(x_2)\right)\dfrac{dr_2}{dx_2}+k\lambda(x_2)r_1\right] \end{cases} \tag{2.68}$$

The function $\zeta(x_2)$ in Equation 2.66 is defined in terms of Lamé's parameters[05] $\lambda(x_2)$ and $\mu(x_2)$ as follows

$$\zeta\left(x_2\right)=4\mu\left(x_2\right)\cdot\frac{\left[\lambda\left(x_2\right)+\mu\left(x_2\right)\right]}{\left[\lambda\left(x_2\right)+2\mu\left(x_2\right)\right]} \tag{2.69}$$

Defining vectors $\mathbf{g}(x_2) = [l_1\ l_2]^T$ and $\mathbf{f}(x_2) = [r_1\ r_2\ r_3\ r_4]^T$, and denoting matrices $\mathbf{B}(x_2)$ and $\mathbf{A}(x_2)$, the 2×2 and 4×4 arrays of Equations 2.65 and 2.66, respectively, the latter equations can be rewritten in a compact form

For *Love waves*: $\dfrac{d\,\mathbf{g}\left(x_2\right)}{dx_2}=\mathbf{B}\left(x_2\right)\cdot\mathbf{g}\left(x_2\right)$ \hfill (2.70)

For *Rayleigh waves*: $\dfrac{d\,\mathbf{f}\left(x_2\right)}{dx_2}=\mathbf{A}\left(x_2\right)\cdot\mathbf{f}\left(x_2\right)$ \hfill (2.71)

Relations 2.70 and 2.71 are two sets of first-order, linear, ordinary differential equations with variable coefficients. They define two *differential eigenvalue problems* with *linear operator* d/dx_2. The displacement and stress eigenfunctions are $l_1(x_2, k, \omega)$ and $l_2(x_2, k, \omega)$, respectively, for *Love waves*, whereas for *Rayleigh waves* they are the pair $[r_1(x_2, k, \omega)$, $r_2(x_2, k, \omega)]$ and $[r_3(x_2, k, \omega)$, $r_4(x_2, k, \omega)]$, respectively.

The boundary conditions associated with the eigenproblems (Equations 2.70 and 2.71) can be immediately deduced from Equation 2.59 and Equations 2.63 and 2.64

For *Love waves*:
$$\begin{cases} l_2\left(x_2,k,\omega\right)=0 & \text{at } x_2 = 0 \\[2mm] l_1\left(x_2,k,\omega\right)\to 0 & \text{as } x_2 \to \infty \end{cases} \tag{2.72}$$

For *Rayleigh waves*:
$$\begin{cases} r_3\left(x_2,k,\omega\right)=r_4\left(x_2,k,\omega\right)=0 & \text{at } x_2 = 0 \\[2mm] r_1\left(x_2,k,\omega\right)\to 0,\ \ r_2\left(x_2,k,\omega\right)\to 0 & \text{as } x_2 \to \infty \end{cases} \tag{2.73}$$

---

[05] It would be more appropriate to denote *Lamé parameters* λ and μ as *Lamé functions* in light of their dependence on x_2.

For a given frequency ω, nontrivial solutions of the two eigenproblems (Equations 2.70 and 2.71) subjected to boundary conditions (Equations 2.72 and 2.73) exist only for particular values of the wavenumber $k_j = k_j(\omega)$, $j = 1, M$.[06] These special values k_j are called the *eigenvalues* associated with the eigenproblem, and the corresponding solutions $l_m^{(j)}(x_2, k_j, \omega)$, $r_n^{(j)}(x_2, k_j, \omega)$, $m = 1,2$; $n = 1,4$ are called the *eigenfunctions*. It can be shown that, for a given frequency ω, the set of eigenfunctions satisfies a series of orthogonality conditions with appropriate weighting functions (Keilis-Borok 1989).

The systems of linear, first-order, ordinary differential Equations 2.70 and 2.71 have variable coefficients because the Lamé parameters λ and μ and the mass density ρ are functions of x_2. This characteristic makes their solution nontrivial. Several techniques have been developed in this regard, and some of the most important ones will be briefly reviewed in the next section. They share the common goal of constructing the Love and Rayleigh dispersion relations from which to compute the eigenvalues $k_j = k_j(\omega)$, $j = 1, M$. However, the dispersion relations can only be defined implicitly and may be written in the form

$$\Phi_{L/R}[\lambda(x_2), \mu(x_2), \rho(x_2), k, \omega] = 0 \qquad (2.74)$$

Equation 2.74 is called either the *Love* or the *Rayleigh dispersion equation* depending on whether it is obtained from Equations 2.70 and 2.72 or from Equations 2.71 and 2.73. In the most general case, Equation 2.74 is a highly nonlinear, transcendental function of the arguments. It states that, in vertically inhomogeneous media, the velocity of propagation of surface Love and Rayleigh waves is a multivalued function of frequency.

Each set, $\{k_j, l_m^{(j)}(x_2, k_j, \omega)\}$ or $\{k_j, r_n^{(j)}(x_2, k_j, \omega)\}$, defines a mode of propagation and, in general, there are M normal modes of propagation at any given frequency. The number M can be finite or infinite, depending upon the x_2 dependence of the medium properties and on the frequency of excitation. Furthermore, the distribution of the modes, called the *mode spectrum*, can be continuous or discrete, and in some cases both (Keilis-Borok 1989). It can be shown that in a medium composed of a finite number of homogeneous layers overlaying a homogeneous half-space, the total number of surface wave modes of propagation is always finite (Ewing et al. 1957).

From a physical point of view, the existence of different modes of propagation at a given frequency is due to constructive interference phenomena occurring among waves. In continuously varying heterogeneous media, the ray paths are curved (as a result of Snell's law); hence, they interfere with each other. In stratified media, the ray paths are rectilinear, and interference

---

[06] To simplify the notation, the same letter M has been used to denote the number of modes resulting from the solution of either the *Love* or the *Rayleigh* eigenproblem. In the most general case, they are characterized by different eigenvalues.

phenomena occur among waves undergoing multiple reflections at the layer interfaces. In either case, the dispersion equation is the mathematical statement of this condition of constructive interference (Achenbach 1984).

2.4.1.1 Solutions by numerical techniques

This section briefly reviews some of the most important methods used to solve the Love and Rayleigh eigenproblems, Equations 2.70 and 2.71, and their associated boundary conditions, Equations 2.72 and 2.73. Although several techniques are available for the solution of linear systems of first order, ordinary differential equations with variable coefficients, including numerical integration, finite difference, finite element, boundary element, and more recently spectral element methods, are the first that have been specifically developed for surface wave applications belonging to the family of the *propagator matrix algorithms* (Pestel and Leckie 1963). The Thomson–Haskell procedure (Thomson 1950; Haskell 1953), also called the *transfer matrix method*, is probably the most notorious of these algorithms because of its simplicity and its ease of computer implementation. Even recently, it has been the subject of updates and efficient reformulations (Rokhlin and Wang 2002; Liu 2010; Ganpan Ke et al. 2011). The application of this algorithm, however, is limited to vertically inhomogeneous media that can be idealized by a stack of homogeneous strata overlying a homogeneous half-space.

In the Thomson–Haskell algorithm, the nontrivial solutions of the Love and Rayleigh eigenproblems are found from the roots of dispersion relations (Equation 2.74) constructed through a sequence of matrix multiplications after imposing the continuity of stress and displacement fields at each layer interface. Each matrix is formed by elements built from transcendental functions of material properties of an individual layer, and for this reason, it is named a *layer matrix*. The roots of the dispersion equations are the wavenumbers corresponding to the modes of propagation of surface waves at different frequencies. Once the roots have been found, it is easy to determine the eigenfunctions for each mode of propagation by straightforward algebraic operations. The eigenfunctions provide the depth-dependence of stress and displacement fields.

An important contribution to the transfer matrix method originally developed by Thomson (1950) was made by Haskell (1953), who provided asymptotic expressions for the dispersion relations for the important limiting cases of short and long wavelengths. Because the initial formulation of the Thomson–Haskell algorithm suffered from problems of numerical instability at high frequencies (Knopoff 1964), this method has been modified and improved throughout the years by numerous researchers (Pestel and Leckie 1963; Dunkin 1965; Schwab and Knopoff 1970; Abo-Zena 1979; Harvey 1981). It is also worth mentioning that the matrix propagator method has been recently generalized to layered porous media (Jocker et al. 2004).

Turning their attention to different techniques, Kausel and Roesset (1981) derived a finite element formulation of the Thomson–Haskell algorithm, which was then called the *dynamic stiffness matrix method*. The main feature of this method is the replacement of the Thomson–Haskell transfer matrices with layer stiffness matrices that are similar to conventional stiffness matrices used in structural analysis. The advantage of this formulation is the ability to use standard structural analysis techniques such as condensation and substructuring to solve the Love and Rayleigh eigenproblems and the inhomogeneous elastodynamic problem of layered media subjected to dynamic loads (Kausel 1981). The first attempts at using finite element techniques to solve wave propagation problems in seismology and earthquake engineering date back to the early 1970s with the work of Lysmer and Waas (1972) and Lysmer and Drake (1972).

Another important class of algorithms for solving the surface waves' eigenvalue problems is the method of *reflection and transmission coefficients* developed by Kennett and his coworkers (Kennett 1974, 1983; Kennett and Kerry 1979) and modified and/or improved by other researchers (Luco and Apsel 1983; Chen 1993; Hisada 1994, 1995; Chapman 2003; Pei et al. 2008). This method, like the Thomson–Haskell algorithm, is only suitable for applications in multilayered media. It is based on the use of reflection and transmission coefficients to construct reflection and transmission matrices for each layer and then combining them through a recursive algorithm into the global reflection and transmission matrices of the whole layered system. The result is a very efficient iterative algorithm for constructing the dispersion equations (Equation 2.74). The method of reflection and transmission coefficients also offers an interesting physical interpretation because it explicitly models the constructive interference that leads to formation of the surface waves modes (Kennett 1983). Earlier versions of this algorithm were numerically unstable at high frequencies because of the presence of certain frequency-dependent terms that have been eliminated in more recent formulations (Chen 1993).

Independent from a specific family of methods used, most of the computational efforts of the algorithms adopted for the solution of surface wave eigenproblems (Equations 2.70 and 2.71) are devoted to the implementation of the following two tasks:

- Building the Love/Rayleigh secular functions $\Phi_{L/R}$ [$\lambda(x_2)$, $\mu(x_2)$, $\rho(x_2)$, k, ω]
- Computation of the roots of $\Phi_{L/R}[\cdot]$ as a function of frequency or wavenumber.[07]

---

[07] The choice between frequency and wavenumber as independent variables in the computation of the roots of Equation 2.74 is not as equivalent and perfectly symmetric as it may seem. One way of proceeding may reveal unexpected difficulties and results (Wilmanski 2005).

For an elastic medium, the use of complex arithmetic in constructing the Love and Rayleigh secular functions can be avoided (Haskell 1953; Schwab and Knopoff 1971) and the roots of the dispersion relations can be obtained by means of root-bracketing techniques combined with bisection (Hisada 1995). The use of these slow converging root-finding techniques is recommended by the rapidly oscillating behavior of the Love and Rayleigh secular functions, particularly at high frequencies, which requires the use of robust methods (Press et al. 1992).

Figure 2.20 shows the absolute value of the Rayleigh secular function $\Phi_R[\cdot]$ plotted against the wavenumber for a prescribed frequency of excitation. The function $\Phi_R[\cdot]$ has been constructed using the Thomson–Haskell algorithm modified by Schwab and Knopoff (1972). The material properties of the layered medium considered for the analysis are reported in Table 2.4.

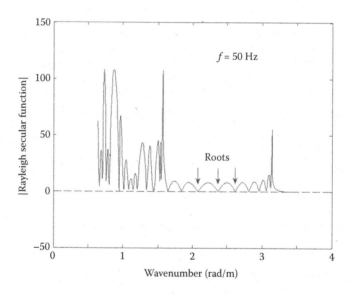

Figure 2.20 Behavior of the Rayleigh secular function. The plot shows the function $\Phi_R[\cdot]$ constructed using the Thomson–Haskell algorithm as a function of the wavenumber at a given frequency of excitation.

Table 2.4 Properties of layered system used to build Rayleigh secular function plotted in Figure 2.20

Layer	Thickness (m)	V_P (m/s)	V_S (m/s)	ρ (Mg/m³)
1	10	200	100	1.9
2	10	400	200	1.9
3	10	600	300	1.9
Half-space	∞	1000	500	1.9

The closely spaced roots of the Rayleigh secular function are also identifiable from Figure 2.20. The number of roots, at a given frequency, defines the eigenvalues associated with the eigenproblem (Equation 2.71) and identifies the number of Rayleigh modes that can exist for a particular layered medium. Their precise and reliable computation is not a trivial task, especially at high frequencies (Strobbia 2003).

Determination of the roots of the Rayleigh (or Love) secular function in viscoelastic continua is even more difficult because in this case $\Phi_{L/R}[\cdot]$ is a complex-valued mapping of the complex-valued wavenumber (Lai and Rix 2002).

Figure 2.21 shows a graph of the roots of the Rayleigh dispersion equation (Equation 2.74) for the layered medium of Table 2.4 where the phase velocity rather than the wavenumber has been plotted against frequency. Each dispersion curve is associated with a particular mode of propagation. In general, there are several modes of propagation at a given frequency, with the higher modes characterized by a higher speed of propagation.

Also, as the frequency increases, the number of modes associated with that frequency increases as well and the modes appear more closely spaced. As $\omega \to \infty$ each of the modes tends to an asymptote, which is the Rayleigh phase velocity of the outcropping layer of the vertically inhomogeneous

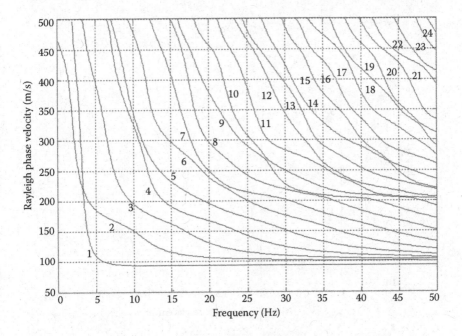

Figure 2.21 Rayleigh dispersion curves for the layered system of Table 2.4. Each curve denotes a single mode of propagation and corresponds to an eigenvalue of Equation 2.71 expressed as a function of frequency.

half-space and that corresponds to a 10 m thick stratum with $V_S = 100$ m/s. Conversely, as the frequency decreases, the higher modes cease to exist below their *cutoff* frequencies, and the dispersion curve associated with the fundamental mode tends to approach the Rayleigh phase velocity of the half-space with $V_S = 500$ m/s.

Once the roots of the Rayleigh secular function have been obtained, computation of the eigenfunctions is a straightforward task requiring standard algebraic operations. Figures 2.22 and 2.23 illustrate the mode shapes of the displacement and stress eigenfunctions (components τ_{21} and σ_{22}), respectively, for the layered medium of Table 2.4 at $f = 16$ Hz. The eigenfunctions have been normalized with respect to the maximum of the absolute value of the vertical displacement (Figure 2.22) and the maximum value of the stress component σ_{22} (Figure 2.23).

A common feature of multimode Rayleigh wave propagation is that higher modes have a greater penetration depth than lower modes. This property, which is clearly shown in Figures 2.22 and 2.23, is very important in the solution of the inverse problem because the ability to resolve deeper layers can be directly related to the ability to explicitly recognize higher modes of propagation.

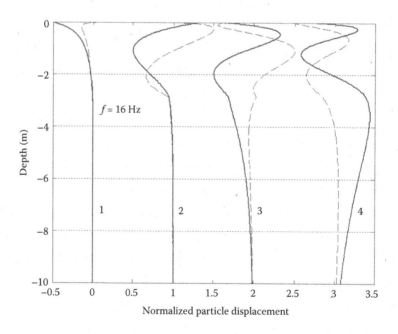

Figure 2.22 Horizontal (dashed lines) and vertical (bold lines) components of Rayleigh *displacement eigenfunctions* for the layered medium of Table 2.4 at $f = 16$ Hz. Each pair of curves denotes a single mode.

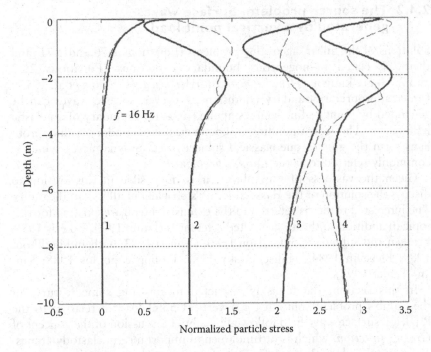

Figure 2.23 Components τ_{21} (dashed lines) and σ_{22} (bold lines) of Rayleigh *stress eigenfunctions* for the layered medium of Table 2.4 at f = 16 Hz. Each pair of curves denotes a single mode.

As mentioned at the beginning of this section, other numerical techniques can be used to solve the systems of linear, first-order, ordinary differential equations (Equations 2.70 and 2.71), including the finite difference method (Boore 1972), numerical integration (Takeuchi and Saito 1972), the boundary element method (Manolis and Beskos 1988), and the spectral element method (Faccioli et al. 1996; Komatitsch and Vilotte 1998). Although these techniques are less popular than the propagator matrix methods briefly described in this section, they have several advantages. Numerical integration, for example, can be used in vertically inhomogeneous media where the medium properties vary continuously with depth; therefore, they are more general than techniques belonging to the class of propagator matrix methods. Boundary element methods are best suited for modeling unbounded or semi-infinite media because they only require discretization of the boundary. As a result, they reduce the dimensionality of the problem by one. Moreover, boundary element methods eliminate the need, required by finite element–based methods, of using fictitious or nonreflecting boundaries to simulate the Sommerfeld condition at infinity.

2.4.2 The source problem: Surface waves generated by a vertical point load

Solutions of the linear eigenvalue problems (Equations 2.70 and 2.71) and their associated homogeneous boundary conditions (Equations 2.72 and 2.73) are known as *free Love and Rayleigh waves* (Ewing et al. 1957). Under appropriate boundary conditions, however, surface waves can be generated by point or line sources applied at a free boundary of an elastic half-space. The corresponding inhomogeneous boundary value problems form the guts of one classical subject of elastodynamics,[08] which is commonly referred to as the *Lamb's problem*.[09]

Given the vastness of the subject, it is impossible in this section to discuss the solution of the classical Lamb's problem in all its ramifications. The interested reader is referred to the extensive literature existing for this topic in addition to the already cited references (Lamb 1904; Pekeris 1955; Eringen and Suhubi 1975; Hudson 1980; Kennett 1983; Båth and Berkhout 1984; Bleistein 1984; Malischewsky 1987; Philippacopoulos 1988; Sato and Fehler 1997).

In this section, the focus is restricted to showing some features of the Lamb's problem solution for a vertical point load in relation to the Rayleigh surface wave field followed by a brief discussion of the concept of *Green's function*, which is of fundamental importance in elastodynamics. The section ends with the definition of *apparent phase velocity*, a concept arising in connection with mode superposition of surface Rayleigh waves generated by harmonic sources.

2.4.2.1 Lamb's problem for time-harmonic, vertical point load

The problem of determining the displacement field induced by a vertical, time-harmonic point load applied at the free boundary of a homogeneous, isotropic, linear elastic half-space was first solved by Lamb (1904) in his classical paper entitled "On the propagation of tremors over the surface of an elastic solid." Lamb used complex variable theory and contour integration to find the solution to what can be considered as the dynamic version of the *Boussinesq's problem*, another classical theoretical problem of linear elasticity—that of determining the stress–strain fields induced in a linear elastic half-space by a concentrated, vertical point load.

---

[08] In reality, the wave field generated by shallow or buried sources at the free boundary of an elastic half-space includes P-, S-, and head waves, in addition to surface waves.

[09] Although a somewhat more general meaning is currently attributed to the *Lamb's problem* (Achenbach 1984).

In his famous paper, Lamb actually solved other boundary value problems of elastodynamics including the calculation of the displacement field caused by a vertical line load applied normally to the free surface or inside the half-space. For both the vertical point and line loads, he considered harmonic and impulsive time variations. Figure 2.24 shows Lamb's computations for the horizontal and vertical components of the displacement generated at the free boundary (in the far-field approximation) by an impulse vertical line loading function. This plot has been considered by many as the first ever computed *synthetic seismogram* (Aki and Richards 2002).

The wave field in Figure 2.24 has been computed by Lamb at a particular spatial location, far from the source. The first disturbance is due to the arrival of longitudinal P-wave (P) followed by the transversal S-wave (S). The greatest disturbance corresponds to the arrival of the Rayleigh wave, which is referred to in the original Lamb's paper as the "major tremor" whereas body P- and S-waves are named "minor tremors."

For the case of a time-harmonic, vertical point source applied at the free boundary of a linear elastic, homogeneous half-space (Figure 2.25), the solution given by Lamb for the Rayleigh vertical displacement field $u_2(r, \omega)$ in the *far-field approximation* can be written as follows

$$u_2(r,\omega) = \frac{Fe^{i\omega t}}{2i\mu} \cdot k_R \cdot \Psi(k_R) \cdot H_0^{(2)}(k_R r) \tag{2.75}$$

where $r = \sqrt{x_1^2 + x_3^2}$ is the distance from the vertically oscillating harmonic force $F \cdot e^{i\omega t}$, μ is the shear modulus of the elastic medium, k_R is the Rayleigh

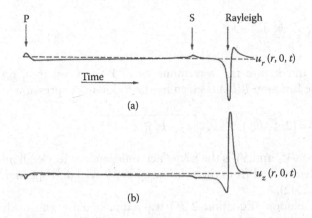

(a)

(b)

Figure 2.24 (a) Horizontal and (b) vertical components of surface displacement obtained in the far-field by Lamb (1904) for an impulse-like vertical line loading function. In panel a, the time of arrival at a specific location of longitudinal (P), transversal (S), and Rayleigh waves are indicated.

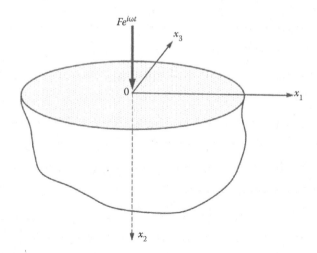

Figure 2.25 Coordinate system for the solution of Lamb's problem for the case of time-harmonic, vertical point source applied at the free boundary of a linear elastic, homogeneous half-space.

Table 2.5 Medium properties used to compute Lamb's solution

Layer	Thickness (m)	V_P (m/s)	V_S (m/s)	ρ (Mg/m³)
Half-space	∞	600	300	1.9

wavenumber, and $H_0^{(2)}(\cdot)$ denotes the Hankel function of the second kind of zero order.[10] The function $\Psi(k_R)$ is defined as follows

$$\Psi(k_R) = -\frac{k_S^2 \sqrt{k_R^2 - k_P^2}}{R'(k_R)} \tag{2.76}$$

where k_P and k_S are the wavenumbers of P- and S-waves, respectively. Finally, the function $R(k_R)$ is given by the following expression

$$R(k_R) = \left(2k_R^2 - k_S^2\right)^2 - 4k_R^2 \sqrt{\left(k_R^2 - k_P^2\right)\left(k_R^2 - k_S^2\right)} \tag{2.77}$$

where $k_R = \omega/V_R$ and V_R is the frequency-independent Rayleigh phase velocity to be determined from the solution of the Rayleigh dispersion equation (Equation 2.42).

Lamb's solution (Equation 2.75) was computed for the medium properties illustrated in Table 2.5 and for a vertical force of unit magnitude.

---

[10] The Hankel function $H_0^{(2)}(z)$ is defined as $H_0^{(2)}(z) = J_0(z) - iY_0(z)$ where $J_0(z)$ and $Y_0(z)$ with $z \in \mathbb{C}$ are the Bessel functions of the first kind and second kind of zero order, respectively.

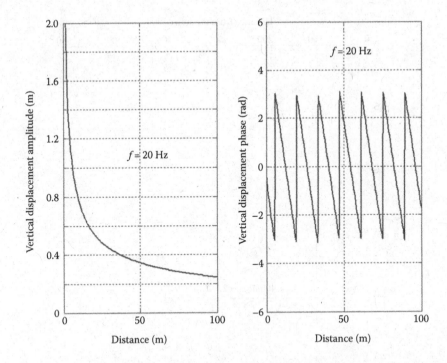

Figure 2.26 Vertical components of displacement amplitude and phase as functions of distance from the source as computed from Lamb's solution (Equation 2.75). The calculation has been carried out at a frequency of 20 Hz.

Figure 2.26 shows the variation of displacement amplitude and phase with distance from the source at a frequency of 20 Hz. As expected, the displacement amplitude decreases with distance from the source with a rate of decay that decreases as the distance grows. On the contrary, the displacement phase exhibits jump discontinuities at regular intervals corresponding to the wavelength of the propagating Rayleigh wave ($\lambda_R = V_R/f = 279.76/20 = 13.99$m).

Figure 2.27 shows the synthetic seismograms obtained from the inverse Fourier transform of Equation 2.75 calculated at five distances from the source: 100, 200, 300, 400, and 500 m. The vertical component of the displacement field $u_2 (r, \omega)$ was computed for a frequency range from 0.1 to 20 Hz. The seismograms shown in Figure 2.26 illustrate the response of the halfspace to a vertical point load with time variation given by a Dirac δ function. It is important to emphasize that waveforms computed from Equation 2.75 represent *only* the contribution of the Rayleigh surface wave field. Other contributions (mainly the body wave field) are not accounted for in Equation 2.75.

Figure 2.27 also shows an approximate calculation of the Rayleigh phase velocity from the slope of a straight line joining points of equal phase in the seismograms (in this case, the peaks of displacement amplitudes).

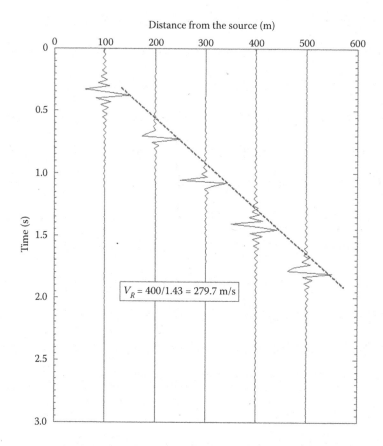

Figure 2.27 Synthetic seismograms obtained from the inverse Fourier transform of Lamb's solution (Equation 2.75) computed at distances of 100, 200, 300, 400, and 500 m from the source and for $f = 0.1–20$ Hz. The signals represent the medium response to a vertical point load varying with time as a Dirac δ function.

The amplitudes in each seismogram have been normalized with respect to the maximum of the absolute value of vertical displacement.

The rate of decay of displacement amplitude with distance from the source may be better appreciated by considering the *asymptotic expansion* of the Hankel function of the second kind of order κ (Lebedev 1972):

$$H_\kappa^{(2)}(z) = \left(\frac{2}{\pi \cdot z}\right)^{0.5} \cdot e^{-i \cdot (z - 0.5 \cdot \kappa \cdot \pi - 0.25 \cdot \pi)} \cdot \left[\sum_{j=0}^{n}(\kappa, j)(2iz)^{-j} + O(|z|^{-n-1})\right] \quad (2.78)$$

for $|\arg z| \leq \pi - \varepsilon$

where ε is an arbitrarily small positive number and

$$(\kappa, j) = \frac{(-1)^j}{j!}\left(\tfrac{1}{2}-\kappa\right)_j\left(\tfrac{1}{2}+\kappa\right)_j = \frac{\left(4\kappa^2-1\right)\left(4\kappa^2-3^2\right)\cdots\left(4\kappa^2-(2j-1)^2\right)}{2^{2j}j!}, \quad (2.79)$$

$$(\kappa, 0) = 1$$

In Lamb's solution, represented by Equation 2.75, the argument is $z = k_R \cdot r$ and $\kappa = 0$. For sufficiently large values of $|z|$,[11] the Hankel function $H_0^{(2)}(z)$ can satisfactorily be approximated by the product of a complex exponential (i.e., the phase factor) by a term proportional to $z^{-0.5}$. Thus, in the far-field approximation, Rayleigh waves generated by a vertical, time-harmonic point load applied at the free boundary of a linear elastic, homogeneous half-space attenuate geometrically with the increase of the distance r from the source as $r^{-0.5}$; hence, they propagate along *cylindrical* wave fronts (Figure 2.12).

Using the same argument of asymptotic expansions, Lamb (1904) showed that, for the same problem, P and S body waves suffer a geometric decay proportional to r^{-2} at the free surface. Thus, body waves propagate along hemispherical wave fronts. On the contrary, it can be shown that Rayleigh waves generated by a surface time-harmonic, vertical line source do not undergo geometric attenuation, whereas cylindrical body waves decay with increasing distance from the source as $r^{-1.5}$ (Achenbach 1984). It is important to remark, however, that these laws of spatial attenuation are not applicable, in general, to transient wave forms because for the latter, the spatial attenuation of the wave results from a combination of geometric spreading and decaying of the signal with time (Keilis-Borok 1989).

2.4.2.2 Features of wave propagation in two dimensions

A remarkable result that comes from the theory of wave propagation is revealed when the phenomenon takes place in spaces having *even* rather than *odd* dimensions. For instance, the radiation field in two dimensions has features that have no counterpart in a three-dimensional (3D) Euclidean space. In the latter, a space–time localized impulse emitted by a source is reproduced at all other positions throughout the space after a finite time t^* has elapsed. This travel time is a function of the distance from the source and of the speed of propagation of a disturbance in that space. The original signal of finite duration emitted from the source will be reproduced identically at later times at different points of the space. The only difference will be the amplitude decay of the signal due to geometric and possibly material (i.e., dissipative) spreading. This phenomenon is known as the *Huyghens's principle*.

---

[11] On the order of one half of the wavelength λ_R.

In 2D wave propagation, Huyghens's principle does not hold, and an external disturbance of finite duration generates a signal that does not cease to persist even after the external disturbance has ended its action. Geometric (and possibly material) attenuation will still yield its effects. However, in a 2D space, a single impulse is perceived at a particular position as a blurred signal that lives on, although decaying in time, beyond the duration of the impulse emitted at the source.[12]

Surface waves propagating along the free boundary of a half-space are essentially a 2D phenomenon, and as such they share all the features described here with regard to the radiation in two dimensions. Lamb's solution reflects these features as well. It can be shown by taking the inverse Fourier transform of Equation 2.75 and then convolving it with a source function of finite time duration. The result would be a signal that, at a particular distance from the source, will have all the described characteristics of a blurred waveform despite being generated by a Dirac δ impulse function.

Figure 2.28 shows the half-space, free-surface response at a distance of 100 m from a vertical point source, the time variation of which is a *smooth square*[13] wave function *of 0.996 s duration* (top of Figure 2.28). The medium response (bottom of Figure 2.28) was computed from the inverse Fourier transform of Lamb's solution (Equation 2.75) convolved with the smooth square wave source function. To superimpose on the same chart, the source function and the response of the virtual receiver, the latter has been normalized with respect to the maximum of the absolute value of displacement amplitude.

The material properties adopted for the half-space are those reported in Table 2.5. It is readily apparent that the time window of *0.996 s* of the source function is spread in a response of almost *7s*. This effect is also visible in Figure 2.27, where the seismograms of finite time duration represent the half-space response to a source function varying as a Dirac δ function (i.e., an impulse of infinitesimal time window).

2.4.2.3 Geometric spreading function for surface Rayleigh waves

As discussed at the end of Section 2.4.2, surface Rayleigh waves generated by a time-harmonic, vertical point load applied at the free surface of a half-space propagate along cylindrical wave fronts; thus, they attenuate

---

[12] If humans were to live like cartoons in a 2D world, they could hardly communicate because their conversations would overlie in time, thereby producing unclear, indistinct dialogs. This feature of wave propagation in *two dimensions* can be proved to hold in any Euclidean space of even dimensions. Likewise, the validity of Huyghens's principle in a 3D world can be shown to hold in any Euclidean space of odd dimensions (Courant and Hilbert 2004).

[13] A smooth square function is a square function where the discontinuities of the derivative at the four corners of the function have been eliminated by smooth transitions to avoid the *Gibbs's effect*.

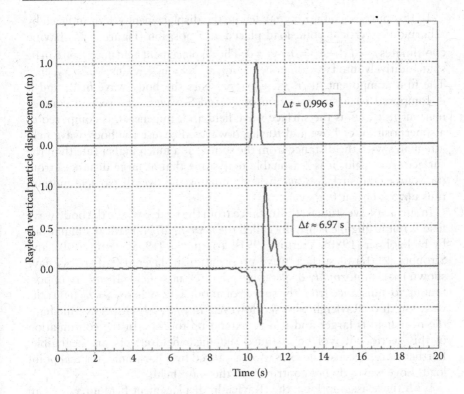

Figure 2.28 (Top) Source function representing a smooth square wave triggered after 10 s from origin and of 0.996 s duration; (bottom) normalized response at a virtual receiver located 100 m distance from the source at the free surface. Result obtained by taking the inverse Fourier transform of Lamb's solution (Equation 2.75) and convolving it with the source function. Medium properties are reported in Table 2.5.

geometrically with a factor proportional to $r^{-0.5}$, if r is the distance from the source. Body waves on the contrary attenuate at the free surface with a factor proportional to r^{-2} and thus much faster.

These laws of geometric spreading are valid for a linear elastic, homogeneous half-space and change with the geometry of the source. A vertically inhomogeneous medium modifies the geometric spreading function controlling the spatial attenuation of Rayleigh waves. This happens because of interference phenomena occurring among waves undergoing multiple reflections at the layer interfaces. The aim of this section is to derive an analytical expression of the geometric spreading law of Rayleigh waves produced by a time-harmonic, vertical point load applied at the free surface of an elastic, inhomogeneous half-space.

To this purpose, let $\mathbf{u}(r, x_2, \omega)$ denote the displacement vector induced by a harmonic, vertical point load placed at a position (Figure 2.25) having coordinates $r = \sqrt{x_1^2 + x_2^2} = 0$, $x_2 = 0$. The displacement field $\mathbf{u}(r, x_2, \omega)$ can be split additively into two components: $\mathbf{u}(r, x_2, \omega) = \mathbf{u}_B(r, x_2, \omega) + \mathbf{u}_S(r, x_2, \omega)$.[14] The first component, $\mathbf{u}_B(r, x_2, \omega)$, represents the body wave field, and it is composed by a superposition of P- and S-waves. The second component, $\mathbf{u}_S(r, x_2, \omega)$, is the surface wave field and, in general, is composed of a superposition of Love and Rayleigh waves. Because the body wave field attenuates with the distance from the source at a much higher rate than the surface wave field, it is reasonable to assume that at large distances from the source, the surface wave field dominates the overall particle motion, thus $\mathbf{u}(r, x_2, \omega) \approx \mathbf{u}_S(r, x_2, \omega)$.

In surface wave studies, the distance from the source where the body wave field is not negligible is usually called the *near-field*. Numerical simulations by Holzlohner (1980), Vrettos (1991), Tokimatsu (1995), Foti (2000), and Strobbia (2003) of surface wave propagation in layered half-spaces have shown that, in *normally* dispersive media, the near-field effects are important up to a distance from the source of about $\lambda_R/2$ [where $\lambda_R = \lambda_R(\omega)$ is the wavelength of Rayleigh waves]. However, in *inversely* dispersive media,[15] the near-field is larger and it may extend up to $2\lambda_R$. In the continuation of this section, it will be assumed that near-field effects are negligible. Furthermore, because the excitation is caused by a harmonic, vertical point load, Love waves do not contribute to the wave field.

With these assumptions, the (Rayleigh) displacement field $\mathbf{u}(r, x_2, \omega)$ can be computed from the solution of the Rayleigh eigenproblem using modal superposition.[16] At large distances, the wave field induced by a harmonic, vertical point load $F \cdot e^{i\omega t}$ acting in a direction perpendicular to the free surface of the half-space can be expanded, in the radial direction, in a series of κth order Hankel functions.[17] For large values of r, the κth order Hankel functions can be approximated by their asymptotic expansions involving complex exponential functions as shown by Equations 2.78 and 2.79. Thus, the particle displacement $\mathbf{u}(r, x_2, \omega) = u_r(r, x_2, \omega)\, \mathbf{e}_r + u_2(r, x_2, \omega)\, \mathbf{e}_2$

---

[14] This additive decomposition of the wave field in body and surface wave contributions is possible only in the *far-field*. A precise definition of this term is provided in the next paragraph.

[15] By *normally* dispersive media are denoted half-spaces where the *mechanical impedance* ρV_S is constant or increases smoothly with depth. Conversely, *inversely* dispersive media are vertically inhomogeneous half-spaces where the mechanical impedance ρV_S varies irregularly with depth.

[16] One way of solving a boundary value problem in linear elastodynamics (e.g., a *source problem*) is through a weighted "sum" (discrete or continuous) of eigenfunctions. Each term of the expansion is a measure of the contribution given by a particular mode of propagation to the wave field.

[17] This expansion is in a sense *natural* when considering the form of the Lamb's problem solution in a homogeneous half-space given by Equation 2.75.

resulting from the superposition of M distinct Rayleigh modes can be written in *cylindrical coordinates* $\{r, x_2, \vartheta\}$ as follows (Ben-Menahem and Singh 2000; Aki and Richards 2002)

$$u_l(r, x_2, \omega) = \sum_{j=1}^{M} [A_l(r, x_2, \omega)]_j \cdot e^{i(\omega \cdot t - k_j \cdot r + \varphi_l)} \tag{2.80}$$

where $l = r, 2$ denotes the radial and the vertical directions, respectively; $[A_l (r, x_2, \omega)]_j$ and $k_j(\omega)$ are the Rayleigh displacement amplitudes and wavenumber, respectively, associated with the jth mode of propagation. Finally, $\varphi_l = -\pi/4$ for $l = r$ and $\varphi_l = \pi/4$ for $l = 2$.

Equation 2.80 shows that $u_l(r, x_2, \omega)$ is independent of the azimuthal angle ϑ. This was expected due to the cylindrical symmetry of the problem. The actual displacement field is obtained by taking either the *real* or *imaginary* part of Equation 2.80. By choosing the latter, this equation may be rewritten as follows (Lai 1998)

$$\Im[u_l(r, x_2, \omega)] = \Im \left\{ \sum_{j=1}^{M} [A_l(r, x_2, \omega)]_j \cdot e^{i(\omega \cdot t - k_j \cdot r + \varphi_l)} \right\} =$$

$$= \sum_{j=1}^{M} [(C_l)_j \sin(\omega t) - (D_l)_j \cos(\omega t)], \quad l = r, 2 \tag{2.81}$$

where $\Im(\cdot)$ denotes the imaginary part of the argument, $(C_l)_j = (A_l)_j \cdot \cos(k_j r + \varphi_l)$ and $(D_l)_j = (A_l)_j \cdot \sin(k_j r + \varphi_l)$. Now, using simple trigonometric identities, Equation 2.81 becomes

$$\Im[u_l(r, x_2, \omega)] = U_l(r, x_2, \omega) \cdot \sin[\omega t - \psi_l(r, x_2, \omega)], \quad l = r, 2 \tag{2.82}$$

where

$$\begin{cases} U_l(r, x_2, \omega) = \left\{ \sum_{i=1}^{M} \sum_{j=1}^{M} [A_l(r, x_2, \omega)]_i \cdot [A_l(r, x_2, \omega)]_j \cdot \cos[r \cdot (k_i - k_j)] \right\}^{0.5} \\[4mm] \psi_l(r, x_2, \omega) = \tan^{-1} \left\{ \dfrac{\sum_{i=1}^{M} [A_l(r, x_2, \omega)]_i \cdot \sin(k_i \cdot r + \varphi_l)}{\sum_{j=1}^{M} [A_l(r, x_2, \omega)]_j \cdot \cos(k_j \cdot r + \varphi_l)} \right\}, \quad l = r, 2 \end{cases} \tag{2.83}$$

Taking the real part of Equation 2.80, repeating the procedure that lead to Equation 2.82, and combining the results yields

$$u_l(r,x_2,\omega) = U_l(r,x_2,\omega) \cdot e^{i\left[\omega t - \psi_l(r,x_2,\omega)\right]}, \qquad l = r, 2 \qquad (2.84)$$

The expressions for $U_l(r, x_2, \omega)$ and $\psi_l(r, x_2, \omega)$ in Equation 2.83 are functions of the modal amplitudes $[A_l(r, x_2, \omega)]_j$ of Rayleigh waves. For a harmonic, vertical point load $F \cdot e^{i\omega t}$ placed at $r = \sqrt{x_1^2 + x_2^2} = 0$, $x_2 = 0$ (Figure 2.25), the Rayleigh displacement amplitudes $[A_l(r, x_2, \omega)]_j$ of the individual modes of propagation are related to the displacement eigenfunctions $r_1(x_2, k, \omega)$ and $r_2(x_2, k, \omega)$ and to other modal parameters (Section 2.4.1) by the following relationship (Aki and Richards 2002)

$$\left[A_l(r,x_2,\omega)\right]_j = \begin{bmatrix} A_r(r,x_2,\omega) \\ A_2(r,x_2,\omega) \end{bmatrix}_j$$

$$= \frac{F \cdot r_2(0,k_j,\omega)}{4(V_R)_j \cdot (U_R)_j \cdot (I_R)_j \cdot \sqrt{2\pi r \cdot k_j}} \cdot \begin{bmatrix} r_1(x_2,k_j,\omega) \\ r_2(x_2,k_j,\omega) \end{bmatrix} \qquad (2.85)$$

with $l = r, 2$ and $(V_R)_j$, $(U_R)_j$, k_j are the phase velocity, group velocity, and wavenumber of the Rayleigh jth mode of propagation, respectively ($j = 1, M$). In Equation 2.85, the notation $r_2(0, k_j, \omega)$ has been used to indicate that the eigenfunction has to be evaluated at $x_2 = 0$, which is the position of the source at the free surface of the half-space. The term $(I_R)_j$ is the first Rayleigh energy integral and is defined by

$$(I_R)_j = \frac{1}{2} \int_0^\infty \rho(x_2) \left[(r_1^2)_j + (r_2^2)_j \right] dx_2, \qquad j = 1, M \qquad (2.86)$$

The substitution of Equation 2.85 into Equation 2.83 yields (Lai 1998)

$$\begin{cases} U_r(r,x_2,\omega) = \dfrac{F}{4\sqrt{2\pi} \cdot r} \left\{ \displaystyle\sum_{i=1}^{M} \sum_{j=1}^{M} \dfrac{r_1(x_2,k_i) r_1(x_2,k_j) r_2(0,k_i) r_2(0,k_j) \cos\left[r(k_i - k_j)\right]}{\sqrt{k_i k_j} \left[(V_R)_i (U_R)_i (I_R)_i\right] \left[(V_R)_j (U_R)_j (I_R)_j\right]} \right\}^{0.5} \\[20pt] U_2(r,x_2,\omega) = \dfrac{F}{4\sqrt{2\pi} \cdot r} \left\{ \displaystyle\sum_{i=1}^{M} \sum_{j=1}^{M} \dfrac{r_2(x_2,k_i) r_2(x_2,k_j) r_2(0,k_i) r_2(0,k_j) \cos\left[r(k_i - k_j)\right]}{\sqrt{k_i k_j} \left[(V_R)_i (U_R)_i (I_R)_i\right] \left[(V_R)_j (U_R)_j (I_R)_j\right]} \right\}^{0.5} \end{cases}$$

$$(2.87)$$

and

$$
\begin{cases}
\psi_r(r,x_2,\omega) = \tan^{-1}\left[\dfrac{\displaystyle\sum_{i=1}^{M}\dfrac{r_1(x_2,k_i)r_2(0,k_i)}{\sqrt{k_i}\cdot\left[(V_R)_i(U_R)_i(I_R)_i\right]}\cdot\sin\left(k_i\cdot r-\dfrac{\pi}{4}\right)}{\displaystyle\sum_{j=1}^{M}\dfrac{r_1(x_2,k_j)r_2(0,k_j)}{\sqrt{k_j}\cdot\left[(V_R)_j(U_R)_j(I_R)_j\right]}\cdot\cos\left(k_j\cdot r-\dfrac{\pi}{4}\right)}\right] \\[40pt]
\psi_2(r,x_2,\omega) = \tan^{-1}\left[\dfrac{\displaystyle\sum_{i=1}^{M}\dfrac{r_2(x_2,k_i)r_2(0,k_i)}{\sqrt{k_i}\cdot\left[(V_R)_i(U_R)_i(I_R)_i\right]}\cdot\sin\left(k_i\cdot r+\dfrac{\pi}{4}\right)}{\displaystyle\sum_{j=1}^{M}\dfrac{r_2(x_2,k_j)r_2(0,k_j)}{\sqrt{k_j}\cdot\left[(V_R)_j(U_R)_j(I_R)_j\right]}\cdot\cos\left(k_j\cdot r+\dfrac{\pi}{4}\right)}\right]
\end{cases}
$$

$$(2.88)$$

To simplify the notation, the frequency dependence of the eigenfunctions $r_1(x_2, k_j, \omega)$ and $r_2(x_2, k_j, \omega)$ has been omitted in Equations 2.87 and 2.88. Obviously, if the displacement is calculated at the free surface of the half-space, the eigenfunctions in Equations 2.87 and 2.88 are evaluated at $x_2 = 0$.

Equation 2.84 is instructive because it shows that a multiplicative decomposition of the displacement field of the type $u_l = |u_l|\cdot e^{i\cdot\arg(u_l)}$ ($l = r, 2$) is possible even in a vertically inhomogeneous half-space where surface wave propagation is multimodal. However, because the wavenumber $k_l(r, x_2, \omega)$ is no longer a constant, as happens in a homogeneous medium, the spatial variation of the displacement field is no longer harmonic even though the temporal variation of the source is sinusoidal. Equations 2.84, 2.87, and 2.88 also show the remarkable result: the three main variables controlling the outcome of $\mathbf{u}(r, x_2, \omega)$, namely the source depth ($x_2 = 0$), the receiver depth (x_2), and the distance from the source (r), are uncoupled in the sense that their contribution is independent. The final step is to rewrite Equation 2.84 as follows

$$
u_l(r,x_2,\omega) = F\cdot\Upsilon_l(r,x_2,\omega)\cdot e^{i\left[\omega t-\psi_l(r,x_2,\omega)\right]}, \quad l = r, 2
$$

$$(2.89)$$

from a comparison with Equation 2.87

$$
\begin{cases}
\Upsilon_r(r,x_2,\omega) = \dfrac{1}{4\sqrt{2\pi}\cdot r}\left\{\displaystyle\sum_{i=1}^{M}\sum_{j=1}^{M}\dfrac{r_1(x_2,k_i)r_1(x_2,k_j)r_2(0,k_i)r_2(0,k_j)\cos\left[r(k_i-k_j)\right]}{\sqrt{k_ik_j}\left[(V_R)_i(U_R)_i(I_R)_i\right]\left[(V_R)_j(U_R)_j(I_R)_j\right]}\right\}^{0.5} \\[40pt]
\Upsilon_2(r,x_2,\omega) = \dfrac{1}{4\sqrt{2\pi}\cdot r}\left\{\displaystyle\sum_{i=1}^{M}\sum_{j=1}^{M}\dfrac{r_2(x_2,k_i)r_2(x_2,k_j)r_2(0,k_i)r_2(0,k_j)\cos\left[r(k_i-k_j)\right]}{\sqrt{k_ik_j}\left[(V_R)_i(U_R)_i(I_R)_i\right]\left[(V_R)_j(U_R)_j(I_R)_j\right]}\right\}^{0.5}
\end{cases}
$$

$$(2.90)$$

The function $\Upsilon_l(r, x_2, \omega)$ $(l = r, 2)$ has been named the *Rayleigh geometric attenuation function* (Lai 1998). It has the important physical interpretation of modeling the geometric attenuation of Rayleigh surface waves in vertically inhomogeneous half-spaces. Whereas in a homogeneous half-space, Rayleigh waves generated by a vertical, harmonic point load attenuate with distance by a factor proportional to $r^{-0.5}$ as a result of geometric spreading, this simple attenuation law (which follows directly from the principle of conservation of energy) does not hold in vertically inhomogeneous continua. The reason is that in stratified media, the Rayleigh wave field originates from the superposition of several modes of propagation that are caused by constructive interference among waves undergoing multiple reflections and refractions at the layer interfaces. As a result of this phenomenon, the geometric spreading law $r^{-0.5}$, valid in homogeneous media, is altered, and it is replaced by the function $\Upsilon_l(r, x_2, \omega)$ $(l = r, 2)$.

From Equation 2.90, it is easy to verify that, if the medium is homogeneous, then $M = 1$, $k_i = k_j = k$, $(U_R)_i = (V_R)_i V_R$, and the geometric spreading function $\Upsilon_l(r, x_2, \omega)$ $(l = r, 2)$ reduces, as expected, to the relation $E_l/\sqrt{r}$, where $E_r = r_1(x_2)r_2(0)/4\sqrt{2} \cdot V_R^2 \cdot I_R \cdot \pi k$ and $E_2 = r_2(x_2)r_2(0)/4\sqrt{2} \cdot V_R^2 \cdot I_R \cdot \pi k$ are two constants. The relevance of the explicit factorization of the Rayleigh displacement field into the product of load magnitude, geometric spreading function, and phase factor exhibited by Equation 2.89 will become apparent in Section 2.5.3, wherein the attenuation of Rayleigh waves in weakly dissipative media is discussed.

In Figure 2.29, a graphical representation of $\Upsilon_2(r, x_2, \omega)$ at the free surface $(x_2 = 0)$ is shown, which has been obtained after performing a series of numerical simulations for three types of systems: a homogeneous half-space, a three-layer normally dispersive medium, and a three-layer inversely dispersive medium. The simulations were carried out at the frequencies of 7, 15, 40, and 90 Hz, using the material properties illustrated in Tables 2.6 through 2.8.

Figure 2.29 shows that the geometric attenuation law $r^{-0.5}$ valid in a homogeneous half-space does not hold in normally and inversely dispersive media, particularly as the frequency increases. This happens because by increasing the frequency, it increases the number of modes of propagation caused by interference phenomena. These findings have also been observed in other independent numerical studies (Gucunski and Woods 1991; Tokimatsu et al. 1992; Strobbia 2003).

2.4.2.4 Apparent phase velocity of surface waves

In a vertically inhomogeneous half-space, surface Rayleigh waves generated by harmonic sources propagate at a speed that is called *apparent* or *effective* Rayleigh phase velocity (Tokimatsu 1995; Lai 1998; Foti 2000; Strobbia 2003). This kinematic quantity describes the speed of propagation

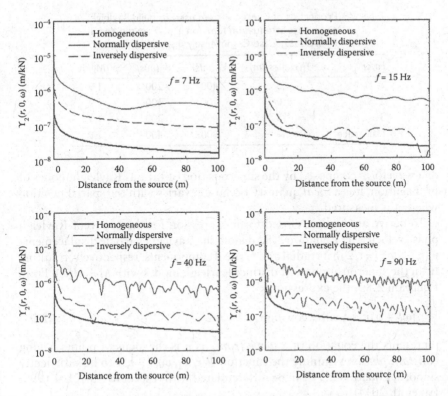

Figure 2.29 Geometric attenuation function $\Upsilon_2(r, 0, \omega)$ in three types of soil media and at different frequencies. The material properties used in the numerical simulations are reported in Tables 2.6 through 2.8.

Table 2.6 Properties of layered system used to build Rayleigh geometric attenuation function $\Upsilon_2(r, 0, \omega)$ plotted in Figure 2.29; Case A—homogeneous half-space

Layer	Thickness (m)	V_P (m/s)	V_S (m/s)	ρ (Mg/m³)
Half-space	∞	600	300	1.9

Table 2.7 Properties of layered system used to build Rayleigh geometric attenuation function $\Upsilon_2(r, 0, \omega)$ plotted in Figure 2.29; Case B—normally dispersive medium

Layer	Thickness (m)	V_P (m/s)	V_S (m/s)	ρ (Mg/m³)
1	5	200	100	1.9
2	10	400	200	1.9
3	15	600	300	1.9
Half-space	∞	800	400	1.9

Table 2.8 Properties of layered system used to build Rayleigh geometric attenuation function $\Upsilon_2(r, 0, \omega)$ plotted in Figure 2.29; Case C—inversely dispersive medium

Layer	Thickness (m)	V_P (m/s)	V_S (m/s)	ρ (Mg/m³)
1	5	400	200	1.9
2	10	200	100	1.9
3	15	600	300	1.9
Half-space	∞	800	400	1.9

of a waveform composed by the superposition of various Rayleigh modes of propagation; it is a *local* quantity because it varies with the spatial position where it is measured.

To derive an explicit, closed-form expression for the apparent Rayleigh phase velocity, let Equation 2.82 denote the Rayleigh particle displacements $u_l(r, x_2, \omega), l = r, 2$ in its radial and vertical components, respectively, resulting from the superposition of M distinct Rayleigh modes with $M = M(\omega)$. From Equation 2.82, the expression

$$[\omega t - \psi_l(r, x_2, \omega)] = \text{constant}, \quad l = r, 2 \tag{2.91}$$

represents the equation of a *wave front*, as it is the locus of points having *constant phase*. Assuming the function $\psi_l(r, x_2, \omega), l = r, 2$ to be sufficiently smooth, Equation 2.91 can be differentiated with respect to time (Lai 1998; Lai et al. 2014)

$$\omega - \frac{\partial \psi_l}{\partial r}(r, x_2, \omega) \cdot \frac{dr}{dt} = 0, \quad l = r, 2 \tag{2.92}$$

to yield

$$V_l^{app}(r, x_2, \omega) = \frac{\omega}{\dfrac{\partial \psi_l(r, x_2, \omega)}{\partial r}}, \quad l = r, 2 \tag{2.93}$$

where the symbol $V_l^{app}(r, x_2, \omega), l = r, 2$ denotes the *apparent or effective* Rayleigh phase velocity. Equation 2.93 shows that the apparent Rayleigh phase velocity is a local quantity; that is, its magnitude is a function of the spatial position where it is calculated.

At a fixed position in space, say at $x_2 = 0$, the function $V_l^{app}(r, 0, \omega)$, $l = r, 2$ describes a 2D surface, called the Rayleigh *dispersion surface*, showing the variation of the apparent Rayleigh phase velocity with frequency and distance from the source (Lai 1998). Equation 2.93 also shows that because the apparent Rayleigh phase velocity is a vector quantity, the two components of $V_l^{app}(r, 0, \omega), l = r, 2$ will, in general, travel at different apparent phase velocities.

The denominator of Equation 2.93 can be interpreted as an effective Rayleigh wavenumber and denoted as $k_l^{app}(r, 0, \omega)$, $l = r, 2$. However, a decomposition of the argument of Equation 2.82 in the form $\left(\omega t - k_l^{app} \cdot r\right)$, which is standard for monochromatic waves, is no longer possible because the *apparent wavenumber* $k_l^{app}(r, 0, \omega)$, being a *local* quantity, must be integrated over r to yield the phase $\psi_l^{app}(r, 0, \omega)$.

From Equation 2.93, it is now possible to obtain an explicit definition of the apparent Rayleigh phase velocity expressed in terms of quantities derived from the solution of the Rayleigh eigenproblem (i.e., modal phase and group velocities, eigenfunctions, etc.). Substitution of Equation 2.83 into Equation 2.93 yields (Lai et al. 2014)

$$V_l^{app}(r, x_2, \omega) = 2\omega \cdot \frac{\displaystyle\sum_{i=1}^{M}\sum_{j=1}^{M}\left\{(A_l)_i (A_l)_j \cdot \cos\left[r(k_i - k_j)\right]\right\}}{\displaystyle\sum_{n=1}^{M}\sum_{m=1}^{M}\left\{(A_l)_n (A_l)_m (k_n + k_m) \cdot \cos\left[r(k_n - k_m)\right]\right\}} \qquad (2.94)$$

Finally, substitution of Equation 2.85 into Equation 2.94 gives

$$\begin{cases} V_r^{app}(r, x_2, \omega) = \dfrac{2\omega \displaystyle\sum_{i=1}^{M}\sum_{j=1}^{M}\left\{\dfrac{r_1(x_2, k_i) r_1(x_2, k_j) r_2(0, k_i) r_2(0, k_j) \cos\left[r(k_i - k_j)\right]}{\left[(V_R)_i (U_R)_i (I_R)_i\right]\left[(V_R)_j (U_R)_j (I_R)_j\right]\sqrt{k_i k_j}}\right\}}{\displaystyle\sum_{n=1}^{M}\sum_{m=1}^{M}\left\{\dfrac{r_1(x_2, k_n) r_1(x_2, k_m) r_2(0, k_n) r_2(0, k_m)(k_n + k_m)\cos\left[r(k_n - k_m)\right]}{\left[(V_R)_n (U_R)_n (I_R)_n\right]\left[(V_R)_m (U_R)_m (I_R)_m\right]\sqrt{k_n k_m}}\right\}} \\[40pt] V_2^{app}(r, x_2, \omega) = \dfrac{2\omega \displaystyle\sum_{i=1}^{M}\sum_{j=1}^{M}\left\{\dfrac{r_2(x_2, k_i) r_2(x_2, k_j) r_2(0, k_i) r_2(0, k_j) \cos\left[r(k_i - k_j)\right]}{\left[(V_R)_i (U_R)_i (I_R)_i\right]\left[(V_R)_j (U_R)_j (I_R)_j\right]\sqrt{k_i k_j}}\right\}}{\displaystyle\sum_{n=1}^{M}\sum_{m=1}^{M}\left\{\dfrac{r_2(x_2, k_n) r_2(x_2, k_m) r_2(0, k_n) r_2(0, k_m)(k_n + k_m)\cos\left[r(k_n - k_m)\right]}{\left[(V_R)_n (U_R)_n (I_R)_n\right]\left[(V_R)_m (U_R)_m (I_R)_m\right]\sqrt{k_n k_m}}\right\}} \end{cases}$$

$$(2.95)$$

where $V_r^{app}(r, x_2, \omega)$ and $V_2^{app}(r, x_2, \omega)$ denote the components of the apparent Rayleigh phase velocity along the radial and vertical directions, respectively. Again, to reduce the length of Equation 2.95, the frequency dependence of the eigenfunctions $r_1(x_2, k, \omega)$ and $r_2(x_2, k, \omega)$ has been left unspecified. As shown by Equation 2.95, the apparent Rayleigh phase velocity is completely determined from the solution of the Rayleigh eigenvalue problem (Section 2.4.1). Indeed, all the modal quantities appearing in Equation 2.95, including $(V_R)_j = \omega/k_j$, $(U_R)_j = d\omega/dk_j$, and $(I_R)_j$, $j = 1,M$, can be computed from the set $\left\{ k_j, r_n^{(j)}(x_2, k_j, \omega) \right\}$, $n = 1,4$.

An example of application of the concept of apparent Rayleigh phase velocity is illustrated in Figures 2.30 and 2.31, where the modal dispersion curves associated with two different layered half-spaces—one normally dispersive, the other inversely dispersive—have been superimposed to the corresponding apparent dispersion curves (dotted lines). The material properties used in the numerical simulations are those reported in Tables 2.7 and 2.8. The dependence of $V_2^{app}(r,0,\omega)$ on the distance from the source has been eliminated via an averaging procedure, which also eliminates the near-field effects (Lai et al. 2014). In real testing, the adopted experimental configuration of the receivers will affect the shape of the apparent dispersion curve that can be extracted (see Chapter 3). Details on the experimental determination of the apparent Rayleigh dispersion curve are given in Chapter 4.

Figure 2.30 is instructive because it shows that even in a normally dispersive half-space, the apparent dispersion curve does not coincide with the dispersion curve associated with the fundamental mode of propagation in the whole frequency range. It is often assumed that in normally dispersive media most of the surface wave energy travels with the group velocity of the first

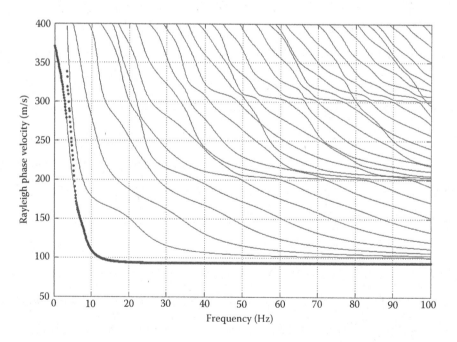

Figure 2.30 Comparison between modal and apparent Rayleigh dispersion curves in a *normally* dispersive half-space. Material properties are shown in Table 2.7.

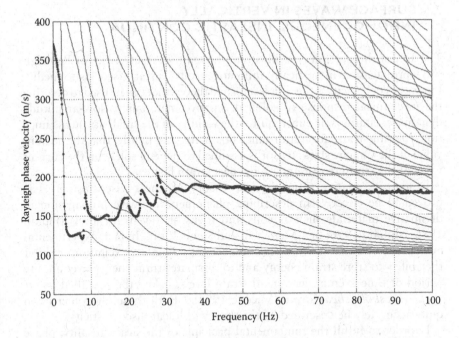

Figure 2.31 Comparison between modal and apparent Rayleigh dispersion curves in a *inversely* dispersive half-space. Material properties are shown in Table 2.8.

mode of propagation because the contribution of higher modes is relatively small. Even though this is a rather common situation (Gucunski and Woods 1991; Tokimatsu et al. 1992; Foti 2000), it cannot be assumed as a conclusion of general validity because there are examples where multimode wave propagation is important even in normally dispersive sites (Strobbia 2003). In the example shown in Figure 2.30, the relative larger importance of higher modes in the low-frequency range causes a deviation of the apparent dispersion curve from that associated with the fundamental mode. A further discussion about this issue is reported in Section 7.4, where some examples are provided.

On the contrary, in inversely dispersive media, the contribution of higher modes of propagation is important—as shown in Figure 2.31. The apparent dispersion curve separates from the fundamental mode of propagation at a frequency of about 8 Hz and never turns back as the frequency increases. Thus, surface wave propagation in inversely dispersive media is a multimodal phenomenon, and, as such, any attempt to successfully determine the medium parameters (e.g., the V_S-profile) from the inversion of experimentally determined Rayleigh dispersion curves should take into account modes of propagation higher than the first. See Chapter 6 for an in-depth discussion of this issue in relation to the solution of the inverse problem associated with surface waves.

2.5 SURFACE WAVES IN VERTICALLY INHOMOGENEOUS, INELASTIC CONTINUA

The aim of this section is to illustrate the main characteristics of Rayleigh wave propagation in vertically inhomogeneous, linear viscoelastic media. Linear viscoelasticity is the simplest formal theory that can be used to describe the mechanical response of solid, dissipative materials to low-amplitude dynamic excitations. It only requires the validity of the small-strain assumption, the time-translation invariance postulate, and finally, that the current value of the Cauchy stress tensor is assumed to depend solely on the current value of the strain tensor and on the past strain history.

Despite its relative simplicity, the theory of linear viscoelasticity has proved to be effective in describing phenomena of wave propagation in dissipative materials such as soils and rocks at low-strain levels (Pipkin 1986; Ishihara 1996; Ben-Menahem and Singh 2000). In fact, experimental evidence shows that geomaterials subjected to dynamic excitations exhibit the ability to store strain energy and to dissipate strain energy over a finite period of time even at very small strain levels, below the so-called *linear cyclic threshold shear strain* (Vucetic 1994).[18] Both these phenomena can quite accurately be described by the theory of linear viscoelasticity.

In order to fulfill the fundamental principle of physical causality, phase velocity and attenuation of a mechanical disturbance propagating in a linear viscoelastic medium cannot be assigned independently. They must satisfy the *Kramers–Krönig equation*, which establishes a relation between the material damping ratio and the frequency-dependent speed of propagation of a viscoelastic pulse. Thus, dissipative materials are inherently dispersive. Illustration of these and other features concerned with the propagation of viscoelastic bulk and surface Rayleigh waves in vertically inhomogeneous continua is preceded by a brief section dedicated to a review of viscoelastic constitutive modeling with definition of the associated model parameters.

2.5.1 Constitutive modeling of linear dissipative materials

Experimental evidence shows that under dynamic excitation, most geomaterials such as soils and rocks exhibit a mechanical response that is strongly dependent upon the *norm* of the deviatoric strain tensor.[19] Figure 2.32 illustrates

---

[18] This is a value of shear strain below which particulate materials, such as soils, exhibit a linear response. This may be inelastic, however, no appreciable phenomena of instantaneous energy dissipation take place which would be typical of an elastoplastic response. Energy losses occur only over a finite period of time.

[19] The norm of a second-order tensor T is defined by $\|T\| = \max\left\{\left|\lambda^{(1)}\right|, \left|\lambda^{(2)}\right|, \left|\lambda^{(3)}\right|\right\}$, where $\lambda^{(i)} j = 1, 2, 3$ are its eigenvalues, which may be computed from the equation $\det(T - \lambda^{(i)} 1) = 0$. If T coincides with the small-strain tensor ε, the eigenvalues $\lambda^{(i)}$ represent the principal stretches.

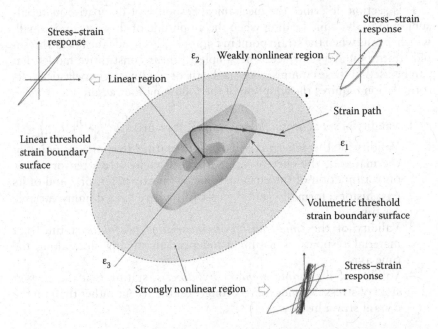

Figure 2.32 Conceptual representation of the mechanical response of geomaterials to dynamic excitations. The dependence of the response from the magnitude of the deviatoric strain tensor is illustrated through the notions of linear and volumetric cyclic threshold strain boundary surfaces in the principal strain space.

this statement using a 3D conceptual representation of the *linear* and *volumetric cyclic threshold strain boundary surfaces* in the principal strain space. These are generalizations of the uniaxial notions of cyclic threshold shear strains (Vucetic 1994) for the case of multiaxial loading.

For states of strain in the interior of the *linear* threshold strain boundary surface, geomaterials tend to exhibit linear response under static-monotonic and dynamic loading when the phenomenon of energy dissipation cannot be neglected (even though it is small in magnitude). In the weakly nonlinear region, which is bounded by the *volumetric* threshold strain boundary surface (Figure 2.32), the cyclic stress–strain response is characterized by hysteretic loops having a larger area (greater energy dissipation) that remains stable with an increase in the number of cycles. In this region, dilatancy phenomena are minor and stiffness degradation is limited. With the increase in the strain level, the strain point trespasses the *volumetric* threshold strain boundary surface and enters into the strongly nonlinear region where the hysteretic loops become unstable as the number of cycles increases (Figure 2.32). In this phase, soil response exhibit marked nonlinearities with severe stiffness and strength degradation.

This section describes the mechanical response of dissipative materials subjected to dynamic loading when the amplitude of shear strain is small, which means when the strain point in Figure 2.32 is inside the linear threshold strain boundary surface. The simplest, linear constitutive model that can be used in these circumstances is the theory of linear viscoelasticity. Its formulation requires the adoption of the following four assumptions:

1. Validity of *small-strain theory*, namely $\|\varepsilon\| = \max\left\{\left|\lambda^{(1)}\right|, \left|\lambda^{(2)}\right|, \left|\lambda^{(3)}\right|\right\} \ll 1$.
2. Validity of the *inheritance postulate* stating that at any point of the material, the current value of the Cauchy stress tensor $\sigma(t)$ is only a function of the current value of the strain tensor ε and of its past history; formally $\left\{\varepsilon(\tau)\right\}_{-\infty}^{t} \mapsto \sigma(t)$, where τ is a dummy variable for time.
3. Validity of the *time-translation invariance postulate* stating that material response is assumed independent of any shift along the time axis.
4. Validity of the *fading memory hypothesis*, stating that the current state of stress depends more strongly on the recent rather than on the distant strain history.

Continuity of strain history is desirable but not strictly required because discontinuous strain histories can be easily handled by means of integral operators intended in the *Stieltjes sense* (Fung 1965). Under these assumptions, the *Riesz representation theorem* of functional analysis (Christensen 1971) guarantees the existence of a unique relationship between the Cauchy stress tensor $\sigma(t)$ and the strain history $\left\{\varepsilon(\tau)\right\}_{-\infty}^{t}$ via the following linear functional

$$\sigma(t) = \int_{-\infty}^{t} G(t-\tau) : \frac{d\varepsilon(\tau)}{d\tau} d\tau \tag{2.96}$$

where $G(t)$ is a fourth-order tensor-valued function of the material called the *relaxation tensor function*. In deriving Equation 2.96, it was assumed that the strain history is continuous; however, as mentioned earlier, discontinuities in the strain history may be handled as well, if the integral appearing in Equation 2.96 is intended in the *Stieltjes sense*.

The constitutive relationship described by Equation 2.96 is sometimes called the *Boltzmann's equation* because it can also be derived by applying Boltzmann's superposition principle. The relaxation tensor function $G(t)$ has 81 components; however, only 21 are independent due to the symmetry of the stress and strain history tensors of a general viscoelastic anisotropic material.

For an isotropic, linear, viscoelastic material, the relaxation tensor function has only two independent components, and they are sufficient to completely describe the mechanical response of the material. In this case, Equation 2.96 can be rewritten as

$$
\begin{cases}
s(t) = \int\limits_{-\infty}^{t} 2G_S(t-\tau)\dfrac{de(\tau)}{d\tau}\,d\tau \\[4mm]
tr\big[\sigma(t)\big] = \int\limits_{-\infty}^{t} 3G_B(t-\tau)\dfrac{d\,tr\big[\varepsilon(\tau)\big]}{d\tau}\,d\tau
\end{cases}
\tag{2.97}
$$

where $s = \sigma - \dfrac{1}{3}tr(\sigma)1$ and $e = \varepsilon - \dfrac{1}{3}tr(\varepsilon)1$ are the components of the deviatoric stress and strain tensors, respectively. The scalar functions, $G_S(t)$ and $G_B(t)$, are the shear and bulk relaxation functions, respectively. From Equation 2.97, shear and volume deformations of viscoelastic isotropic materials are uncoupled, mimicking a well-known fact of linear isotropic elasticity.

$G_S(t)$ and $G_B(t)$ are material response functions, and they are analogous of the shear and bulk moduli of linear elasticity with the important difference that the relaxation functions are not constants but are time dependent. By definition, $G_S(t)$ is the shear stress response of a material subjected to a shear strain history specified as a Heaviside or step function.[20] A similar interpretation applies to the bulk relaxation function $G_B(t)$.

Although viscoelastic constitutive relations often are given a physical interpretation in terms of rheological models made up by various combinations of springs and viscous dashpots, this is not necessary; far more general models for viscoelastic materials may be constructed without resorting to networks of springs and dashpots.

Once $G_S(t)$ and $G_B(t)$ have been specified, Equation 2.97 may be used to the compute stress response of the material to a prescribed strain history. However, this operation may be not trivial. Instead, if the strain history is specified as a harmonic function of time, the viscoelastic constitutive relationships assume a very simple form. For example, let us assume that the strain history in Equation 2.96 is specified by the function $\varepsilon(t) = \varepsilon_0 \cdot e^{i\omega t}$, where ε_0 is the strain amplitude tensor. Then, the integral equation degenerates into the following algebraic equation

$$
\sigma(t) = G^*(\omega) : \varepsilon_0 \cdot e^{i\omega t}
\tag{2.98}
$$

---

[20] Alternatively to the shear relaxation function, the response of a viscoelastic material could be specified in terms of the *shear creep function* $J_s(t)$, which is the shear strain response of a material subjected to a shear stress history specified as a Heaviside function.

where $G^*(\omega) = G_{(1)}(\omega) + i \cdot G_{(2)}(\omega)$ is the fourth-order *complex-valued tensor modulus*.

The real and imaginary components $G_{(1)}(\omega)$ and $G_{(2)}(\omega)$ of $G^*(\omega)$ are not independent. Their relationship can be easily found using Fourier integral theorem (Christensen 1971). The result is

$$G_{(1)}(\omega) = G_{(e)} + \frac{2}{\pi} \cdot \int_0^\infty \frac{G_{(2)}(\tau) \cdot \omega^2}{\tau \cdot (\omega^2 - \tau^2)} \cdot d\tau \qquad (2.99)$$

where $G_{(e)} = G(t \to \infty)$ is known as the *equilibrium response* of the relaxation function. Equation 2.99 represents one form of the *Kramers–Krönig relations*. They are important because they state that viscoelastic materials are inherently *dispersive*; thus, the speed of propagation of a mechanical disturbance is frequency dependent (see Section 2.5.2). If the material is isotropic, Equation 2.98 simplifies as follows

$$\begin{cases} s(t) = 2G_S^*(\omega) \cdot \mathbf{e}_0 \cdot e^{i\omega t} \\ \mathrm{tr}[\sigma(t)] = 3G_B^*(\omega) \cdot \mathrm{tr}(\varepsilon_0) \cdot e^{i\omega t} \end{cases} \qquad (2.100)$$

where $\mathbf{e}_0$ is the amplitude of the deviatoric strain tensor and $G_S^*(\omega)$ and $G_B^*(\omega)$ are the complex shear and bulk moduli, respectively.[21] A relevant feature predicted by Equation 2.100 is that the stress components of viscoelastic materials undergoing steady state harmonic oscillations are, in general, out of phase with the corresponding strain components. The amount by which the stress lags behind the strain is measured by the argument of the complex modulus, which is also a measure of the amount of energy dissipated by the viscoelastic material during harmonic oscillations. To demonstrate this, let Equation 2.100 be rewritten for the simple case of shear, uniaxial oscillation

$$\sigma_S(\omega) = G_S^*(\omega) \cdot \varepsilon_S(\omega) \qquad (2.101)$$

where $\varepsilon_S(\omega) = \varepsilon_{S0}(\omega) \cdot e^{i\omega t}$ with $\varepsilon_{S0} \in \mathbb{R}$. By taking the real part of $\varepsilon_S(\omega)$, Equation 2.101 can be rewritten as

$$\sigma_S(\omega) = \left| G_S^*(\omega) \right| \cdot \varepsilon_{S0} \cdot \cos\left[\omega t - \phi_S(\omega) \right] \qquad (2.102)$$

---

[21] Similarly to the relation between the *relaxation* and *creep* functions in the time domain, the response of a viscoelastic material in the frequency domain could be specified by the shear and bulk complex *compliances* $J_S^*(\omega)$ and $J_B^*(\omega)$ as alternatives of $G_S^*(\omega)$ and $G_B^*(\omega)$.

where

$$\tan\left[\phi_S(\omega)\right] = \arg\left[G_S^*(\omega)\right] = \frac{G_{(2)S}}{G_{(1)S}} \tag{2.103}$$

is the *loss shear angle* or the *loss shear tangent*. The real and imaginary parts $G_{(1)S}$ and $G_{(2)S}$ of the complex modulus $G_S^*(\omega)$ are often referred to as the *storage* and *loss shear modulus*, respectively (Pipkin 1986). Using trigonometric identities, Equation 2.102 can be rewritten as

$$\sigma_S(\omega) - \left|G_S^*(\omega)\right| \cdot \varepsilon_{S0}(\omega) \cdot \cos\left[\phi_S(\omega)\right] = \left|G_S^*(\omega)\right| \cdot \varepsilon_{S0} \cdot \sin\left[\phi_S(\omega)\right] \cdot \sin(\omega t)$$

$$\tag{2.104}$$

which, combined with the relation $\varepsilon_S(\omega) = \varepsilon_{S0} \cdot \cos(\omega t)$, yields

$$\left(\frac{\varepsilon_S}{\varepsilon_{S0}}\right)^2 + \left(\frac{\sigma_S - G_{(1)S} \cdot \varepsilon_S}{G_{(2)S} \cdot \varepsilon_{S0}}\right)^2 = 1 \tag{2.105}$$

This is the equation of an ellipse rotated by an angle $\psi_S(\omega)$ with respect to the strain axis assumed to coincide with the x-axis. This ellipse represents the stress–strain *hysteretic loop* exhibited by a general linear viscoelastic material subjected to harmonic oscillations. It can be shown that the inclination of the principal axes of the ellipse $\psi_1 = \psi_S(\omega)$ and $\psi_2 = (\pi/2 + \psi_1)$ is given by the following relation (Lai 1998)

$$\tan\psi_{1,2}(\omega) = \frac{\left[|G_S^*|^2 - 1\right] \mp \sqrt{\left[|G_S^*|^2 + 1\right]^2 - 4G_{(2)S}^2}}{2 \cdot G_{(1)S}} \tag{2.106}$$

The energy dissipated by the material (per unit volume) during a cycle of harmonic oscillation can be computed from the area enclosed by the elliptic hysteretic loop. In a stress-controlled test, the following relation defines the area ΔW_S^{dissip}:

$$\Delta W_S^{dissip}(\omega) = \oint_l dW_S = \oint_l \Re(\sigma_S) \cdot \Re(d\varepsilon_S) \tag{2.107}$$

where the symbol $\Re(\cdot)$ denotes the real part of a complex quantity, and l is length of the hysteretic loop. In Equation 2.107, the term dW_S represents the work done by the stress (per unit volume of material) for an infinitesimal variation of strain. Considering Equation 2.102 and the fact that $\Re(d\varepsilon_S) = \omega \cdot \varepsilon_{S0} \cdot \cos(\omega t + \pi/2) \, dt$, Equation 2.107 can be rewritten as follows

$$\Delta W_S^{dissip}(\omega) = \int_0^{2\pi/\omega} \omega \cdot \varepsilon_{S0}^2 \cdot \left|G_S^*\right| \cdot \cos(\omega t - \phi_S) \cdot \cos(\omega t + \pi/2) \cdot dt \tag{2.108}$$

Finally, making use of trigonometric identities, this integral can be easily solved to yield

$$\Delta W_S^{dissip}(\omega) = \pi \cdot G_{(2)S} \cdot |\varepsilon_S|^2 \tag{2.109}$$

Thus, the quantity ΔW_S^{dissip} is directly proportional to the loss modulus $G_{(2)S}$. Although these results have been obtained for the simple case of shear uniaxial oscillation, they can be easily generalized for *multi axial* loading. Figure 2.33 shows an experimental stress–strain hysteretic loop measured during a cyclic torsional shear test performed on a cylindrical clay specimen. Despite the fact that the maximum shear strain ampli-tude is below the cyclic linear threshold shear strain[22] (Figure 2.32), the stress–strain loop encloses a nonvanishing area, which is evidence that in geomaterials energy dissipation takes place even at very low amplitude dynamic excitations. In this particular experiment, the damping ratio was measured to be about 2%.

At the microscopic level, different mechanisms have been proposed to explain the process of energy dissipation at very small strain levels that occurs in geomaterials subjected to dynamic excitations (Biot 1956; Stoll 1974; Johnston et al. 1979; White 1983; Leurer 1997). These studies indicate that an interactive combination of several individual mechanisms is

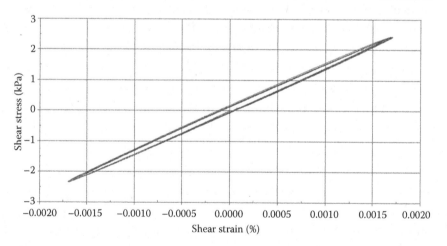

Figure 2.33 Experimental stress–strain hysteretic loop measured during a cyclic torsional shear test on a cylindrical clay specimen. Maximum shear strain amplitude $1.7 \cdot 10^{-5}$ at $f = 0.01$ Hz. (From Lai, C. G. et al., *Misura simultanea del modulo di taglio e dello smorzamento intrinseco dei terreni a piccole deformazioni* (in Italian), ANIDIS, Torino, 1999.)

---

[22] For normally consolidated clays with plasticity index of 50, the cyclic linear threshold shear strain is in the order of 10^{-4} (Vucetic 1994).

responsible for most of the phenomena macroscopically called *energy dissipation*. For coarse-grained soils, the two mechanisms that have been postulated are frictional losses between soil particles and fluid flow losses due to the relative movement between the solid and fluid phases. Fine-grained soils exhibit more complex phenomena, which are controlled by electromagnetic interactions between water dipoles and microscopic solid particles.

Another important feature illustrated by Figure 2.33 is that the elliptical shape of cyclic stress–strain loops predicted by the theory of linear viscoelasticity is matched fairly well by geomaterials in experimental measurements performed at very small strains, which has also been shown by other researchers (Dobry 1970).

Based on Equation 2.109, several definitions have been proposed in the literature as measures of energy dissipation for geomaterials (O'Connell and Budiansky 1978; Ishihara 1996; Aki and Richards 2002). All of them are consistent with each other only when applied to *weakly* dissipative viscoelastic materials.[23] Seismologists and geophysicists have adopted definitions of energy losses borrowed from disciplines such as electrical engineering (Cole and Cole 1941). Despite different approaches, most of these definitions involve dimensionless parameters proportional to the ratio between the energy dissipated during a cycle of harmonic oscillation ΔW_S^{dissip} and some measures of the stored energy per unit volume. In soil dynamics, the parameter traditionally used as a measure of energy dissipation during harmonic excitation is the so-called *material damping ratio*

$$D_\chi^{geo}(\omega) = \frac{\Delta W_\chi^{dissip}(\omega)}{4\pi \cdot W_\chi^{max}(\omega)} \tag{2.110}$$

where $W_\chi^{max}(\omega)$ is the maximum stored energy per unit volume during one cycle of harmonic excitation and the subscript $\chi = P, S$ is used to denote the material damping ratio associated with irrotational (or longitudinal) and equivoluminal (or transversal) wave motion, respectively.

Although Equation 2.110 seems a plausible definition of material damping ratio, it is actually inconvenient to use. The reason is that the maximum stored energy $W_\chi^{max}(\omega)$ per unit volume of a harmonically excited linear viscoelastic material depends not only on the storage modulus $G_{(1)\chi}$ but also on the loss modulus $G_{(2)\chi}$ as well as on their derivatives with respect to frequency. This is caused by the phase lag existing among the various energy storing mechanisms governing the response of linear viscoelastic materials during harmonic excitations (O'Connell and Budiansky 1978;

---

[23] A weakly dissipative (or *low-loss or loss-less*) material is such that sup(D) ≤ 0.05 where sup(·) denotes the least upper bound of the argument and D is *material damping ratio*, which will be defined in the Section 2.5.2.

Tschoegl 1989). As a result, when the definition of material damping ratio given by Equation 2.110 is expressed in terms of the complex modulus $G_\chi^*(\omega)$, the ensuing expression is cumbersome. This is required to correlate material damping ratio with the constitutive parameters of linear viscoelasticity. These difficulties of Equation 2.110 can be overcome by redefining this parameter as follows

$$D_\chi(\omega) = \frac{\Delta W_\chi^{dissip}(\omega)}{8\pi \cdot W_\chi^{ave}(\omega)} \tag{2.111}$$

where $\chi = P,S$ and the term $W_\chi^{ave}(\omega)$ is the average stored energy over one cycle of harmonic oscillation.

An analogous, dimensionless definition of energy dissipation is used by seismologists and geophysicists via a material parameter called the *quality factor* (O'Connell and Budiansky 1978; Aki and Richards 2002) and denoted by $Q_\chi(\omega)$. The relation between the two parameters $D_\chi(\omega)$ and $Q_\chi(\omega)$ is

$$Q_\chi(\omega) = \frac{1}{2D_\chi(\omega)} \tag{2.112}$$

with $\chi = P, S$. It can be shown that $W_\chi^{ave}(\omega)$ in Equation 2.111 can be expressed as (Tschoegl 1989)

$$W_\chi^{ave}(\omega) = \frac{1}{4}G_{(1)\chi} \cdot |\varepsilon_\chi|^2 \tag{2.113}$$

where the complex-valued *constrained modulus*

$$G_P^* = G_B^* + 4/3 G_S^* = G_{(1)P} + i \cdot G_{(2)P} \tag{2.114}$$

has been introduced. Thus, combining Equations 2.113 and 2.109, Equation 2.111 can be rewritten as

$$D_\chi(\omega) = \frac{G_{(2)\chi}}{2G_{(1)\chi}} \tag{2.115}$$

with $\chi = P,S$. For small losses, Equations 2.110 and 2.111 yield identical results. However, Equation 2.111 has the advantage of linking $D_\chi(\omega)$ directly to the constitutive parameters of linear viscoelasticity. Furthermore, Equation 2.115 is independent from the magnitude of the energy losses.

This section ends with a few considerations on how a viscoelastic constitutive model can actually be constructed with specific regard to geomaterials. As shown by Equation 2.97, in the time domain, the response of a viscoelastic material is fully described by the shear $G_S(t)$ and bulk $G_B(t)$ or constrained $G_P(t)$ relaxation functions. In the frequency domain, the fundamental material parameters (in isotropic materials) are the complex moduli $G_S^*(\omega)$ and $G_B^*(\omega)$ or $G_P^*(\omega)$ as shown by Equation 2.100. Working in the time and frequency domain is not equivalent. Comparison of the corresponding constitutive Equations 2.97 and 2.100 shows that, in the frequency domain, the stress–strain relationship is simply given by an algebraic equation. However, in the time domain, it is given by a much more complicated integro-differential equation. Hence, solving a linear viscoelastic boundary value problem with all the field variables specified as harmonic functions of time is the preferred approach.

Because the real and imaginary parts of the complex moduli are not independent, as shown by Equation 2.99, specification of $G_S^*(\omega)$ and $G_P^*(\omega)$ will actually require two and not four material functions—just as it happens with the relaxation (or creep) function in the time domain. Section 2.5.2 will demonstrate that, in geomaterials, these two material functions could be replaced by the dispersion curves $V_P(\omega)$ and $V_S(\omega)$ of P- and S-waves, respectively, or alternatively by the corresponding damping ratio spectra $D_P(\omega)$ and $D_S(\omega)$. However, in general, explicit determination of material functions $G_S^*(\omega)$ and $G_B^*(\omega)$, or alternatively $G_P^*(\omega)$, is accomplished through specific laboratory measurements. Yet, in contrast to linear elasticity where the elastic parameters are determined from the slopes of experimental stress–strain curves associated with a given mode of deformation,[24] in viscoelastic materials, the process of defining a material function is considerably more involved, and there is no unique recipe for actually doing it. Each model has its own features, which may or may not be desirable for modeling a specific material behavior. If, for instance, a given viscoelastic material exhibits *hysteretic*[25] behavior (Visintin 1994) from a series of experimental tests, the corresponding constitutive model should be able to reproduce this important feature.

Typically, the construction of a viscoelastic model can actually be performed following either the *direct* or the *inverse* approach. With the direct method, a particular constitutive model (Kelvin–Voigt, Maxwell, standard linear solid, etc.) is chosen and then an assessment is made a posteriori to check whether the model fits the experimental data and what are the most relevant limitations. An alternative strategy would be to reverse the process and follow an inverse approach where the starting points are the experimental measurements. These will follow a certain pattern, and the objective

---

[24] In linear elasticity, the difference in mechanical response of two materials is simply that one is stiffer than the other, either in bulk or in shear or in both or in other modes of deformation.

[25] This behavior corresponds to a frequency or rate-independent material response. The term hysteretic is often used in physics and other sciences to denote memory effects processes that are scale independent.

is to find a pair of material functions (for isotropic materials), say $G_S^*(\omega)$ and $G_B^*(\omega)$, that capture the observed response. The problem of defining a particular viscoelastic model has then been reduced to a *curve fitting operation*.

The inverse approach, though more involved than the direct method, has the advantage of a greater generality, and it can be applied to any viscoelastic material. Examples of application of this approach to geomaterials include the work of Liu et al. (1976), who assumed a hyperbolic distribution of the relaxation spectrum, and Kjartansson (1979), who adopted a power law time dependence for the creep response function. Both models were able to predict quite accurately several features of geomaterials, including the seismic bandwidth hysteretic behavior as well as the effects of material dispersion.

2.5.2 Viscoelastic waves in unbounded homogeneous media

As discussed in Section 2.5.1, in the frequency domain, the constitutive relations of viscoelasticity become simple and compact algebraic equations, which resemble those of linear elasticity, as is shown by Equations 2.98 or 2.100. This resemblance goes even farther for it can be shown that the Fourier or Laplace-transformed field equations of linear viscoelasticity are formally identical to those associated with linear elasticity except that the elastic shear and bulk moduli μ and B are replaced by the complex moduli $G_S^*(\omega)$ and $G_B^*(\omega)$ in case of Fourier transform, or by $s \cdot G_S^*(s)$ and $s \cdot G_B^*(s)$ in case the field equations are transformed using the Laplace transform[26] (Christensen 1971; Pipkin 1986).

This analogy between the field equations of linear elasticity and viscoelasticity forms the essence of what is known as the *elastic–viscoelastic correspondence principle* (Read 1950; Fung 1965; Christensen 1971; Ben-Menahem and Singh 2000). According to this principle, elastic solutions to steady-state harmonic boundary value problems can be easily converted into viscoelastic solutions for identical boundary conditions by simply replacing the elastic shear and bulk moduli μ and B with the corresponding complex-valued and frequency-dependent parameters $G_S^*(\omega)$ and $G_B^*(\omega)$.[27] The validity of the correspondence principle, however, is restricted to problems where

---

[26] The variable "s" is used here to denote the complex-valued frequency defining the Laplace transform.

[27] An analogous result holds for the solution of viscoelastic initial-boundary value problems using the Laplace transform, which is suitable when the prescribed boundary conditions are arbitrary functions of time. In this case, μ and B must be replaced by the products $s \cdot G_S^*(s)$ and $s \cdot G_B^*(s)$, respectively.

the boundary conditions (e.g., prescribed displacements and tractions at the boundaries of a domain) are time invariant.

Application, in a homogeneous medium, of the elastic–viscoelastic correspondence principle to the Navier's equations (Equation 2.14) of linear elastodynamics with no body forces yields

$$\left(G_B^* + \tfrac{4}{3}G_S^*\right)\text{grad div } \hat{u} - G_S^*\text{curl curl } \hat{u} = -\rho\omega^2\,\hat{u} \tag{2.116}$$

where the vector $\hat{u} = \hat{u}(\mathbf{x},\omega)$ is the Fourier-transformed displacement vector, $\mathbf{x}$ is the position vector, and ρ is the mass density of the homogeneous medium assumed time independent. Applying the *divergence* and *curl* differential operators to Equation 2.116 yields[28]

$$\begin{cases} \left(V_P^*\right)^2 \nabla^2\left(div\ \hat{u}\right) = -\omega^2 \cdot div\ \hat{u} \\[2ex] \left(V_S^*\right)^2 \nabla^2\left(curl\ \hat{u}\right) = -\omega^2 \cdot curl\ \hat{u} \end{cases} \tag{2.117}$$

where $\nabla^2(\cdot)$ denotes the *Laplacian* differential operator, and V_P^* and V_S^* are the *complex-valued* longitudinal (or irrotational) and transversal (or equivoluminal) speeds of propagation of P- and S-waves, respectively. They define *phase velocity* and *spatial attenuation* of monochromatic, bulk waves propagating in a linear, homogeneous, viscoelastic unbounded medium, and they are expressed by the following relations

$$\begin{cases} V_P^*(\omega) = \sqrt{\dfrac{G_B^*(\omega) + \tfrac{4}{3}\cdot G_S^*}{\rho}} \\[3ex] V_S^*(\omega) = \sqrt{\dfrac{G_S^*(\omega)}{\rho}} \end{cases} \tag{2.118}$$

Equation 2.117 shows that *distortional* and *volume* deformations in linear, isotropic viscoelastic materials are uncoupled from each other, such as that occurs in linear elasticity. A general solution of Equation 2.117 may be written as follows

$$\mathbf{u}(\mathbf{x},t) = \mathbf{A}_1 \exp\left[i\left(\omega t - \mathbf{k}_\chi^* \cdot \mathbf{x}\right)\right] + \mathbf{A}_2 \exp\left[i\left(\omega t + \mathbf{k}_\chi^* \cdot \mathbf{x}\right)\right] \tag{2.119}$$

---

[28] An identical procedure would be that of applying the Helmholtz's decomposition theorem to Equation 2.116.

where $\mathbf{A}_1$ and $\mathbf{A}_2$ are two arbitrary constant *bivectors*[29] to be determined from boundary conditions; $\chi = P, S$ is a subscript denoting longitudinal and transversal wave motion, respectively; and $\mathbf{k}_\chi^* = \mathbf{k}_\chi - i\boldsymbol{\alpha}_\chi$ is the bivector wavenumber and defines the *direction of propagation* through the vector $\mathbf{k}_\chi$ and the *direction of attenuation* through the vector $\boldsymbol{\alpha}_\chi$ for the χ-wave.

It can be shown that the vector $\mathbf{k}_\chi$ is normal to planes of constant phase that are defined by the equation (Ben-Menahem and Singh 2000)

$$\mathbf{k}_\chi \cdot \mathbf{x} = \text{constant}, \quad \chi = P, S \tag{2.120}$$

Conversely, the vector $\boldsymbol{\alpha}_\chi$ is normal to planes of constant amplitude that are defined by the equation

$$\boldsymbol{\alpha}_\chi \cdot \mathbf{x} = \text{constant}, \quad \chi = P, S \tag{2.121}$$

The *phase velocity* of the monochromatic χ-wave is equal to $\omega/|\mathbf{k}_\chi|$.

The two vectors $\mathbf{k}_\chi$ and $\boldsymbol{\alpha}_\chi$ do not need to be parallel (Aki and Richards 2002). When the vector $\boldsymbol{\alpha}_\chi$ is parallel to the vector $\boldsymbol{\alpha}_\chi$, the corresponding χ-wave, $\chi = P, S$, is called *simple* or *homogeneous* (Lockett 1962). In a simple χ-wave, the direction of propagation is always coincident with the direction of maximum attenuation such as it happens in 1D wave propagation. *Nonsimple* waves may arise as a result of boundary effects (e.g., reflection and refraction of monochromatic waves at a plane interface) combined with special types of viscoelastic materials (Christensen 1971). All viscoelastic waves considered in this chapter are assumed to be *simple*. If Equation 2.119 is particularized for the case of 1D wave propagation, all vectors and bivectors degenerate into scalars yielding the following relationship:

$$u(x,t) = A_1 \exp\left[i\left(\omega t - k_\chi^* \cdot x\right)\right] + A_2 \exp\left[i\left(\omega t + k_\chi^* \cdot x\right)\right] \tag{2.122}$$

where the complex-valued, scalar wavenumber k_χ^* associated with the propagation of the χ-wave is defined by

$$k_\chi^* = \frac{\omega}{V_\chi^*} = \left(\frac{\omega}{V_\chi} - i\alpha_\chi\right) \tag{2.123}$$

---

[29] A *bivector* can be thought of as the 2D version of an ordinary vector. The need for introducing bivectors in Equation 2.119 originates from the simultaneous complex valuedness and vector nature of $\mathbf{A}_1$, $\mathbf{A}_2$, and $\mathbf{k}_\chi$.

where V_χ and α_χ, $\chi = P, S$, are the real-valued, physical phase velocity and attenuation coefficient of the χ-wave, respectively. This can be easily verified by substituting Equation 2.123 into Equation 2.122.

The attenuation coefficient is a measure of the spatial amplitude decay of the χ-wave as it propagates through a viscoelastic and, hence, dissipative medium. From Equations 2.118 and 2.123, the phase velocities and attenuation factors of bulk P- and S-waves are given by the following relations (Fung 1965)

$$
\begin{cases}
V_\chi(\omega) = \left[\Re\left(\sqrt{\dfrac{\rho}{G_\chi^*(\omega)}} \right) \right]^{-1} \\[2em]
\alpha_\chi(\omega) = \omega \cdot \Im\left(\sqrt{\dfrac{\rho}{G_\chi^*(\omega)}} \right)
\end{cases}
\tag{2.124}
$$

where again $\chi = P, S$ and $G_P^* = G_B^* + 4/3 G_S^*$ is the complex-valued constrained modulus. The symbols $\Re(\cdot)$ and $\Im(\cdot)$ denote the real and the imaginary part of the argument, respectively.

Thus, in linear viscoelastic, unbounded media, the mechanics of wave propagation is completely defined in the frequency domain either by the complex-valued phase velocities V_P^*, V_S^* or by the real-valued phase velocities V_P, V_S and attenuation coefficients α_P, α_S. Whereas V_P and V_S give a measure of the speed at which irrotational and equivoluminal disturbances propagate in a viscoelastic medium, α_P and α_S describe the spatial attenuation of these waves as they propagate through a dissipative material. Hence, the attenuation factors are directly related to the physical mechanisms responsible for the energy losses. However, since only two material functions are required to specify the constitutive parameters of a linear, isotropic, viscoelastic solid, V_χ and α_χ ($\chi = P, S$) in Equation 2.124 cannot be prescribed arbitrarily, just like it was for the real and the imaginary parts of the complex modulus G_χ^* in Equation 2.99. They are linked by another form of the Kramers–Krönig relations that, *translated* into wave propagation parameters, state that the real and the imaginary parts of the complex wavenumber k_χ^* have to be a *Hilbert transform* pair. It can be demonstrated that this is a necessary and sufficient condition to satisfy the fundamental *principle of causality*, which claims that in a physical system the reaction can never precede the action (Bracewell 1978; Tschoegl 1989).[30]

---

[30] Applied to wave propagation, the principle of physical causality postulates that a disturbance originated at a point in a medium (the source) is not allowed to arrive at a different point in the same medium (the observer) before the time d/c has elapsed, where d is the distance between the source and the observer and c is the speed of propagation of the disturbance in the medium.

One additional note regarding Equation 2.124—because of the frequency dependence of G_χ^*, V_χ and α_χ, $\chi = P, S$ are also frequency dependent, thereby causing a pulse to change its shape as it propagates through a viscoelastic medium. This is due to the phenomenon of *material dispersion*.

Introducing Equation 2.124 in Equation 2.123 after recalling Equation 2.115 and that $G_\chi^* = G_{(1)\chi} + i \cdot G_{(2)\chi}$ yields (Lai and Rix 2002)

$$V_\chi^*(\omega) = \frac{V_\chi(\omega)}{\sqrt{[1+4D_\chi^2(\omega)]}} \cdot \left[\frac{1 + \sqrt{[1+4D_\chi^2(\omega)]}}{2} + i \cdot D_\chi \right] \qquad (2.125)$$

with $\chi = P, S$. Similarly, Equation 2.125 expressed in terms of quality factor $Q_\chi(\omega)$ and defined by Equation 2.112 is

$$V_\chi^*(\omega) = \frac{V_\chi(\omega)}{2 \cdot \sqrt{1+Q_\chi^{-2}(\omega)}} \cdot \left[1 + \sqrt{1 + Q_\chi^{-2}(\omega)} + i \cdot Q_\chi^{-1}(\omega) \right] \qquad (2.126)$$

Equations 2.125 and 2.126 are *exact*; hence, they are valid for *arbitrary* values of material damping ratio $D_\chi(\omega)$ and quality factor and $Q_\chi(\omega)$. Expanding Equation 2.125 in a Mac Laurin series about D_χ and retaining only terms up to second-order yields

$$V_\chi^*(\omega) = \frac{V_\chi(\omega) \cdot [1 + i \cdot D_\chi(\omega)]}{[1 + D_\chi^2(\omega)]} \qquad (2.127)$$

If only first-order terms are retained, the result is

$$V_\chi^*(\omega) = V_\chi(\omega) \cdot [1 + i \cdot D_\chi(\omega)] \qquad (2.128)$$

Equation 2.128 often is adopted in soil dynamics and geotechnical earthquake engineering in applications involving weakly dissipative media (Kramer 1996). Figure 2.34 shows the magnitude of V_χ^* given by the three equations (2.125, 2.127, and 2.128) normalized with respect to the value of phase velocity in a corresponding elastic medium (i.e., for $D_\chi = 0$).

Figure 2.34 shows that the expression that includes up to second-order terms is valid for $D_\chi \leq \sim 0.20$, whereas the relation that accounts for first-order terms only, differs considerably from the other two expressions for $D_\chi \geq \sim 0.05$.

From an experimental viewpoint, measurements in geomaterials show that for states of strain in the interior of the linear cyclic threshold strain boundary surface (Figure 2.32), the following result holds (Vucetic 1994; Ishihara 1996)

$$\sup [D_\chi^{\text{exp}}(\omega)] \leq 0.05 \qquad (2.129)$$

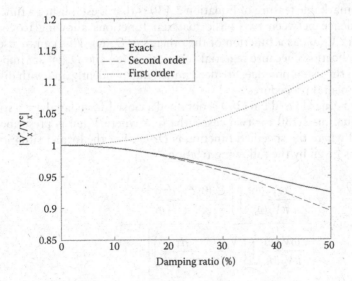

Figure 2.34 Magnitude of the normalized complex-valued phase velocity.

where sup(·) denotes the least upper bound of the argument. Thus, in light of Equation 2.129 and of Figure 2.34, for practical applications within the range of validity of the linear theory of viscoelasticity, Equation 2.128 can be considered an acceptable approximation for the calculation of material function V_χ^*, $\chi = P,S$. This conclusion is important, and in Section 2.5.3, it will be profitably exploited when studying the propagation of surface waves in dissipative media.

By substituting Equations 2.115 and 2.124 into Equation 2.99 written for an isotropic material, it is possible to write the Kramers–Krönig relation (Equation 2.99) in terms of physical parameters $V_\chi (\omega)$ and $D_\chi (\omega)$. The result is

$$V_\chi^2 (\omega) + \omega^2 \cdot \int_0^\infty \frac{4}{\pi} \cdot \left[\frac{D_\chi(\tau)}{\tau \cdot (\tau^2 - \omega^2)} \right] \cdot V_\chi^2(\tau) d\tau = G_{(e)\chi} \cdot \frac{2 \cdot (1+4D_\chi^2)}{1+\sqrt{1+4D_\chi^2}} \quad (2.130)$$

where $\chi = P,S$ and $G_{(e)\chi} = G_\chi (t \to \infty)$ is the *equilibrium response* of the relaxation function for the χ-mode of deformation. Formally, Equation 2.130 is a *singular, Fredholm integral equation of second kind with Cauchy kernel* for the unknown function $V_\chi(\omega)$.[31] This equation defines the mathematical constraint imposed by physical causality on material functions $V_\chi(\omega)$ and $D_\chi(\omega)$, which cannot be assigned independently.

---

[31] Integral Equation 2.130 is linear in $V_\chi^2(\omega)$.

A remarkable feature of Equation 2.130 is that it establishes a functional dependence between two basic material functions allowing to compute one, say $V_\chi(\omega)$, as a function of the other, say $D_\chi(\omega)$. This can be useful in the applications because material functions $V_\chi(\omega)$ and $D_\chi(\omega)$ are important and, in the usual practice, are determined independently and with different experimental procedures.

Meza-Fajardo and Lai (2007) obtained a closed-form, analytical solution for Equation 2.130 for two cases: the first where $V_\chi(\omega)$ is prescribed, the second where the specified function is $D_\chi(\omega)$. For the former situation, the result is given by the following relation

$$
D_\chi(\omega) = \frac{\dfrac{2\omega V_\chi(\omega)}{\pi V_\chi(0)} \displaystyle\int_0^\infty \left(\frac{V_\chi(0)}{V_\chi(\tau)} \cdot \frac{d\tau}{\tau^2 - \omega^2} \right)}{\left[\dfrac{2\omega V_\chi(\omega)}{\pi V_\chi(0)} \displaystyle\int_0^\infty \left(\frac{V_\chi(0)}{V_\chi(\tau)} \cdot \frac{d\tau}{\tau^2 - \omega^2} \right) \right]^2 - 1}
\tag{2.131}
$$

where $\chi = P,S$ and $V_\chi(0) = \lim_{\omega \to 0} V_\chi(\omega)$. Equation 2.131 represents an explicit form of the material dispersion relation for arbitrary dissipative, linear viscoelastic materials. By measuring the frequency dependence of body waves, $V_\chi(\omega)$, this equation allows the calculation of damping ratio spectra $D_\chi(\omega)$.

An explicit, particular solution of Equation 2.130, well known in seismology, is that obtained under the assumption that material damping ratio $D_\chi(\omega)$ is frequency independent (i.e., *hysteretic*) over the seismic bandwidth (~0.001–10 Hz). It is given by the following expression (Ben-Menahem and Singh 2000; Aki and Richards 2002)

$$
V_\chi(\omega) = \frac{V_\chi(\omega_{ref})}{\left[1 + \dfrac{2D_\chi}{\pi} \ln\left(\dfrac{\omega_{ref}}{\omega} \right) \right]}
\tag{2.132}
$$

where ω_{ref} denotes a reference angular frequency usually assumed in seismology equal to 2π. Equation 2.132 predicts values of phase velocity $V_\chi(\omega)$ that increase monotonically with material damping ratio for a prescribed frequency. Conversely, for a particular value of damping ratio, Equation 2.132 predicts an asymptotic increase of $V_\chi(\omega)$ with frequency. This particular type of dispersion relation is often invoked in seismology and in geotechnical earthquake engineering in light of the fact that a large number of experimental data on geomaterials seem to support the frequency independence of material damping ratio over the seismic band (Shibuya et al. 1995; Lo Presti and Pallara 1997; Aki and Richards 2002).

The subject of wave propagation in viscoelastic media is broad and has several ramifications. The interested reader can find a detailed treatment of some of the aspects discussed in this section or other topics in Borcherdt (1971, 1973, 2009) and Buchen (1971).

2.5.3 Surface Rayleigh waves in dissipative half-spaces

The procedure illustrated in Section 2.4.1 to define and solve the Love and Rayleigh eigenvalue problems in elastic, isotropic, layered half-spaces can be extended to viscoelastic media using the Laplace and Fourier transform methods. This approach, as mentioned in Section 2.5.2, naturally leads to the exploitation of the *elastic–viscoelastic correspondence principle*, which allows us to elegantly solve virtually any type of initial-boundary value problems with time-independent boundary conditions.

When this association between the field equations of elasticity and viscoelasticity is invoked for the eigenvalue problem associated with free surface waves (Section 2.4.1), the result is that Equations 2.70 and 2.71 and corresponding boundary conditions (Equations 2.72, 2.73) are still valid provided that arrays $f(x_2)$, $g(x_2)$, $A(x_2)$, $B(x_2)$, and Lamé parameters $\lambda(x_2)$, $\mu(x_2)$ are intended to be complex valued.[32] Consequently, most of the features described in Section 2.4.1 for the elastic eigenproblems carry over to their viscoelastic counterparts with the important difference that nontrivial solutions of the viscoelastic eigenvalue problems are now complex-valued wavenumbers and eigenfunctions.[33] Even the numerical techniques used to solve the elastic and viscoelastic eigenproblems are essentially the same, the main difference being that in the latter case the use of complex arithmetic can no longer be avoided and algorithms such as root-finding techniques must be properly fit to remain applicable for complex values of the arguments. This turns out to be a nontrivial exercise (Lai and Rix 2002).

However, surface wave propagation in linear viscoelastic media includes a rather interesting *special case*. If the complex-valued Lamé's parameters λ^* and μ^* are specified in such a way that the corresponding Poisson's ratio is a frequency-independent, real-valued parameter, it can be demonstrated that the roots of Love and Rayleigh dispersion equations are real valued (Christensen 1971). Thus, in this special circumstance, the solution of the complex-valued eigenproblem can be obtained using the same procedures employed for the solution of the corresponding

---

[32] Mass density $\rho(x_2)$ and circular frequency ω are still real valued.

[33] It is observed, however, that certain properties of the real-valued eigenproblem require careful consideration before they can be generalized to the complex-valued case. One such example is represented by the definition of orthogonality between eigenfunctions.

elastic eigenproblem, even though the resulting wavenumbers will still be complex valued.

In the most general situation, Poisson's ratio is a complex-valued, frequency-dependent function; therefore, the solution of Love and Rayleigh eigenproblems is not trivial, particularly with regards to the computation of the eigenvalues, which are the zeros of Love and Rayleigh secular functions $\Phi_{L/R}[\cdot]$. A technique for the solution of the Rayleigh eigenproblem in arbitrarily dissipative linear viscoelastic half-spaces has been proposed by Lai and Rix (2002) based on the application of Cauchy's theorem of complex variable theory.

One of the consequences of material dispersion on the propagation of Rayleigh waves in viscoelastic media is that the phase difference between the horizontal and the vertical components of the displacement field is no longer equal to $\pi/2$, such as in the elastic case. This happens because the Rayleigh eigenfunctions are now complex valued; thus, the principal axes of the ellipse describing the orbit of the Rayleigh particle motion are sloping. The degree by which they are rotated forward or backward with respect to the free boundary of the half-space depends on the mechanical properties of the medium (Båth and Berkhout 1984).

Figure 2.35 shows the orbits of Rayleigh particle motion calculated in a linear viscoelastic, layered half-space with material properties listed in Table 2.9.[34]

The numerical simulations were conducted using a time-harmonic, vertical point source applied at the free boundary of the half-space. Horizontal and vertical Rayleigh displacement was computed from the superposition of different modes of propagation of Rayleigh waves. Figure 2.35 shows the results obtained at two frequencies and distances from the source.[35] They clearly show the inclination of the principal axes of the ellipse with respect to the free boundary of the half-space and also to a variable semi-axes ratio.[36]

The degree of the sloping and the semi-axes ratio varies independently with frequency and distance from the source. It is also noted from the Figure that the amplitude of particle motion decreases with the increase

---

[34] To properly account for material dispersion and satisfy causality, the Rayleigh secular function $\Phi_{L/R}[\cdot]$ must be constructed using phase velocities $V_P(x_2,\omega)$, $V_S(x_2,\omega)$ and material damping ratios $D_P(x_2,\omega)$, $D_S(x_2,\omega)$ that cannot be assigned independently, but must satisfy Equation 2.130 and thus an appropriate (causal) frequency-dependence law. The properties of Table 2.9 have been chosen using material dispersion Equation 2.132. This corresponds with assuming hysteretic, that is, frequency-independent damping ratio over the seismic band. The phase velocities of body waves were specified at a reference frequency of $\omega_{ref} = 2\pi$.

[35] The particle orbits were computed at distances from the source outside the near-field where the influence of body waves was considered negligible.

[36] At the free surface of a vertically inhomogeneous half-space, the particle orbit of a Rayleigh wave is not necessarily retrograde even in a perfectly elastic medium (Haskell 1953).

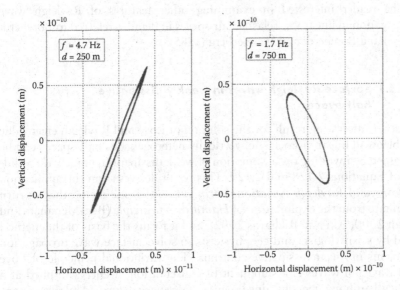

Figure 2.35 Rayleigh particle orbits at the free boundary of a linear viscoelastic, layered half-space at variable frequencies (f) and distances (d) induced by a time-harmonic, vertical point source. Material properties used in the calculations are reported in Table 2.9.

Table 2.9 Properties of layered system used for computation of Rayleigh particle orbits of Figure 2.35

Layer	Thickness (m)	V_P (m/s)	V_S (m/s)	ρ (Mg/m³)	D_P (m/s)	D_S (m/s)
I	5.0	300	150	1.7	0.03	0.04
2	5.0	400	200	1.7	0.03	0.05
3	5.0	500	250	1.8	0.02	0.04
4	7.5	600	300	1.8	0.02	0.02
5	7.5	700	350	1.9	0.03	0.04
Half-space	∞	800	400	1.9	0.03	0.05

Note: Medium parameters correspond to table I of Lai and Rix (2002) for weakly dissipative half-space.

of distance from the source, as expected.[37] The rotation at the free surface of the axes of the elliptical orbit of Rayleigh particle motion predicted by the theory has also been observed experimentally (Strobbia 2003).

---

[37] The amplitude of particle motion also decreases with the increase of the excitation frequency. In fact, as the frequency increases, the wavelength of the disturbance decreases; thus, it increases the number of cycles traveled by the signal to cover the same distance. Because the medium is inelastic, this yields greater energy dissipation. The analogy in the time domain would be to consider a fixed point in space and observe the attenuation in the same time window of two low- and high-period signals.

The reader interested in examining other features of Rayleigh wave propagation in linear viscoelastic half-spaces in detail is referred to Borcherdt (1973) and Borcherdt and Wennerberg (1985).

2.5.3.1 Surface Rayleigh waves in weakly dissipative half-spaces

A rather interesting result for the solution of Love and Rayleigh eigenvalue problems in linear viscoelastic, vertically heterogeneous half-spaces can be obtained by invoking the assumption of weak dissipation, namely the validity of Equations 2.128 and 2.129. This result derives from the application of *Love and Rayleigh variational principles* in elastic media that in turn originate from the exploitation of *Hamilton's principle* (Ben-Menahem and Singh 2000; Aki and Richards 2002), and it forms the basis of the method used by seismologists and geophysicists to solve surface wave propagation problems in linear dissipative continua. The variational principle of Love and Rayleigh waves is a statement of conservation of energy applied at a vertically inhomogeneous, linear elastic half-space deformed by the passage of free plane, Love, and Rayleigh waves: the average kinetic energy associated with a given mode of propagation of surface wave equals the average elastic strain energy.

From a computational viewpoint, the peculiar feature of the method is that the attenuation coefficients $\alpha_L(\omega)$ and $\alpha_R(\omega)$, which are the parameters characterizing the response of a dissipative half-space to the propagation of surface waves, are calculated from the solution of the Love and Rayleigh eigenproblems in the corresponding elastic medium. Although the solution is approximated, the accuracy of the results is indistinguishable in weakly dissipative media (Lai and Rix 2002).

With reference to Figure 2.19, the formalism of variational calculus associated with the assumption of weak dissipation can be used to obtain the following important relations (Aki and Richards 2002)

$$\begin{cases} V_R(\omega) = V_R^e + \left\{ \int_0^\infty V_S \left[\frac{\partial V_R}{\partial V_S} \right]_{\omega, V_P} \left[1 - \frac{V_S^e}{V_S} \right] dx_2 + \int_0^\infty V_P \left[\frac{\partial V_R}{\partial V_P} \right]_{\omega, V_S} \left[1 - \frac{V_P^e}{V_P} \right] dx_2 \right\} \\ \alpha_R(\omega) = \frac{\omega}{\left[V_R(\omega) \right]^2} \cdot \left\{ \int_0^\infty V_S D_S \left[\frac{\partial V_R}{\partial V_S} \right]_{\omega, V_P} dx_2 + \int_0^\infty V_P D_P \left[\frac{\partial V_R}{\partial V_P} \right]_{\omega, V_S} dx_2 \right\} \end{cases}$$

$$(2.133)$$

where $V_P = V_P(x_2, \omega)$ and $V_S = V_S(x_2, \omega)$ should be intended as frequency-dependent phase velocities of body waves propagating in a weakly dissipative medium, whereas V_χ^e, $\chi = P,S,R$, is the phase velocity of the corresponding wave propagating in the same medium after setting $D_\chi = 0$, that is, in the associated elastic half-space (Figure 2.19). Because some of the quantities appearing in Equation 2.133, for example, $V_R(\omega)$, are referred to a specific mode of propagation, Equation 2.133 should be intended in the *modal sense*. The expression at the right-hand side of this equation

$$\left[1 - \frac{V_\chi^e(x_2)}{V_\chi(x_2,\omega)}\right] \quad \text{with } \chi = P, S \qquad (2.134)$$

specifies the frequency dependence law of the speed of propagation of longitudinal P- and transversal S-waves. In other words, Equation 2.134 defines a material dispersion relation to be specified on a case-by-case basis. As illustrated in Section 2.5.2, for a waveform propagating in a dissipative medium to be *causal*, this relation cannot be assigned arbitrarily but must satisfy the Kramers–Krönig equation (Equation 2.130).

Equation 2.133[38] constitutes an important result conceptually and from the practical viewpoint because it shows that in vertically heterogeneous, weakly dissipative half-spaces, Rayleigh phase velocity $V_R(\omega)$ and the associated attenuation coefficient $\alpha_R(\omega)$ can be computed from the solution of the corresponding elastic eigenvalue problem including the partial derivatives of $V_R(\omega)$ with respect to medium parameters V_P and V_S.

Indeed, Equation 2.133 forms the basis of the procedure used by seismologists and geophysicists to solve surface wave propagation problems in the inelastic Earth, which is based on the assumption of weak dissipation (Lee and Solomon 1979; Keilis-Borok 1989; Ben-Menahem and Singh 2000; Aki and Richards 2002; Herrmann 2007). In recent years, the same approach has also been used in geotechnical engineering for near-surface site characterization (Rix et al. 2000; Foti 2003).

These procedures can be used to solve the *forward* surface and the *inverse* problem associated with surface wave motion. In the forward or direct problem, the objective is to determine Rayleigh (or Love) dispersion and attenuation curves $V_R(\omega)$ and $\alpha_R(\omega)$, given a model of a vertically inhomogeneous half-space specified by a set of medium parameters $\{\rho(x_2), V_P(x_2), V_S(x_2)\}$. Equation 2.133 is written in a form already suitable for the solution of the forward problem. Conversely, the inverse or backward problem associated with Equation 2.133 is a problem where the goal is to determine which set or subset of model parameters $\{\rho(x_2), V_P(x_2), V_S(x_2)\}$ corresponds to a given collection of dispersion and attenuation curves $V_R(\omega)$ and $\alpha_R(\omega)$. As will be illustrated in Chapter 6, the

---

[38] A perfectly analogous result can also be obtained for Love waves.

solution of the Rayleigh inverse problem is far more involved mathematically than the solution of the corresponding forward problem.

Figures 2.36 and 2.37[39] show a plot of the Rayleigh modal attenuation curves for the layered systems of Tables 2.7 (normally dispersive) and 2.8 (inversely dispersive). A constant value of $D_P = 0.01$ and $D_S = 0.02$ for all layers and the half-space has been assumed for the hysteretic damping ratio. Equation 2.132 has been adopted as material dispersion law. The attenuation curves $\alpha_R(\omega)$ shown in Figures 2.36 and 2.37 are associated with the modal dispersion curves $V_R(\omega)$ of Figures 2.30 and 2.31. They appear more irregular, although the attenuation coefficient $\alpha_R(\omega)$ exhibits a natural tendency to increase with frequency. This feature is expected after inspecting the second part of Equation 2.133. The cutoff frequencies below which higher modes cease to exist are the same for Rayleigh dispersion and for attenuation curves. This result is inherent to Equation 2.133 given the procedure used to compute the Rayleigh attenuation coefficients $\alpha_R(\omega)$ under the assumption of weak dissipation. In strongly dissipative half-spaces,

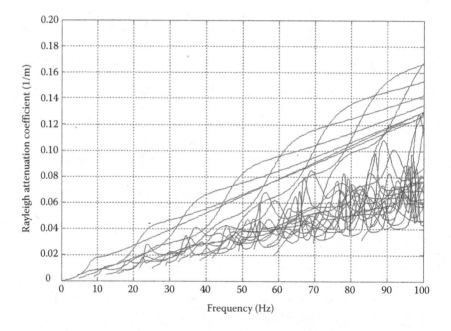

Figure 2.36 Rayleigh attenuation curves for the normally dispersive system of Table 2.7. Each curve denotes a single mode of propagation.

---

[39] Figures 2.36 and 2.37 were calculated by Dr. Maria-Daphne Mangriotis from the Institute of Petroleum Engineering at Heriot-Watt University, Edinburgh, UK.

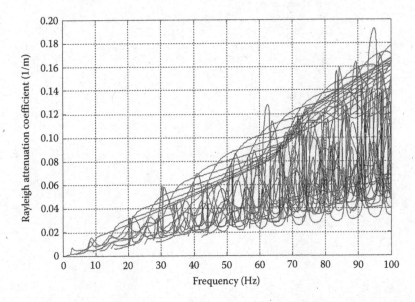

Figure 2.37 Rayleigh attenuation curves for the inversely dispersive system of Table 2.8. Each curve denotes a single mode of propagation.

cutoff frequencies for dispersion and attenuation curves are not necessarily coincident (Lai and Rix 2002).

When using Equation 2.133 to solve the Rayleigh inverse problem, a feature that is particularly relevant is that this equation uncouples the two problems of inverting dispersion and attenuation data. The procedure, the details of which will be described in Chapter 6, involves *three* major steps. The *first* is the experimental determination of the dispersion and attenuation curves, namely $V_R(\omega)$ and $\alpha_R(\omega)$, from surface wave measurements.[40] In the *second* step, the experimental dispersion curve $V_R(\omega)$ is inverted to obtain the profile $V_S(x_2)$ of shear wave velocity of an unknown elastic vertically inhomogeneous half-space (Figure 2.19). The *third* and final step involves the use of Equation 2.133 as the basis of the inversion of the experimentally measured attenuation curve $\alpha_R(\omega)$ to obtain the material damping ratio profile $D_S(x_2)$ or quality factor $Q_S(x_2)$. An important aspect of this procedure is that whereas the inversion of $\alpha_R(\omega)$ to obtain $D_S(x_2)$ is *linear*, the inversion of $V_R(\omega)$ to determine the $V_S(x_2)$ profile is highly nonlinear. Some case histories are reported in Section 7.3.

---

[40] In seismological applications, this is done through observed earthquake ground motion recordings. In seismic prospecting and near-surface geophysics, the dispersion and attenuation curves are obtained via nondestructive active or passive testing methods (see Chapters 4 and 5 for more details).

Chapter 3

Measurement of surface waves

This chapter illustrates the theory and practice of surface wave data acquisition. The measurement is an experimental process involving the generation of surface waves and the observation of their effects in the medium to investigate. The final aim is the estimation of the wave propagation parameters, which are then used for subsurface characterization. In most applications, the focus is on the wave velocity and its dependence on frequency (see also Chapter 4). The attenuation of surface waves and its dependence on frequency is of interest for the estimation of the material damping (see also Chapter 5). In some applications, the polarization (i.e., the ratio between different components of the particle motion) is also analyzed.

The experimental procedures and the equipment have a primary importance in the whole process of surface wave testing. The acquired data have limitations that affect the available and usable information. Most limitations of the measured data cannot be overcome by data processing.

The general principles of seismic data acquisition will be introduced in the chapter, and then the measurement of surface waves will be presented as sampling of a multidimensional signal. In order to explain limitations and trade-off of surface wave acquisition, some basic principles of signal processing will be introduced even if they are a key element of Chapter 4, where they are presented in more detail. This sequence has been chosen to follow the actual workflow of surface wave tests: acquisition—processing—inversion. The correct planning of field activities requires an understanding of subsequent processing. In turn, the latter is influenced by limitations and constrains introduced in acquisition.

Elements of signal processing are discussed from a conceptual point of view to outline the main issues related to acquisition and processing of surface wave data. A formal treatment of signal processing is outside the scope of the present book, and the reader is referred to specific textbooks on the matter (e.g., Bendat and Piersol 2010; Santamarina and Fratta 2010).

Finally, the typical acquisition parameters and procedures will be discussed, and seismic equipment of common use will be described.

3.1 SEISMIC DATA ACQUISITION

3.1.1 Seismic data

Data acquisition is the first step of any seismic characterization method. The acquisition of seismic data is the generation and observation of the effects of the propagation of seismic waves, in time and in space. A seismic source, a set of receivers, and an acquisition system are deployed with an appropriate geometrical configuration to record the wave field. The seismic wave induces vibrations (i.e., motion of the particles from their position of equilibrium). The motion involves, for a point in the medium, a variation of stress and strain in time. Each receiver detects the effect in time (the particle motion or the associated pressure variations) of the propagating wave at a specific location. Particle motion is typically sampled in terms of velocity or acceleration, which are easier to observe than displacements.

The seismic *trace* (e.g., a time signal) is the elementary unit of seismic data. It describes the medium response to a certain source in a given position in space. Seismic traces are usually represented as wiggles or as density plots (Figure 3.1).

The identification of the different wave types in a single trace is not straightforward, even in global seismology applications where propagation paths can be very long and different events are well separated in time.

In engineering and exploration geophysics, multichannel recording is the standard practice. Recording simultaneously with a plurality of receivers at different locations allows the wave propagation to be observed in space and time. The acquired set of multiple traces, often called seismic record or seismic gather or multichannel seismogram, represents the effects of the propagating wave field at different locations. The different events, or wave types, that constitute the wave field can be identified in a record that measures the vibrations in time at different distances. In a simple seismic acquisition, surface waves and body waves with direct, refracted, and reflected paths are propagated. They appear in different regions of the seismograms with different properties. The example of Figure 3.2 shows schematically refracted, reflected, and surface waves in a simple laterally homogeneous model.

Seismic records are analyzed and processed to estimate the properties of different wave types. For instance, the velocity of events can be obtained extracting the travel time at different distances along the propagation path, although the attenuation can be estimated extracting the amplitude variations. In actual practice, it is not possible to perform a single complete acquisition in which, ideally, all wave types are recorded. The wave field is complex even in rather simple media, with multiple wave types and complex propagation paths. Optimal recording of each event has its

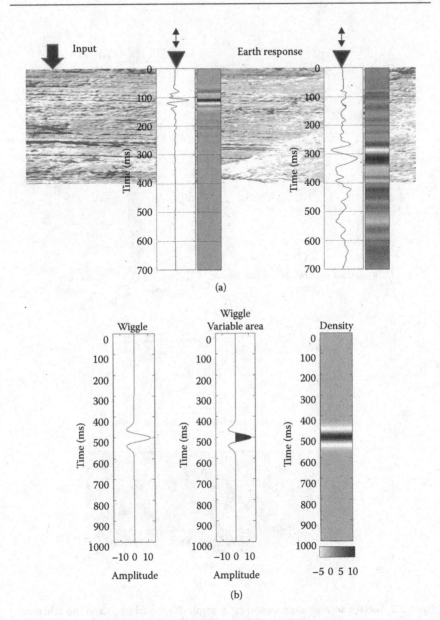

Figure 3.1 Seismic acquisition as generation of waves and observation of their effects:
(a) seismic traces describe the effect of the wave at a specific location;
(b) each trace can be represented as a wiggle or as a density plot.

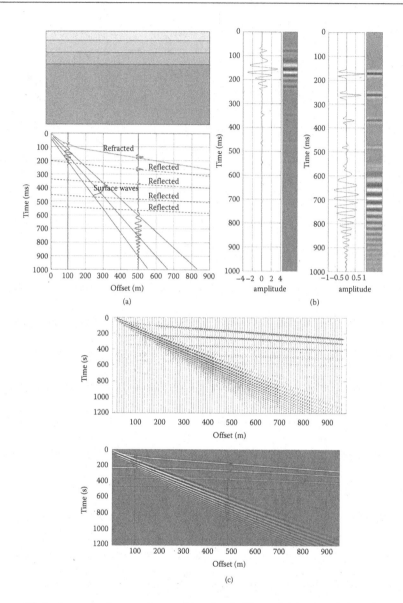

Figure 3.2 Surface seismic experiment: (a) a simple 1D model (top) and the schematic representation of the position of the different wave types in time and offset (bottom); (b) two sample traces are plotted as wiggle and density plots; (c) a wiggle multichannel seismogram (top) and a density grayscale representation of the same seismogram (bottom).

own requirements, for instance, in terms of offsets. The ideal measurement would require an infinite number of ideal receivers. In reality, a limited number of receivers with specific characteristics and limitations are available.

The seismic data acquisition requires a careful design and control, as with any laboratory experiment or field measurement. The survey design shall consider the objective of extracting the *propagation parameters* of some *wave types*, which will be used to characterize the medium under investigation, given a *processing and inversion method*, a desired *resolution*, and an estimated target *depth*, also considering an expected *level of noise*.

In seismic refraction applications, for example, the acquisition aims at extracting the travel times of refracted body waves as a function of the source–detector distance. Even if a seismic source will not generate only refracted waves, they will be identified in recorded data for their kinematic properties because they are associated with the first arrivals in the wave train. The maximum offset and the receiver spacing have to be designed according to the depth of the targets.

In surface wave methods, the acquisition scope is the observation and measurement of different propagation properties of the surface waves: the phase or group velocity, the attenuation, the amplitude, and the polarization, as a function of frequency. Therefore, measuring surface waves requires recording the effects of the surface wave propagation—reducing or controlling the presence of other wave phenomena in the acquired data. The crucial difference with respect to most other seismic acquisitions is that, despite the fact that data are acquired in time and space, the completeness and accuracy of the data depend on the property distribution in the *frequency domain*. The need for evaluating the data in the frequency domain is one of the peculiar aspects of surface wave measurement.

3.1.2 Surface wave acquisition

For most surface wave methods, data analysis is commonly performed in the frequency domain. The characterization of the subsoil is then carried out by solving an inverse problem (see Chapter 6), in which the properties of different wavelengths are mapped into the parameters of the subsurface at different depths. Generating and recording a wide wavelength range of surface waves is therefore crucial in order to collect information about different layers.

The wavelength depends on the frequency and on the velocity of the propagating wave. The latter is a function of the mechanical properties of the subsurface, which are unknown (because they are the objective of the experiment) and cannot be controlled. Therefore, the key parameter is the frequency: the lower the frequency, the longer the wavelength.

Surface waves are a low-velocity event and are often very dispersive in shallow applications. The duration of the wave train, even at relatively short offset, can be large, exceeding some seconds at less than 100 m in distance. It is not possible to generate only surface waves; therefore, due to the needed long duration of records, seismograms acquired to observe surface waves typically contain all other seismic events. Their identification and separation is not straightforward, but surface waves are, in general, high-energy dominant events in active seismic records. In shallow engineering applications, the surface wave acquisition is less challenging than refraction or reflection data acquisition.

Surface wave measurement requires the accurate estimation of the wave propagation properties in the frequency domain, over a wide frequency range, but the data are contaminated by incoherent and coherent noise. The limitations of the space and time sampling affect the accuracy and the bandwidth of the estimated properties, which are the final objectives of the survey design and of the data acquisition, in a way that is not intuitive. The surface wave measurement can be presented as a problem of multidimensional signal sampling. To explain the survey design and the data acquisition, the concepts of signals in time and space and of coherent and incoherent noise have to be introduced. The relationships between different domains (time and frequency, and space and wavenumber) and the corresponding transforms are indeed essential to understand the limitations of the data acquisition.

3.2 THE WAVE FIELD AS A SIGNAL IN TIME AND SPACE

The propagation of the wave field in a continuous medium induces vibrations, which are motion of the particles from their position of equilibrium. The wave field can be completely described with a continuous 4D function of space and time representing, for example, the vector particle velocity. Considering the plane surface waves at the free surface in a 1D medium, the wave field is a continuous vector function of distance x and time t. If we consider a single component of the particle displacement, it is a continuous scalar function of offset and time, $s(x,t)$. In Figure 3.3, a snapshot of a vertical section, strained by a propagating plane Rayleigh wave at constant frequency, is shown at a fixed time instant. The bottom continuous image depicts, in grayscale, the vertical component of the particle velocity at the surface, as a function of offset and time. A slice at constant time represents the particle velocity versus offset at a given time, while a section at constant offset represents the particle velocity of a specific point (such as point A) in time.

We define a *signal* as a physical quantity, measurable over time and space; in this case, as in communication theory, the signal is also a dataset carrying information. The ideal acquisition experiment should generate a

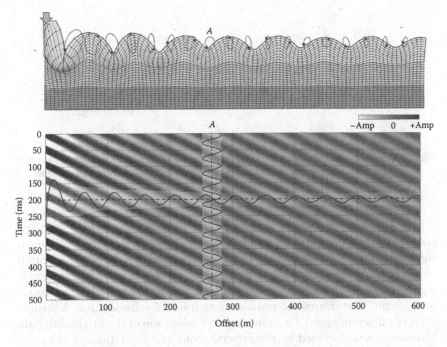

Figure 3.3 The wave field is a continuous function of space and time. Considering a single component of the particle motion of the points along a straight line at the free surface, the wave field is a 2D continuous function of *x* and *t*. A seismic trace is the value at constant *x* (for instance, at point A).

wave field with only surface waves in a noise-free environment, over the required wide frequency band. The ideal signal, that is, the effect of the wave propagation, should be observed continuously and infinitely, and its true value should be recorded without any distortion induced by the acquisition system and equipment.

In reality, this is not possible; the ideal wave field cannot be generated and the signal cannot be ideally recorded. A seismic source will generate a band-limited wave field, with different wave types superimposed on each other in different domains. The presence of lateral variations, of near-field effects, and so on, is not an acquisition limitation, strictly speaking. Yet, it affects the nature of the propagating wave field and has to be considered in the acquisition design. Other nonsource-generated events will superimpose onto the vibration field, constituting incoherent or pseudorandom noise. It is not possible to generate and observe only the desired signal, rather a composition of signal $\bar{s}$ and noise n

$$w(x,t) = \bar{s}(x,t) + n(x,t) \qquad (3.1)$$

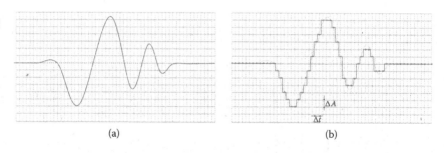

Figure 3.4 Schematic representations of (a) a continuous signal and its digitized version (b) the sampling and quantization transform a continuous real variable into a discrete array with a limited set of possible values.

The wave field is a physical quantity, for instance, the particle velocity, continuous in time, in the continuous medium to be investigated. The measurement of the wave field at a specific location involves the transformation of the physical quantity into an analog electric signal using a transducer, its transmission, conditioning, digitization, and recording. This "measurement chain" transforms the ground motion into a seismic trace. A seismic trace is a discrete signal (i.e., a limited representation of the original signal). Moreover, it is distorted by the imperfections and limitations of the measuring system and is affected by other types of noise.

An important aspect of the digitization is that the physical quantity is not recorded continuously in time and amplitude; sampling and quantization are schematically represented in Figure 3.4. A continuous vibration signal, defined at any time instant as a real-valued number, is transformed into a discrete series of points that can assume a limited discrete set of amplitude values.

The fidelity of the digital data is the similarity of the recorded version of the signal to the true physical quantity. It depends mainly on the density of sampling (i.e., the time interval between samples) and on the resolution of the analog-to-digital conversion, as will be discussed in Section 3.3.6.

In seismic data acquisition, the sampling of the signal *in space* is even more important than the time sampling. The effect of the wave propagation is indeed detected by a limited number of receivers. The wave field w is recorded on a discrete, finite set of points, in time and space. The effects of the sampling in time and space are added onto the physical noise (other events, coherent and incoherent).

Figure 3.5 represents the difference between the ideal experiment and the reality. Ideally, surface waves should be measured continuously in time and space. In reality, the gathered data are limited by the time and space sampling, are affected by the instrument, and are contaminated by noise:

$$\text{data} = \text{sampling (instrument } (\bar{s} + n)) \tag{3.2}$$

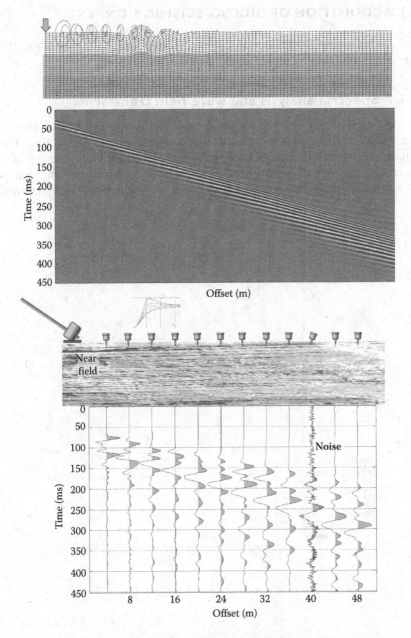

Figure 3.5 An ideal experiment with a continuous wave field consisting only of surface waves and a real acquisition with discrete sampling, different types of noise, and the equipment effects.

3.3 ACQUISITION OF DIGITAL SEISMIC SIGNALS

Before discussing the effects and limitations of sampling and noise, the formal relationships of signal analysis have to be introduced.

3.3.1 Spectral analysis and wave field transforms

The concept of *frequency* is intuitive for cyclic processes; in general, frequency indicates the number of occurrences of an event per unit of time. For a harmonic signal, the frequency is the reciprocal of the time duration of a cycle, called a *period*. Noncyclic signals can be decomposed into the sum of cyclic functions. In Figure 3.6, a signal (e.g., a seismic trace) is represented

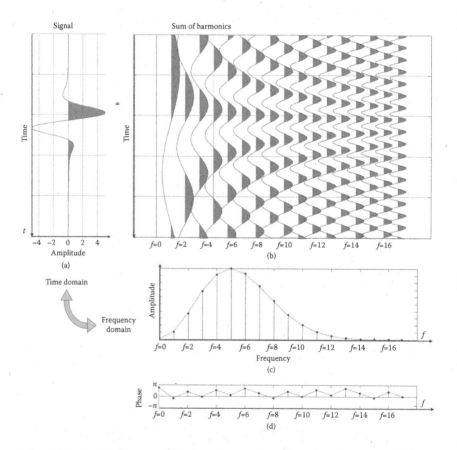

Figure 3.6 (a) A generic nonperiodic signal can be decomposed in (b) the sum of simple cyclic functions. The amplitude and phase of the elementary cyclic signal are the frequency-domain representation of the signal, or its spectrum, consisting of the (c) amplitude and (d) phase.

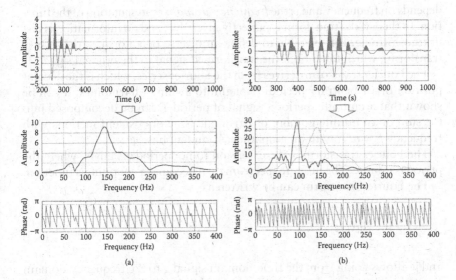

Figure 3.7 Example of spectral analysis of two seismic signals with different frequency content. The amplitude spectrum of the signal (a) is overimposed on the amplitude of the spectrum of the signal (b).

in time domain in Figure 3.6a. It can be exactly obtained by summing the cyclic signals plotted in Figure 3.6b. Each cyclic signal has a different frequency, amplitude, and phase. Each frequency is a multiple of the minimum frequency corresponding to the inverse of the total duration of the signal. The series of amplitudes and phases of these elementary signals (Figure 3.6c and d) represent the trace in frequency domain, also known as the spectrum of the trace.

Often the word *spectrum* is used to indicate the amplitude spectrum of a signal, and the process of decomposing a signal in its frequency components is called *spectral analysis*. The spectral analysis, in practice, consists of finding the amplitude and phase of the elementary cyclic signals. Two real seismic traces are analyzed in Figure 3.7. Their amplitude spectra show different frequency content. The signal on the left is dominated by higher frequency components than the signal on the right.

The mathematical operators used to decompose a signal in its cyclic components are called *transforms*.

3.3.2 Fourier series and Fourier transform

Integral transform are used to transform a function of one variable (e.g., time) into a related function of a different variable (e.g., frequency). The two variables specify two *domains*. The original signal depends on time, and it is addressed as the *time-domain* representation of the function. Its transform depends on frequency and is the *frequency-domain* representation of the function. In theory, the transform is perfectly invertible, and no information is lost in transforming. It is possible to define an inverse operator (i.e., the inverse transform), which takes back the transformed signal to the original signal.

The Fourier transform derives from the Fourier series, which is the decomposition of an arbitrary periodic signal into a sum of harmonics. It can be shown that an infinite, periodic signal of period T can be decomposed into the sum of an infinite number of harmonic (sine or cosine) functions with frequency $f_n = n/T$, with an amplitude A_n and a phase φ_n.

The Fourier transform is the extension to nonperiodic signals, where the period is infinite. The discrete spectrum then becomes a continuous function.

The Fourier transform can be written as

$$G(f) = \int_{-\infty}^{+\infty} g(t)e^{-j2\pi ft}dt \tag{3.3}$$

and it allows going from the time-domain signal g to the frequency-domain signal G. The original independent variable t is eliminated via integration over the infinite interval. The inverse transform operates on the frequency-domain signal G and allows going back, from frequency domain to time domain. The inverse transform can be written as

$$g(t) = \int_{-\infty}^{+\infty} G(f)e^{j2\pi ft}df \tag{3.4}$$

The two representations of the signal are g and G, and the relationship is written as

$$g(t) \xleftrightarrow{\ F\ } G(f) \tag{3.5}$$

The original time signal and its frequency spectrum constitute a *Fourier transform pair*. The two sides of the Fourier transform pair are complementary views of the same signal. It is possible to go from one domain to the other without any loss of information.

The same concept applies to signals in different original domains. For instance, the signals for which the independent variable is *space* are transformed into the *wavenumber* domain. The wavenumber k indicates the cycles per unit distance and is the reciprocal of the wavelength. The Fourier pair in this case can be indicated as

$$g(x) \xleftrightarrow{\ F\ } G(k) \tag{3.6}$$

In some applications, it is common to use the circular wavenumber, representing the spatial frequency in radians per meter: $\kappa = 2\pi k$. The relationship between wavenumber and circular wavenumber is equivalent to the relationship between frequency f and circular frequency, $\omega = 2\pi \cdot f$.

3.3.2.1 Properties of the Fourier transform

Some basic properties of the Fourier transform are of paramount importance for seismic signal processing. In particular, the following are recalled in this section: linearity, scaling, shifting, and the convolution theorem. The properties are hereafter described for time signals, but they are valid for any Fourier pair. For example, for signals in space, the wavenumber would substitute for the frequency in the following sentences.

The linearity of the transform implies that scaling a function scales its transform pair

$$g(t) \xleftrightarrow{\ F\ } G(f) \quad \text{then} \quad ag(t) \xleftrightarrow{\ F\ } aG(f) \tag{3.7}$$

and adding two functions corresponds to adding the two spectra

$$\begin{aligned} g(t) &\xleftrightarrow{\ F\ } G(f) \\ h(t) &\xleftrightarrow{\ F\ } H(f) \end{aligned} \quad \text{then} \quad g(t)+h(f) \xleftrightarrow{\ F\ } G(f)+H(f) \tag{3.8}$$

The scaling property states that a multiplication of the scale factor s to the independent variable (time) changes inversely the frequency axis of the spectrum of the signal. If a signal is shrunk in time, its spectrum is stretched in frequency (and vice versa)

$$g(t) \xleftrightarrow{\ F\ } G(f) \quad \text{then} \quad g(t/s) \xleftrightarrow{\ F\ } sG(s \cdot f) \tag{3.9}$$

Another relevant property is related to time shift, which induces a phase shift proportional to the time shift and to the frequency; that is, the same time shift corresponds to more cycles for a high-frequency (wavenumber) than for a low-frequency (wavenumber) function

$$g(t) \xleftrightarrow{\ F\ } G(f) \quad \text{then} \quad g(t-t_0) \xleftrightarrow{\ F\ } e^{-j\omega t_0} G(f) \tag{3.10}$$

Many physical phenomena can be described by their effects on signals. One of the operators that combine two signals to produce a new signal is the convolution, formally denoted by an asterisk. By definition, the convolution between two signals h and g is

$$p = h * g \quad \text{then} \quad p(t) = \int_{-\infty}^{\infty} g(u)h(t-u)\,du \tag{3.11}$$

The convolution theorem states that the convolution between two signals in time domain is equivalent to the multiplication of their spectra in the frequency domain, and vice versa

$$g(t)\xleftrightarrow{F}G(f) \qquad g(t)*h(f)\xleftrightarrow{F}G(f)\cdot H(f)$$
$$\text{then}$$
$$h(t)\xleftrightarrow{F}H(f) \qquad g(t)\cdot h(f)\xleftrightarrow{F}G(f)*H(f) \qquad (3.12)$$

The aforementioned properties of the Fourier transform are used in the following sections to illustrate the limitations of the sampling and of the acquisition.

3.3.3 Sampling

The data acquisition involves an analog-to-digital conversion, in which the continuous signal is replaced by a discrete series of values at fixed time intervals. The reciprocal of the sampling interval Δt is called sampling frequency F_S. Figure 3.8 illustrates the sampling of a signal. The sampling process implies loss of information.

The process of sampling can be described as a multiplication of the signal y by a sampling *comb* function. The *comb* function consists of an infinite set of regularly spaced impulses, or Dirac's deltas. The spectrum of the sampled function is therefore equal to the spectrum of comb·y, which is the convolution of the two spectra, the one of the signal with the one of the comb function

$$F(\text{comb}\cdot y) = F(\text{comb})*F(y) \qquad (3.13)$$

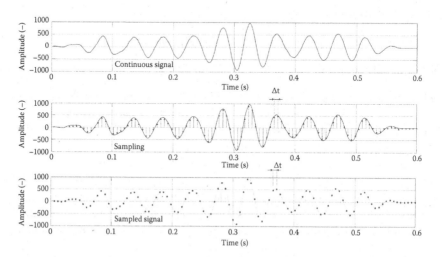

Figure 3.8 Sampling of a continuous signal involves reading the values of the signal at a finite discrete set of points. The sampled signal is defined only at a series of (usually evenly spaced) time instants.

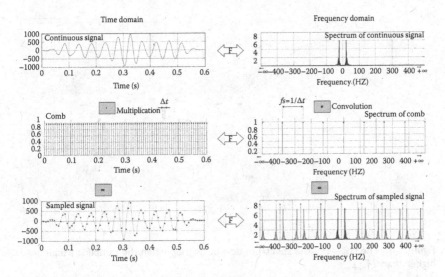

Figure 3.9 Sampling of a continuous function, in time domain (left) and in frequency domain (right); the spectrum of the sampled signal is a repetition of the original spectrum, with a periodicity equal to the inverse of the sampling.

The time-and frequency-domain representations of the sampling are illustrated in Figure 3.9. The top row shows the continuous signal with its symmetric (even) amplitude spectrum. The middle row shows the sampling function, with impulses spaced by Δt, and its amplitude spectrum, with impulses spaced by $f_s = 1/\Delta t$. The comb function is multiplied by the continuous signal, and its spectrum is convolved with the spectrum of the continuous signal. The results are shown in the third row, where the sampled signal is represented on the left and its spectrum is depicted on the right. An infinite set of copies of the original spectrum is obtained, and the spacing between these replicas is equal to the sampling frequency, f_s.

These replicas are separated (i.e., not overlapping each other) when the maximum frequency f_{max} in the signal is lower than half of the sampling frequency, as in Figure 3.10a (top).

Given a certain sampling frequency, f_S, if the maximum frequency exceeds the limit of $0.5f_S$, the replicas of the spectrum overlap and their values are summed up (as in Figure 3.10b (bottom)).

The Nyquist frequency F_{Nyq} is the maximum observable frequency in sampled data. It is equal to half the sampling frequency.

From an acquisition planning perspective, given a maximum frequency in the signal f_{max}, it is possible to evaluate the sampling frequency needed to sample without loss of information as

$$F_S > 2 \cdot f_{max} \tag{3.14}$$

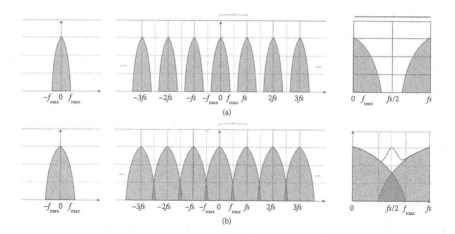

Figure 3.10 The copies of the spectrum do not overlap when the sampling frequency is higher than $2 \cdot f_{max}$.

This criterion is known as the Nyquist–Shannon theorem; no information is lost by regular sampling provided that the sampling frequency is greater than twice the highest frequency in the sampled signal.

3.3.4 Interpolation and aliasing

When the sampling frequency is sufficient, it is possible to recover the spectrum of the continuous signal by multiplying the transform of the sampled function by a boxcar, that is, a box window equal to 1 between $-f_S/2$ and $f_S/2$ and zero outside. This is equivalent to the convolution in time domain with the boxcar spectrum, which is a sinc function. Therefore, the complete sinc function can be used to interpolate a sampled function without any information loss, when the sampling criterion is verified.

If the spectrum has frequencies higher than half of the sampling frequency, the actual spectrum is summed up to its replicas. In this case, the original signal cannot be recovered. Any frequency component above $f_S/2$ is represented as a fictitious component at a lower frequency. This error is known as aliasing. In Figure 3.11, examples of sufficient and insufficient samplings are provided.

Aliasing makes different signals indistinguishable when sampled and creates distortions and artifacts in the reconstruction. A well-known example of distortion is the apparent reverse spinning direction of wheels in movies. When the time sampling is not sufficient to sample the forward moving frequency, related to the distance between the spikes, an apparent negative frequency appears.

To avoid aliasing, the sampling rate has to be increased to satisfy the requirements of the sampling theorem. Alternatively, antialias filters can

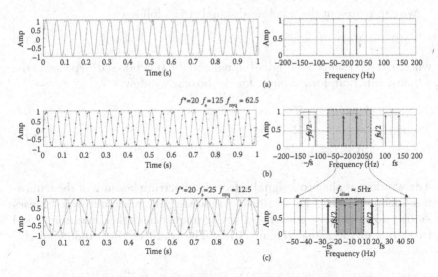

Figure 3.11 Example of aliasing— (a) a continuous signal with the frequency of 20 Hz is plotted with its spectrum. (b) When sampled at 125 Hz with a Nyquist frequency of 62.5 Hz, the first copies of the spectrum appear centered at +125 Hz and −125 Hz, and they fall outside of the range (−62.5 to +62.5), allowing a correct reconstruction of the signal. (c) The signal is sampled at 25 Hz, hence with a usable range (−12.5 to +12.5 Hz). The aliased copies at 5 Hz (25 − 20 Hz) fall in the range, and the signal appears as aliased.

be implemented to remove all components above the highest frequency of interest for the specific application from the signal. These filters restrict the bandwidth of the signal to satisfy the condition of proper sampling by removing frequency components above the Nyquist frequency. The most effective option is represented by electronic analog filters that remove the high-frequency components before the digitization of the signals (hardware filter). Alternatively, the filter can be implemented through digital filtering after oversampling of the signal (software filter). The antialias protection for spatial sampling can be done via spatial oversampling followed by digital filtering or by using arrays of analog receivers that are summed into a single trace.

3.3.5 Windowing

The signals described in the previous sections are still considered indefinite in length, although any real data are limited to a finite time interval. The beginning and end of the acquisition can be described as windowing, or multiplication of an infinite signal by a finite time window. A window function, or tapering function, is a mathematical function that is zero-valued outside a chosen interval. A rectangular window, or boxcar window, is constant inside the interval and zero elsewhere. When the signal s is multiplied by the window, the product is zero-valued outside the interval.

We can represent the windowed signal $z(t)$ as follows

$$z(t) = w(t)s(t) \tag{3.15}$$

where $s(t)$ represents the original signal and the function $w(t)$ represents the window function, for example, for the boxcar window

$$w(t) = \begin{cases} 1 - \dfrac{T}{2} < t < \dfrac{T}{2} \\ 0 \quad \text{elsewhere} \end{cases} \tag{3.16}$$

Any window applied to a signal affects its spectrum because of the trunca-tion spectral leakage. For example, windowing of a harmonic signal causes its transform to have nonzero values at frequencies other than the actual frequency of the signal (Figure 3.12). The result of the windowing is a

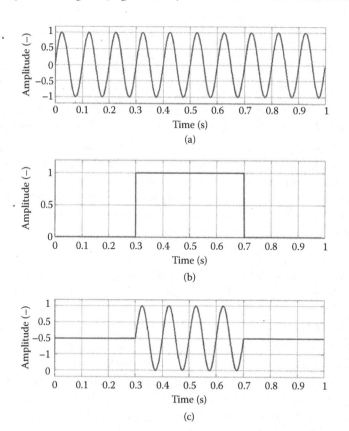

Figure 3.12 Windowing and spectral leakage: (a) harmonic function; (b) boxcar window; (c) windowed signal.

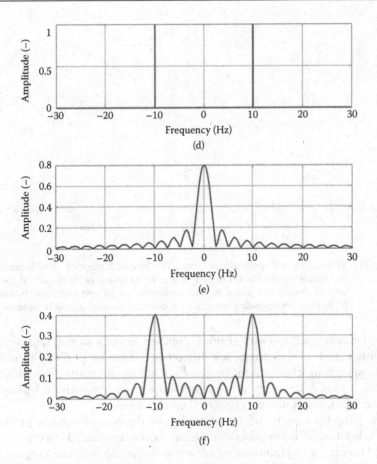

Figure 3.12 (Continued) Windowing and spectral leakage: (d) spectrum of the harmonic function; (e) spectrum of the boxcar window; (f) spectrum of the windowed signal.

truncated signal, zero-valued outside of the window length, and a spectrum replicating the window spectrum. The amplitude of the actual frequency component is altered, and side lobes associated with the spectrum of the box window appear. Windowing a signal corresponds to the multiplication of the signal (a) by the chosen window (c) and is equivalent to the convolution, that is, to the multiplication of the signal spectrum (b) with the window spectrum (d). The spectral leakage spreads the energy out of its real position and affects the resolution of the spectrum (f).

Leakage reduces the spectral resolution and the ability to distinguish multiple events. Figure 3.13 illustrates two examples of ambiguities in the spectral representation of finite signals.

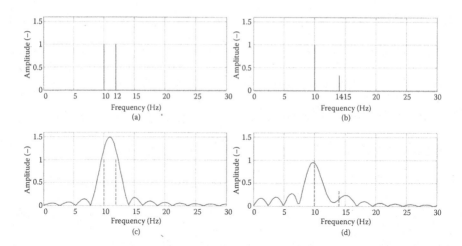

Figure 3.13 Windowing and spectral resolution. The limited acquisition introduces spectral leakage and different events cannot be resolved: (a, b) the actual spectra of two signals composed by the superposition of two harmonic functions; (c, d) the corresponding spectra after convolution with a boxcar window.

Two events, such as two harmonic components of a signal, with similar amplitude and a small frequency difference cannot be properly identified in the spectrum (they cannot be *resolved*) when the main lobes interfere (Figure 3.13a). A stronger event can mask a weaker one with the leakage of its side lobes, also known as ripples (Figure 3.13b).

The jump between 0 and 1 in the values of the boxcar window produces high side lobes. Window functions with a smooth transition between 0 and 1 (e.g., Hanning and Hamming windows) increase the dynamic range at the price of a loss of resolution (Santamarina and Fratta 2010).

The concept of spectral resolution and its dependence on the window size is extremely important for surface wave acquisition because it affects the possibility of separating multiple modes of propagation. For a signal in space, the window is the length of the observation (i.e., the aperture of the array), and this is a key acquisition parameter. The effect of multidimensional windowing is further discussed later in this chapter.

3.3.6 Quantization and analog-to-digital conversion

As mentioned before, the acquisition of digital data implies analog-to-digital conversion, which not only requires sampling but also implies quantization. A real physical signal consists of a variable that can assume, between its maximum and minimum, any real value (i.e., an infinite set of values). Digital data have a finite number of discrete values. The true value has to be rounded, or truncated, to be represented. Usually, an electronic device converts the input analog signal into the digital output, which is coded as a binary number, proportional to the magnitude of the input.

The resolution of the conversion indicates the number of discrete values that can be produced within the input range. Because the output value is usually in binary form, the resolution is expressed in bits and the number of divisions is measured in a power of two. A 4-bit conversion implies 2^4 levels, corresponding to 16 possible levels in the measurement scale, for example, from −8 to +7 if a signed integer coding is used. Typical values of resolution of analog–digital converters (ADCs) in modern seismic acquisition systems are between 16 and 24 bits. The resolution can also be measured in physical units, as the minimum change of input that creates a change in the output level. Usually, input signals are measured in electrical units (because the particle motion is converted into an electric voltage by the transducer) and the resolution can be expressed in volts. The least significant bit (LSB) voltage is the minimum change in input tension corresponding to a change of output. The LSB voltage is equal to the full range divided by the number of divisions.

The quantization error or quantization distortion is the difference, due to the truncation or rounding, between the actual input analog signal and the output digital signal. In a typical case, with a signal of much higher amplitude than the LSB, the quantization noise is not correlated with the signal. In the case of rounding, the distribution is uniform and is between $-\text{LSB}/2$ and $+\text{LSB}/2$, giving a zero mean and a standard deviation of $\dfrac{\text{LSB}}{\sqrt{12}}$, which is the RMS level of this noise. If the signal is uniformly distributed in the full range, then the quantization noise has a ratio equal to the resolution (i.e., 2^Q) corresponding to a quantization noise of $20\log_{10}(2^Q)$. In Figure 3.14, a 4-bit conversion is shown with the associated quantization noise.

The resolution of the conversion affects the accuracy and data integrity. The resolution is called dynamic range, and typical values are from 16 to 24 bits. Some details are given at the end of this chapter when describing data acquisition systems and the AD converter.

3.3.7 Acquisition of 2D signals

Seismic data are multidimensional functions of space and time. For a linear array of receivers, seismic data are two-dimensional (2D) signals (i.e., function of time and distance from the source). It is therefore important to extend the formulation to multidimensional transforms and spectral analysis. Some examples are reported in the following to discuss the limitations of the acquired data.

The two transforms (from time to frequency and from space to wavenumber) can be applied simultaneously to the seismic dataset. The 2D Fourier transform can be written as

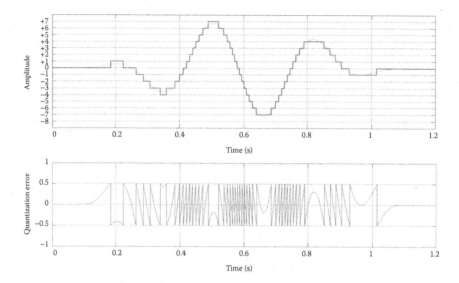

Figure 3.14 Example of signal digital conversion and quantization error. The difference between the real signal and its digitized version represents the inaccuracy introduced by the process. When the signal has large amplitude compared to the unit-amplitude interval, the quantization error is uncorrelated.

$$G(f,k) = \int\limits_{-\infty}^{+\infty} \int\limits_{-\infty}^{+\infty} g(t,x)e^{-j2\pi(kx+ft)}\,dx\,dt \qquad (3.17)$$

The original space and time variables are eliminated with a double integration to obtain the representation of the signal as a function of the two transformed variables, frequency and wavenumber. The two paired domains are the time–space domain (t–x) and the frequency–wavenumber domain (f–k).

The corresponding inverse transform can be written as

$$g(t,x) = \int\limits_{-\infty}^{+\infty} \int\limits_{-\infty}^{+\infty} G(f,k)e^{j2\pi(kx+ft)}\,dk\,df \qquad (3.18)$$

The f–k transform is widely used in seismic data processing, and it is part of the operation called wave field transforms.

The sampling and aliasing issues discussed in the previous sections apply to the 2D domain. In most applications, a regularly spaced array of receivers is deployed to sample the wave field in a regular way in time and offset. In some applications, the spacing between receivers may be not regular, introducing extra complexity in the transforms.

3.3.7.1 Effects of finite sampling

It is useful to discuss the consequences of the windowing in 2D to extend the concept of spectral resolution that was introduced earlier. Considering a unit-amplitude, harmonic plane wave propagating in the x direction with circular wavenumber k_0 and circular frequency ω_0 is

$$s(x,t) = e^{i(\omega_0 t - k_0 x)} \tag{3.19}$$

The frequency–wavenumber spectrum of the signal is

$$S(k,\omega) = 4\pi^2\, \delta(k-k_0)\, \delta(\omega-\omega_0) \tag{3.20}$$

Figure 3.15 shows a graphical representation of this frequency–wavenumber spectrum. Both the circular wavenumber k_0 and circular frequenycy ω_0 of the signal are resolved exactly.

Over space and time, the infinite integrals in Equation 3.17 imply perfect sampling of the signal in both domains. In practice, of course, this is not possible, and the signal is sampled at a limited number of positions and times. Let us initially consider the effect of limited sampling in the time domain. Assume that the signal is sampled with a boxcar window of duration T.

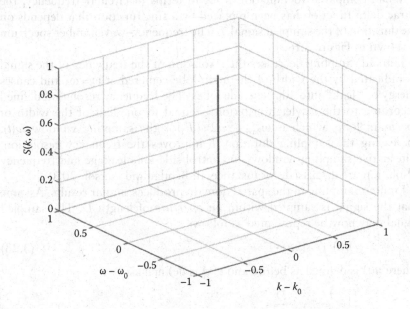

Figure 3.15 Frequency–wavenumber spectrum of a harmonic plane wave.

The frequency–wavenumber spectrum of the sampled signal is

$$Z(k,\omega) = \int\limits_{-\infty}^{\infty} \int\limits_{-\infty}^{\infty} z(x,t)e^{-i(\omega t - kx)}dt\ dx$$

$$= \int\limits_{-\infty}^{\infty} \int\limits_{-\infty}^{\infty} u(t)s(x,t)e^{-i(\omega t - kx)}dt\ dx \tag{3.21}$$

$$= \int\limits_{-\infty}^{\infty} \int\limits_{-\frac{T}{2}}^{\frac{T}{2}} s(x,t)e^{-i(\omega t - kx)}dt\ dx$$

To illustrate the effect of limited temporal sampling, consider the example of a unit-amplitude, harmonic plane wave once again

$$Z(k,\omega) = \int\limits_{-\infty}^{\infty} \int\limits_{-\frac{T}{2}}^{\frac{T}{2}} e^{i(\omega_0 - k_0 x)}\ e^{-i(\omega t - kx)}\ dt\ dx$$

$$= 2\pi|T|\delta(k - k_0)\operatorname{sinc}\left[\frac{T}{2}(\omega - \omega_0)\right] \tag{3.22}$$

When compared to Equation 3.19, in terms of circular frequency, the Dirac delta function has been replaced by a sinc function that depends on the duration of the sampled signal T. The frequency–wavenumber spectrum is shown in Figure 3.16.

Limited sampling decreases the resolution of the frequency of the signal as indicated by the width of the main lobe centered at $\omega = \omega_0$ and causes energy to "leak" into adjacent side lobes. The frequency resolution defined according to the Rayleigh criterion is equal to one-half of the width of the main lobe, which is $\omega_{\text{Rayleigh}} = 2\pi/T$ for the sampling window $w(t)$. Increasing the sampling duration T improves the frequency resolution. Alternative sampling windows to control side-lobe leakage and frequency resolution are discussed, for instance, in Bendat and Piersol (2010).

Limited sampling in the spatial domain produces similar results. Assume that the signal is sampled within an *aperture* of length D. The sampled signal may now be represented as follows

$$z(x,t) = u(t)v(x)s(x,t) \tag{3.23}$$

where $u(t)$ is defined as before and $v(x)$ is defined as

$$v(x) = \begin{cases} 1 - \dfrac{D}{2} \leq x \leq \dfrac{D}{2} \\ 0 \quad \text{elsewhere} \end{cases} \tag{3.24}$$

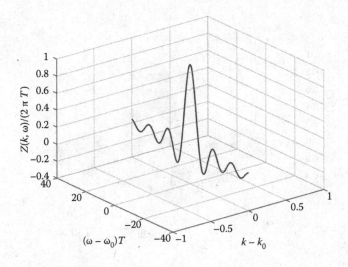

Figure 3.16 Frequency–wavenumber spectrum of a harmonic plane wave with finite temporal sampling.

The frequency–wavenumber spectrum of the sampled signal may be calculated as follows

$$Z(k,\omega) = \int\limits_{-\infty}^{\infty} \int\limits_{-\infty}^{\infty} z(x,t)e^{-i(\omega t - kx)}dt\,dx$$

$$= \int\limits_{-\infty}^{\infty} \int\limits_{-\infty}^{\infty} u(t)v(x)s(x,t)e^{-i(\omega t - kx)}dt\,dx \qquad (3.25)$$

$$= \int\limits_{-\frac{D}{2}}^{\frac{D}{2}} \int\limits_{-\frac{T}{2}}^{\frac{T}{2}} s(x,t)e^{-i(\omega t - kx)}dt\,dx$$

Considering once again the example of a unit-amplitude, harmonic plane wave

$$Z(k,\omega) = \int\limits_{-\frac{D}{2}}^{\frac{D}{2}} \int\limits_{-\frac{T}{2}}^{\frac{T}{2}} e^{i(\omega_0 - k_0 x)}e^{-i(\omega t - kx)}dt\,dx \qquad (3.26)$$

$$= |D||T|\,\mathrm{sinc}\left[\frac{D}{2}(k - k_0)\right]\mathrm{sinc}\left[\frac{T}{2}(\omega - \omega_0)\right]$$

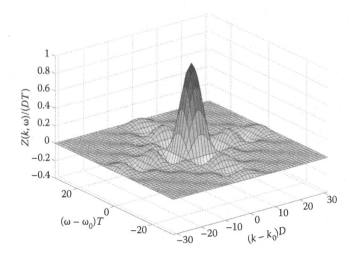

Figure 3.17 Frequency–wavenumber spectrum of a harmonic plane wave with finite temporal and spatial sampling: surface plot.

The frequency–wavenumber spectrum of the temporally and spatially sampled signal is shown in Figure 3.17. The limited resolution and leakage due to finite sampling is present in the wavenumber domain as well.

As before, we can quantify the resolution in the wavenumber domain via the Rayleigh criterion, which is $k_{\text{Rayleigh}} = 2\pi/D$. Increasing the aperture D will improve the resolution of the signal in the wavenumber domain.

Combining the individual functions that define temporal and spatial sampling into a single function

$$w(x,t) = u(t)v(x) \tag{3.27}$$

it is possible to express the frequency–wavenumber spectrum in a compact form through the convolution theorem

$$Z(k,\omega) = \int_{-\infty}^{\infty} \int_{-\infty}^{\infty} w(x,t)s(x,t)e^{-i(\omega t - kx)}dt\,dx$$
$$= W(k,\omega) * S(k,\omega) \tag{3.28}$$

where $*$ is the convolution operator and $W(k,\omega)$ is the Fourier transform of $w(x,t)$

$$W(k,\omega) = \int_{-\infty}^{\infty} \int_{-\infty}^{\infty} w(x,t)e^{-i(\omega t - kx)}\,dt\,dx \tag{3.29}$$

Due to finite sampling in time and space, the limited resolution has important consequences in the context of multimode surface waves. In the example of Figure 3.18, the signal is composed of the first two Rayleigh wave modes at a given frequency with equal amplitudes. In Figure 3.18a, the difference between the wavenumbers of the two modes is sufficiently large given the spatial sampling aperture D; hence, each mode can be identified correctly. In Figure 3.18b, the difference in wavenumbers is less, but the individual peaks in the frequency–wavenumber spectrum can still be resolved. The difference

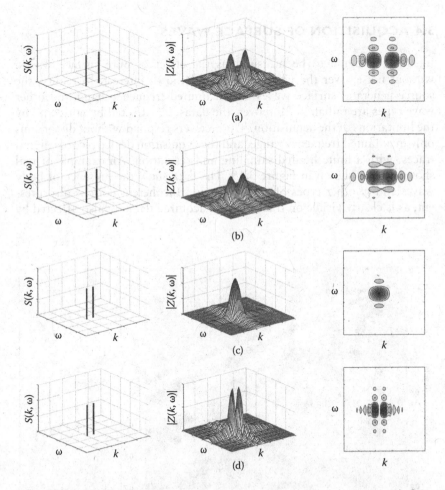

Figure 3.18 Frequency–wavenumber spectrum of a harmonic plane wave with finite temporal and spatial sampling—effect of different spectral resolution related to different apertures. Different distance in wavenumber between the modes are considered for panels a, b, c. In panel d a longer aperture produces a higher spectral resolution and resolves the two modes with the same distance of panel c.

in wavenumbers is further reduced in Figure 3.18c, resulting in a single, "apparent" peak for the combined signal that lies between the "true" peaks associated with the individual modes. Finally, in Figure 3.18d, the difference in wavenumbers is the same as Figure 3.18c, but the spatial sampling aperture is doubled. As a result, the wavenumber resolution is improved and the individual peaks can once again be resolved. In summary, to accurately measure multiple surface wave modes, the spatial sampling aperture has to provide sufficient wavenumber resolution to distinguish between modes.

3.4 ACQUISITION OF SURFACE WAVES

Ideally, the signal to be measured is the effect of pure surface waves, without noise, over the whole frequency band of interest. In reality, the source generates surface waves over a limited frequency band and other wave types are radiated. Moreover, the dataset is affected by noise and by the limitations of the acquisition. The receivers respond without distortions only in a limited frequency range, and the acquisition device records digital traces, with a finite limited sampling and resolution. An example of real shot gather is shown in Figure 3.19. The data show the presence of body waves and of other types of coherent noise. Incoherent noise is also present, as is clearly visible on trace 44. The acquired data are also affected by

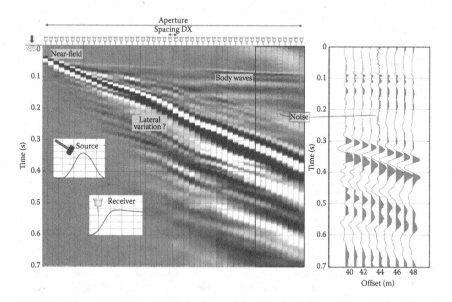

Figure 3.19 Real acquisition with noise, instrument effects, and sampling. Lateral variations appear as changes of dip of the main surface wave event. Body waves are visible as fast early arrivals of P-wave refractions.

the limited source spectrum, the receiver response, and other distortions related to near-field and possibly lateral variations.

These issues cannot be eliminated, but they have to be minimized. The experiment should generate broadband surface waves, with amplitude sufficient to minimize the effect of the incoherent noise. The sampling in time and space should be designed to extract the properties accurately over the desired range of wavelengths, limiting the effects of the coherent noise.

The following issues are discussed in the next sections: coherent and incoherent noise, sampling limitations and survey geometry design, and finally the equipment.

3.4.1 Noise

In general, a *signal* is any function of time and/or space. For seismic tests, a signal is the part of the data carrying meaningful (or useful) information. The unwanted part of the data (i.e., the part that does not carry any useful information) is called *noise*. The latter can mask, distort, and interfere with the signal. The definition of noise is therefore dependent on the objectives of the acquisition.

For instance, in active tests, only the surface waves generated by the controlled seismic source are useful signals. All other recorded events are considered noise. According to this definition, deterministic signals corresponding to an Earth response that cannot be modeled by the assumed Earth model are also considered noise.

Noise is usually classified as *coherent* and *incoherent* noise. Incoherent noise is often referred to as ambient noise, or *random* noise, even if it is statistically not random. Incoherent noise is not deterministically reproducible. It is not a source-generated effect, and it is not correlated with the signal. On the contrary, coherent noise is related to the experiment. It is deterministically reproducible and is correlated to the signal.

The recorded data contain signal and noise: the noise superimposes onto the signal, in different domains, and affects our ability and accuracy in estimating the signal properties. The degree of contamination of the data depends on the relative importance of signal and noise.

The signal-to-noise ratio (SNR) is defined as the ratio between the signal power S and the noise power N. It is often expressed in decibels, as

$$SNR[dB] = 20\log_{10}\frac{S}{N} \tag{3.30}$$

The data acquisition should be designed and performed with the objective of maximizing the SNR to maximize the accuracy of the estimation of the signal properties.

Different strategies are used to tackle the two types of noise. For the incoherent noise, increasing the power of the signal improves the SNR and reduces the distortions of the noise. The coherent noise, due to its deterministic nature and reproducibility, requires different approaches. For example, the acquisition geometry can be designed to minimize the presence of some coherent events, such as the near-field effect.

3.4.1.1 Incoherent noise

Incoherent noise can be the effect of the background vibrations at the site produced by natural and human sources: traffic, vibrating and moving machines, wind, and movements of surface or ground water. It is often dominated by surface waves, which are, however, incoherent with respect to the acquisition experiment. They are generated by sources, the position and time of activation of which are unknown. Passive surface wave methods are based on analysis of these components of the background noise. Specific signal processing techniques are used to extract the propagation properties without knowing the source position and timing.

Other types of incoherent noise are electric or electronic noise in receivers and cables, and in the acquisition system. This noise may be generated by power lines and other external sources and by imperfections in the recording system.

When acquiring active data, recording the ambient vibrations can be useful for understanding the nature and the level of the incoherent noise. Noise records can be collected by recording data from the receiver spread without activating the seismic source. The noise records often are poorly interpretable in time-offset, but their spectrum discloses valuable information. Understanding the nature of the noise can be useful to evaluate the data during the acquisition stage. For example, a strong single-frequency noise can mask the data in time domain without causing relevant distortion in the useful frequency range. Usually narrow-band noise originates from human activities (e.g., industrial facilities or alternate current in the electrical network). When evaluating the ambient noise properties, long records are required. The spectral analysis of single short windows can have peaks and a bandwidth that do not represent the real site conditions. Figure 3.20 shows an example of the noise spectrum at a quiet site; Figure 3.20a shows the averaged amplitude spectrum of 60 min of passive recording on 48 channels. In Figure 3.20b, the spectrum was computed from two single records 2 s long.

The incoherent noise is summed to the signal and affects all estimates made from the data. Improving the data quality means recording data from which the desired propagation properties can be accurately extracted. Increasing the SNR reduces the uncertainty of the estimation. For instance, the uncertainty on the phase shift between two harmonic traces, with a random noise, is a function of the SNR.

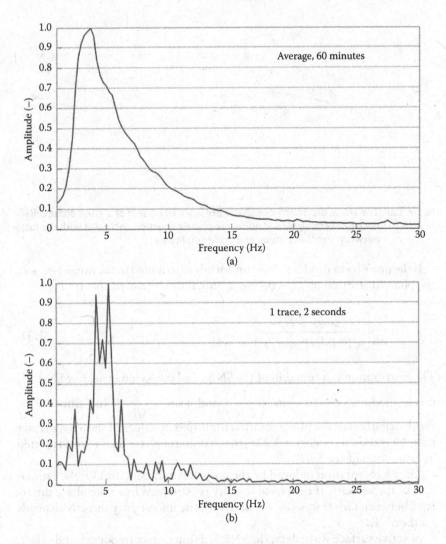

Figure 3.20 Amplitude spectrum of the ambient noise. The spectrum of a single trace, 2 s long, is shown in panel (b). The averaged spectrum is plotted in panel (a).

For a single harmonic trace, the phase distortion of the recorded data $d = s + n$ can be written as (Figure 3.21)

$$\Delta\varphi = \tan^{-1}\left(\frac{N\sin\varphi_n}{N\cos\varphi_n + S}\right) = \tan^{-1}\left(\frac{\sin\varphi_n}{\cos\varphi_n + S/N}\right) \tag{3.31}$$

where φ_n is the phase difference between signal and noise φ_n.

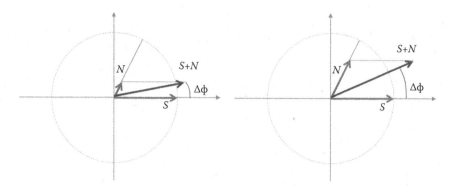

Figure 3.21 The phase distortion due to the presence of noise. For a given phase difference between signal and noise, the phase distortion decreases with the ratio
between signal amplitude and noise amplitude.

If the noise has a random phase uniformly distributed in the range $(-\pi, +\pi)$,
the phase distortion $(\varphi_{S+N} - \varphi_S)$ has a probability density function $P(\varphi)$

$$P = \frac{1}{2\pi}\left(\frac{\cos\varphi \cdot S/N}{\sqrt{1-(S/N)^2 \cdot \sin^2\varphi}} \right) \tag{3.32}$$

The distribution is a function of the SNR, and it is symmetric. For SNR >1,
it is limited to between $-\sin^{-1}\left(\dfrac{1}{S/N}\right)$ and $+\sin^{-1}\left(\dfrac{1}{S/N}\right)$. In Figure 3.22a,
the distribution of the phase rotation for different values of the SNR greater
than 10 is shown. In Figure 3.22b, the maximum phase distortion is plotted
as a function of the SNR.

The phase rotation induced by the noise has a direct effect on the estimation of the velocity. If the phase velocity is estimated from the phase difference between two traces (see Section 4.3), the uncertainty directly depends
on the SNR.

In active surface wave data, the SNR is a function of frequency and offset.
The incoherent noise has an amplitude spectrum that depends on the site. It
is related to the noise sources and to the subsurface response (e.g., see
the spectrum of Figure 3.20). If it is not dominated by near sources, it
is spatially stationary in terms of amplitude. On the contrary, the signal
(the source-generated surface waves) has a frequency- and offset-dependent
amplitude spectrum. With the simplest attenuation model, an exponential
function of offset with constant damping ratio D and constant velocity V,
the signal power in dB takes the form

$$A_{dB}(f,x) = 20 \cdot \log_{10}\left(A_0(f) \cdot e^{-2\pi f/V \cdot D \cdot x} \right) \tag{3.33}$$

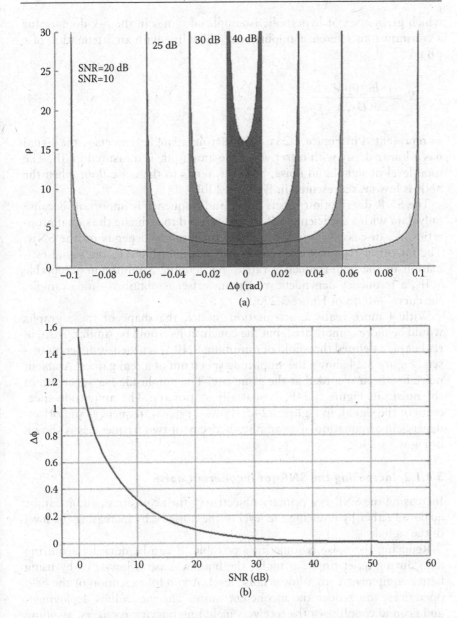

Figure 3.22 Maximum phase distortion induced by noise on a monochromatic signal as a function of the signal-to-noise ratio (SNR): (a) distribution of phase rotation for different values of SNR; (b) maximum phase distortion as a function of SNR.

which gives a set of hyperbolic isoamplitude lines in the f–x domain for a constant unit reference amplitude A_0. The line with an attenuation of c dB is

$$X = \frac{c \cdot ln(10) \cdot V}{20 \cdot D \cdot 2\pi} \cdot \frac{1}{f} \tag{3.34}$$

as represented in Figure 3.23a. Considering a single frequency, the signal has a linear decay with offset when the amplitude is measured in dB. The total level of signal and noise, however, tends to the noise floor when the SNR is low (as represented in Figure 3.23b).

The SNR distribution versus offset and frequency is important because only data with a sufficient SNR should be used to estimate the signal properties. As discussed earlier, the phase uncertainty depends on the SNR. The far-offset region, for example, will not contribute to the signal estimation in the high-frequency range. By defining a minimum acceptable SNR, a frequency-dependent maximum offset is obtained—for example, the curve –40 dB of Figure 3.23a.

With a more realistic attenuation model, the shape of these graphs would be more complicated, but the conclusions would be similar. Also, in real data, a defined threshold of minimum SNR is reached at different offsets. Figure 3.24 shows the amplitude spectrum of a real gather. Ambient noise has been recorded at the same site. The amplitude f–x spectrum of the noise, in Figure 3.24b, is spatially stationary. The amplitude spectrum of the signal, in Figure 3.24c, shows a typical frequency- and offset-dependent attenuation. The amplitude decay of two frequencies is shown in Figure 3.24d.

3.4.1.2 Increasing the SNR for incoherent noise

Increasing the SNR is a primary objective of the acquisition, and it can be achieved either by reducing the level of the noise or by increasing the level of the signal.

Reducing the noise is sometimes possible. It can be done by acquiring data during quiet times (at night, the human noise is lower) or by using better equipment with a lower noise level. A careful execution of the field operations can reduce the incoherent noise. The meticulous deployment and ground coupling of the receivers including burying receivers, avoiding people and vehicle movements, and avoiding vibrations during the acquisition are important practices for minimizing the incoherent noise.

Increasing the signal level can be done by increasing the energy of the source. Possible strategies for increasing the signal level include using more powerful sources or combining different sources for different frequency ranges. An alternative approach is the so-called "vertical stacking" or

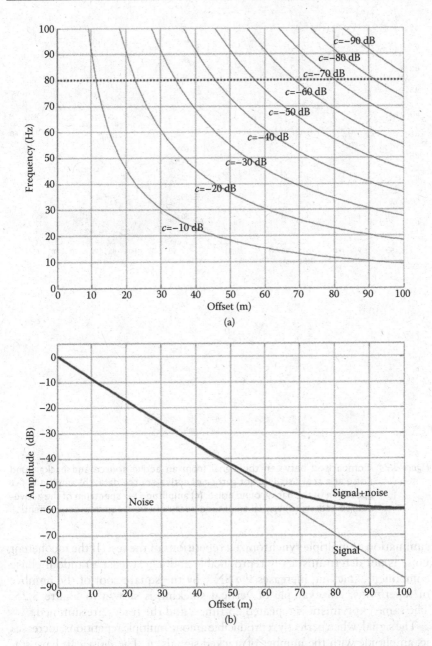

Figure 3.23 (a) The amplitude decay of a signal is plotted. The curves represent the decay of a pure signal with a simple attenuation model (single mode, constant damping, and constant frequency). (b) The amplitude of a recorded frequency with offset is represented schematically as the sum of signal and noise on a dB scale.

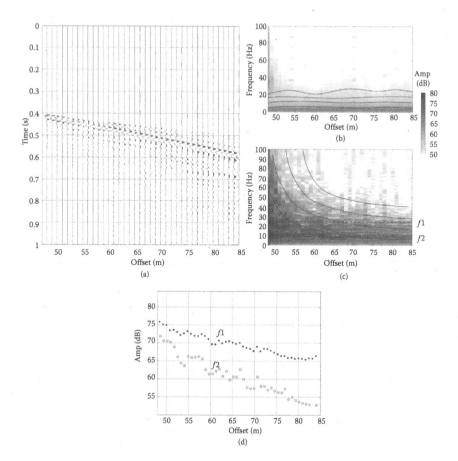

Figure 3.24 Comparison between the signal from an active source and background noise at a real site: (a) shot gather of active-source data; (b) amplitude *f–x* spectrum of the background noise; (c) amplitude *f–x* spectrum of the active-source gather; (d) spatial amplitude decay of two frequency components.

summation of multiple synchronized repetitions of the test. If the incoherent noise is not deterministically reproducible and has at least a random phase component, stacking increases the SNR by the square root of the number of repetitions. An example of vertical stacking is shown in Figure 3.25. The same experiment is repeated 25 times, and the traces are summed.

The signal, when perfectly reproducible among multiple repetitions, increases its amplitude with the number of stacked signals, *n*. The noise can have stationary amplitude but, if incoherent, has a stochastic phase. If the phase is random, as in a frequency component of the random noise, we can assume a uniform distribution in $(-\pi, +\pi)$; the absolute value of the phase difference between two summed components is linearly distributed in $(-\pi/2, +\pi/2)$, and

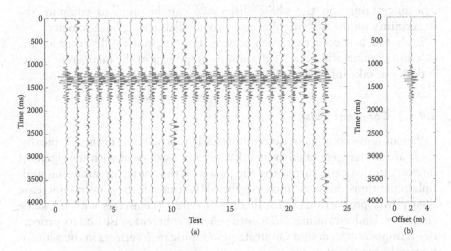

Figure 3.25 The summation of multiple repetitions improves the signal-to-noise ratio because the signal increases more than the noise when the latter is not coherent or deterministically reproducible: (a) ensemble of 25 individual shots; (b) stacked signal.

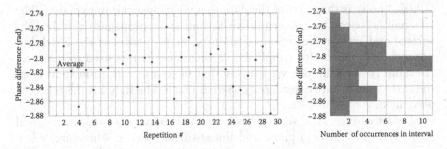

Figure 3.26 The phase difference between two traces estimated for 30 repetitions of the test. The standard deviation in this example is about 1% of the estimated mean value.

this makes the sum lower than its maximum. Considering a constant amplitude, with vertical stacking, the random noise increases as $\sqrt{n}$. The SNR then increases as $n/\sqrt{n} = \sqrt{n}$.

Several repetitions of the experiment are necessary for vertical stacking. If the recorded data are not summed directly in the field but are recorded separately, they provide a way of evaluating the uncertainty of the estimated wave parameters versus the stochastic components of the noise. The phase difference between two traces spaced at 5 m is plotted for 30 repetitions of the test in Figure 3.26. The average phase difference is about 2.8141 radians, and the standard deviation is about 0.0289 radians, or about 1%.

The uncertainty on the phase difference can be used to compute the uncertainty on the wavenumber and propagated to the phase velocity.

It should be noted that such statistical estimations only consider the stochastic uncertainty due to a random component of the incoherent noise. It does not take into account epistemic uncertainty and coherent noise.

3.4.1.3 Coherent noise

Coherent noise consists of deterministically reproducible events. Typically, these are source-generated events that are not associated with the signal of interest. In this case, increasing the power of the source or summing multiple repetitions does not increase the SNR. Indeed, these events increase their power proportionally (in theory) or repeat themselves with the same character. Understanding and identifying the coherent noise and its properties is important to design the strategies to mitigate its effects in the acquisition and processing stages.

The definition of coherent noise is not straightforward. It depends on the objectives of the measurement and on the assumption of the interpretation models. Any part of the data that cannot be explained by the assumed model has to be considered coherent noise. Citing Scales and Snieder (1998), "noise is that part of the data that we choose not to explain." The coherent noise can be a deterministic seismic event propagating in the subsurface and carrying information about some properties of the Earth but not included (maybe for simplicity) in the model that is used to interpret the data. As an example, we can say that in seismic reflection surveying, the surface waves are considered coherent noise. The signal is the reflected wave field.

In most of the cases, surface wave methods assume a one-dimensional (ID) Earth and far-field plane and linear-surface waves. Therefore, when acquiring surface waves, we can consider coherent noise as any other seismic event and, in general, all the effects of the differences between the assumed model and the true Earth response. Whatever produces data that are not explained correctly by the assumed subsoil model is coherent noise. Real data contain then different types of coherent noise: *nonsurface wave events* and *noncompliant surface waves*.

The nonsurface wave events include body waves from direct, refracted, and reflected paths in the subsurface. They also include air blasts (i.e., acoustic waves associated with the source stroke and traveling through the air).

The noncompliant surface wave coherent noise includes all the surface wave energy that does not follow the assumed propagation model for intrinsic and geometric reasons. For example, discontinuities and heterogeneities in the subsurface and in the topography produce diffractions of the surface waves and conversion of body waves into surface waves.

These phenomena, often called scattering, do not propagate directly from the source to the receivers. The associated apparent velocity along the receiver spread can be different from that of the direct surface waves.

The lateral variations of the medium properties affect the wave propagation properties in a deterministic way and distort the extracted velocity. They introduce artifacts in the estimated parameters when a laterally homogenous medium and wave are assumed and can be considered coherent noise. Their presence should be identified and considered in the processing stage.

Surface waves that propagate a short distance from the source cannot be treated as horizontally traveling plane waves. They have different propagation properties and are not correctly interpreted by most inversion approaches. This point has been discussed by many authors and is known as the near-field effect (Stokoe et al. 1994). The near-field effect is deterministic; hence, it can be classified as coherent noise.

Slightly different, from a measurement perspective, is the case of higher modes. The surface wave propagation has a multimodal nature, and the energy associated with higher modes depends on the stratigraphy and on the experimental setup. Especially at inversely dispersive sites, higher modes may dominate the experimental wave field. If the interpretation approach considers only the fundamental mode, or if the acquisition and processing procedures do not allow the identification of multiple events, higher modes have to be considered coherent noise and must be carefully filtered out.

However, in case of higher modes of surface waves, improving the model is recommended; they are not a different, unrelated event, and their properties depend on the same parameters that we try to estimate with the fundamental mode. They can be identified in data and included in the adopted Earth response model; hence, they can be considered as useful signals. Examples of multimodal processing and inversion are reported in the literature (for instance, Gabriels et al. 1987; Maraschini et al. 2010). The advantage is twofold. For one, this adds extra information. Second, upgrading the model allows the acquired data to be correctly interpreted even when the higher modes are dominant and cannot be removed.

3.4.1.4 Body waves

In records, body waves can be P- and S-waves propagating from the source to receivers with different paths. In shallow, small-scale tests, body wave amplitude is often much lower than the surface wave amplitude. They are superimposed onto each other in time-offset, but often they tend to map into different portions of the f–k spectrum. They can be easily identified and removed or ignored. Strong P-waves from refracted paths can be muted in time if necessary.

In large-scale acquisition, for exploration applications, the guided waves can become dominant events in the far offset. Guided modes are surface waves, leaky modes with a dominating pressure component, and interfering total internal reflections of the supercritical energy in a low-velocity waveguide. Because they are faster and often have a smaller damping ratio, they tend to dominate in the far offset. They are usually separated in terms of velocity and frequency, and they can be analyzed for a joint inversion with the Rayleigh waves or Scholte waves. The joint inversion of Rayleigh and guided waves is discussed in Chapter 8.

3.4.1.5 Air blast

The sound emitted by the seismic source propagates as a pressure wave in the atmosphere, sometimes coupling with the ground, and it is detected by seismic receivers. Even in Vibroseis acquisition, the vibrator sound correlates with the pilot and appears as a high-frequency event in the data.

In active data, it is common to observe a linear, high-frequency event propagating with fairly constant velocity at the speed of sound, which varies from 310 to 360 m/s, for a temperature varying from –30 to +50°C. Indeed, pressure and humidity variations can be neglected in normal conditions, and the speed of sound v_{air} can be written as

$$v_{air} = 331.3 \cdot \sqrt{1 + \frac{T}{273.15}} \ [\text{m/s}] \tag{3.35}$$

where T is the temperature.

The presence of a strong air blast is potentially troublesome in surface wave measurement because it may superimpose on the Rayleigh wave signature. Indeed, the velocity of propagation in air may be very close to Rayleigh wave propagation. The air blast can be observed in time-offset, and it is identified in the f–k spectrum as a linear event, with constant velocity and very low attenuation, usually extending to high frequency. It can be muted in time-offset and ignored in the processing stage. In Figure 3.27, the air blast is visible in the shot gather as a high-frequency event preceding the surface wave energy (as annotated).

3.4.1.6 Near-field

Rayleigh waves can be regarded as plane waves only beyond a certain distance from the source (i.e., in the far-field) (Richart et al. 1970). The inner zone is addressed as the zone in the near-field. In the two-station technique (see Section 4.3), the measurements are usually considered as affected by near-field effects if the first receiver is placed at a distance from

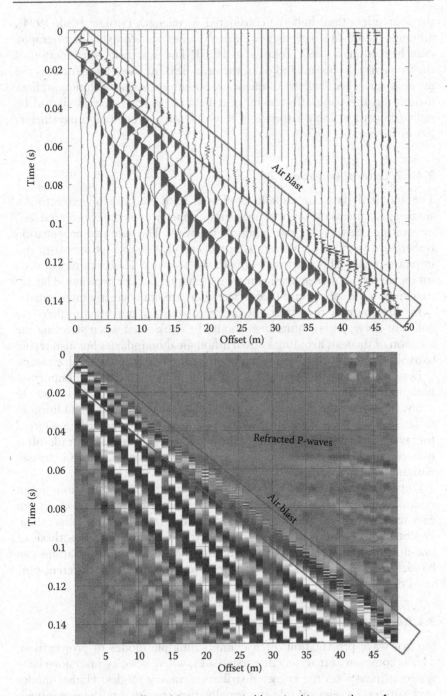

Figure 3.27 Seismogram affected by a strong air blast; in this case, the surface waves arrive after the air blast.

the source less than half the considered wavelength (Stokoe et al. 1994), although some authors suggest larger values for complex stratigraphy (Sànchez-Salinero 1987; Tokimatsu 1995). The near-field is a function of the frequency. In broadband data, the near-field for some frequencies is the far-field for other (higher) frequencies. Some processing techniques allow identifying the zone affected by near-field in which receivers should be muted to get a reliable estimate of Rayleigh wave propagation parameters (see Section 4.4).

3.4.1.7 Lateral variations

The acquired data are usually processed with the aim of extracting a single-space invariant set of propagation properties for the considered site, integrating or averaging over multiple receivers. The assumed propagation is therefore assumed to be laterally homogeneous—linear in time-offset and in phase-offset. This is consistent with the assumed 1D Earth model used for the inversion in most applications (see Chapter 6). The presence of lateral variation in the subsoil produces effects that are not correctly interpreted.

In this context, the effects of lateral variations can be considered as coherent noise. This requirement should be considered when selecting the location of the test, avoiding known lithological boundaries but also trying to avoid the acquisition along the dip direction for expected dipping layers.

In general, it can be observed that the length of the array is an important parameter, as far as the risk of lateral variations is concerned; the longer the array, the higher the chance of significant lateral variations. In addition to different processing techniques, a set of shorter arrays might be preferred for investigating sites where lateral variations are expected. The trade-off is between the lateral resolution and the spectral resolution, which increases with increasing aperture.

Once data are acquired, the identification of lateral variations is an essential step of processing. With the exception of extreme cases, it is not easy to detect the presence of lateral variations from the time-offset data. Processing techniques have been developed to test this basic hypothesis of one-dimensionality using the multichannel data. The lateral variations can be identified by comparison of information extracted from different portions of the array.

3.4.1.8 Higher modes

Surface wave propagation often exhibits multiple modes of propagation. This is common at inversely dispersive sites, where velocity inversions have a strong impact on the energy distribution among modes. Higher modes may also have a relevant role at normally dispersive sites with large impedance contrasts and even with smooth velocity profiles.

Higher modes should be considered as useful information. Their phase velocity, as well as their spatial attenuation and energy distribution, depends on the subsoil properties, and they can be theoretically simulated and inverted. They only become coherent noise when the acquisition, processing, or inversion techniques are not able to deal properly with them.

Therefore, the acquisition should be designed in order to allow the identification of higher modes. Even in cases where the fundamental mode is dominating, the effect of higher modes on the recorded data cannot be neglected. A multichannel acquisition is necessary to properly identify the different modes. However, with some processing technique, the effects of higher modes can be misinterpreted. For instance, jumps of the phase and the oscillating behavior of the amplitude with the offset can be due to the superposition of different modes. The example in Figure 3.28 refers to a site where a higher mode has significant energy in the relevant frequency band.

The shot gather in time-offset and the f–k spectrum show the presence of two modes, with similar energy. The seismogram acquired with an impulsive source and 24 vertical geophones is shown in Figure 3.28a. The presence of two events with different velocities can be recognized in time-offset. The two modes can also be separated in the f–k spectrum (Figure 3.28b). In this case, the spectral resolution allows the separation of the two modes and hence the identification of the properties of both. The effects of mode superposition are clearly illustrated with a representation of the 20 Hz component of the seismogram (Figure 3.29). Phase rotations and minima of the amplitude are observed at the offsets where the two modes are in phase opposition and where they interfere destructively.

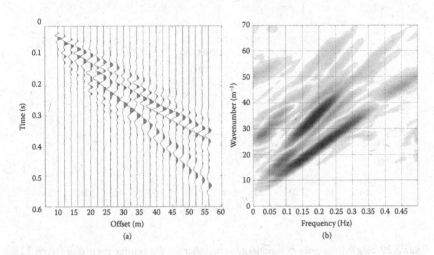

Figure 3.28 Seismogram and f–k spectrum for a case with highly energetic higher modes. The aperture is sufficient to separate the events: (a) seismic gather; (b) f–k spectrum.

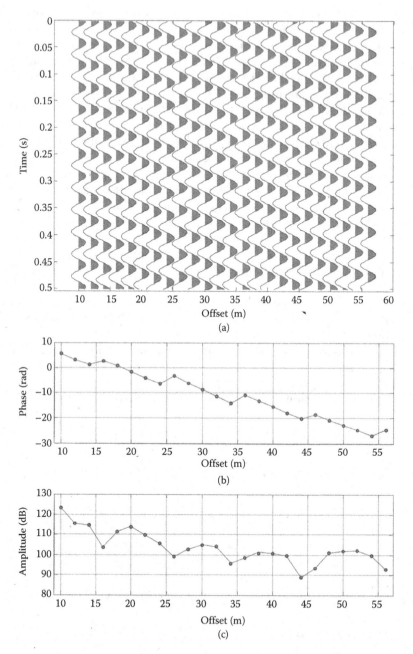

Figure 3.29 Single-frequency component extracted from the seismic record of Figure 3.28: (a) filtered time records; (b) phase versus offset; (c) amplitude versus offset. The amplitude decay shows the interference among multiple modes, and phase rotations are visible as well.

3.4.2 Sampling

The effects of the wave propagation at a specific location are recorded in time as seismic traces, with a fixed time step and a finite length. A finite number of receivers is deployed, and this implies a discrete spatial sampling, with a certain distance between receivers and a total length of the receiver array. The wave field is then sampled using a 2D discrete window in time and space. This induces limitations in terms of accuracy and bandwidth.

The limitations and consequences of the sampling are presented and discussed in time-offset and frequency–wavenumber domains in the following. Yet, the discussion applies independently on the processing method (see Chapter 4) because these limitations are intrinsic. The effects of the sampling are important in any situation and become critical when multiple modes are present.

3.4.2.1 Spatial and temporal discrete and finite sampling

As discussed in Chapter 2, the wave field associated with surface wave propagation in a layered medium can be written in terms of mode superposition, with a discrete set of modal curves. The eigenvalues describe the kinematics: for each frequency f, a set of wavenumbers $k_1(f)$, $k_2(f)$, ... , $k_n(f)$ and phase velocities $v_1 = f/k_1$, $v_2 = f/k_2$, ... , $v_n = f/k_n$ can be found.

The energy propagates only in correspondence with modal wavenumbers. This description of the wave field is valid only in the far-field, neglecting lateral variations, noise, attenuation, and so on, and it allows describing the effects of the sampling on the observed kinematics. The *spatial* sampling is particularly important and directly affects the accuracy of the estimated velocity, the ability to resolve mode, and the ability to identify secondary events with lower amplitudes.

When the wave field is sampled, 2D window w is introduced, and the effects are observed on a discrete set of points

$$r_{obs} = s(x,t) \cdot w(x,t) \tag{3.36}$$

Because the multiplication in time-offset domain is a convolution in frequency–wavenumber domain of the signal spectrum with the spectrum of the 2D window function, we can write

$$R_{obs} = s(f,k) * w(f,k) \tag{3.37}$$

The 2D boxcar window and its spectrum are depicted in Figure 3.30.

A single event—ideally homogeneous, continuous, and infinite—would appear as a spike in f–k. When sampled, it will appear as a spectrum of the acquisition window (Figure 3.31). The effects in k domain are typically more severe than the effects in f domain because the number of sampling

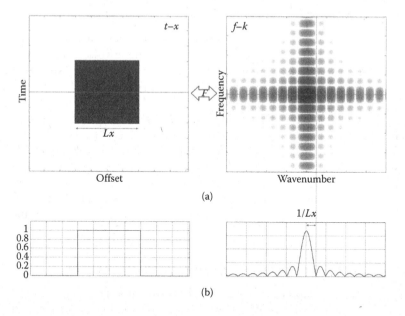

(a)

(b)

Figure 3.30 Time-offset boxcar window: (a) $t-x$ and $f-k$ domains; (b) slides at a constant time and at a constant frequency.

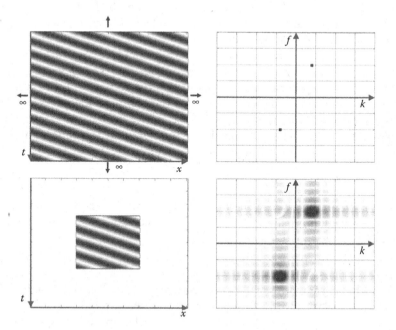

Figure 3.31 Effect of windowing on a harmonic plane wave—the spectral leakage transforms a Dirac's delta into a 2D sinc function.

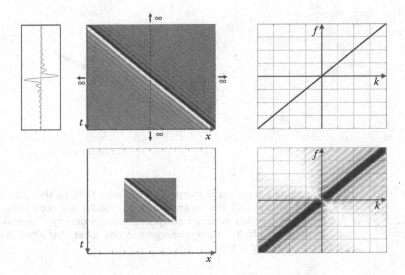

Figure 3.32 Effect of windowing on a broadband wave; the wavelet has a flat spectrum and is not dispersive. The spectral leakage creates a main lobe and side lobes.

points in space is necessarily smaller than in time. The abrupt truncation of the signal in time and space produces side lobes that have to be mitigated.

The sampling, with broadband surface waves, generates the described effects on each point of each mode. In Figure 3.32, the effect of the windowing is shown on a simple, nondispersive signal with a flat amplitude spectrum. The theoretical f–k spectrum is a Dirac's delta line passing through the origin of the f–k plane and with slope V. The finite sampling creates spectral leakage, as shown by the presence of side lobes. They appear in the spectrum of the finite signal as secondary maxima, which are parallel to the main event. Side lobes can be mitigated with tapering during the processing.

The sampling introduces a series of limitations that have important consequences on the inferred dispersion curves. In the following section, the limits introduced by the acquisition are discussed considering the final objective of designing the optimal acquisition to minimize the effects of the sampling over the extracted properties. The effects of the sampling in time and space will be discussed in terms of limitations of the frequency bandwidth, or modal resolution, and of accuracy of the extracted properties.

Windowing introduces a maximum wavenumber, a finite main lobe with leakage, and side lobes. The maximum wavenumber might limit the bandwidth and create spatial aliasing of the data. The spectral resolution affects the accuracy of the velocity estimation and the possibility of discriminating among different modes. A synthetic gather with only surface

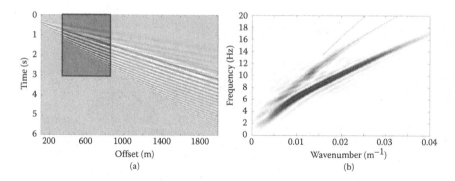

Figure 3.33 Finite sampling of a transient synthetic Rayleigh wave field: (a) shot gather and sampling window; (b) the three theoretical modes are represented in the f–k plane together with the computed spectral density. The truncation in time and offsets creates leakage and side lobes that affect the identification of higher modes.

waves is shown in Figure 3.33a. The spectral analysis of a limited time and space window is shown in Figure 3.33b together with the theoretical modal curves. The fundamental mode corresponds to the energy maximum, but the identification of higher modes is not straightforward.

These sampling issues can lead to quite irregular and misleading behavior of the experimental dispersion curve if they are not correctly assessed in the design of the testing array. Moreover, these limitations also have to be taken into account in the inversion.

Only the modal curves should be considered a characteristic of the site. The experimental dispersion curve is affected by mode superposition effects and also depends on the acquisition parameters. It can then be considered as an apparent dispersion curve (Tokimatsu 1995), which is not an intrinsic property of the site. To compare synthetic and experimental data within an inversion algorithm, it is then necessary to model the experimental test, taking into account all the acquisition parameters and the source position.

The limitations of the temporal sampling are less severe. It is normally possible to sample at a frequency high enough for the proper acquisition of the maximum propagating frequency, and the total duration of the signal can be normally acquired.

3.4.2.2 Maximum wavenumber and spatial aliasing

A broadband dispersion curve is needed for an accurate inversion. The maximum wavenumber that can be identified with a real array depends on the receiver spacing. When the spatial sampling is not sufficient, aliasing occurs, and the short wavelengths cannot be reconstructed properly. With 2D data, a true wavenumber higher than the Nyquist wavenumber $k_{\text{true}} > k_{\text{nyquist}}$ appears in the negative quadrant. The aliased replica falling in the spectrum has an apparent wavenumber of $k_{\text{alias}} = k_{\text{true}} - 2k_{\text{nyquist}}$. The apparent velocity is therefore negative.

The concept can be illustrated with monochromatic plane waves, with the same frequency and decreasing velocity, such as in Figure 3.34. The top row shows the f–k spectra, with a dot indicating the reconstructed position of the event; the bottom row shows the seismograms. The first two cases, with high V_1 and V_2, have a wavenumber smaller than the Nyquist wavenumber and are properly sampled. The apparent velocity in the seismic gather can be identified visually following a phase (e.g., a peak); it is positive with the wave traveling from left to right in the panel. The velocity V_3 corresponds to a wavenumber ($k_3 = f/V_3$) equal to the Nyquist wavenumber. The gather shows two opposite identical apparent velocities; the positive corresponds to the actual velocity.

The cases with V_4 and V_5 are aliased, with $k_{\text{true}} > k_{\text{nyquist}}$ and an apparent negative velocity. In the last case, V_6 has a wavenumber exactly equal to $2k_{\text{nyquist}}$ and the aliased as $k_{\text{alias}} = k_{\text{true}} - 2k_{\text{nyquist}} = 0$. The apparent velocity is therefore infinite, and the event appears flat.

The sign of the apparent velocity is an important element to consider. The event propagates with reference to the gathers in Figure 3.34, from left to right, and the sign of the velocity is known to be positive because of the acquisition geometry. When the acquisition is spatially aliased with a wavenumber smaller than $2k_{\text{nyquist}}$, it appears with a negative velocity (dashed lines in the figure).

In broadband data, all the components with $k_{\text{true}} > k_{\text{nyquist}}$ will be aliased in the negative quadrant (Figure 3.35). However, in end-off gathers, knowing that all the signal energy travels in the positive direction and is associated with positive wavenumbers, we can assume that only noise and aliased events are present in the negative quadrant. It is therefore possible to recover the aliased information in the range ($k_{\text{nyquist}} - 2k_{\text{nyquist}}$), between $-k_{\text{nyquist}}$ and 0, without introducing additional noise. The actual maximum wavenumber with off-end gathers is twice the Nyquist wavenumber; equivalently, the minimum wavelength is equal to the receiver spacing.

3.4.2.3 Spectral resolution and aperture

The actual length of the observation array affects the size of the main lobe of the window transform, which controls the spectral resolution. The leakage of the energy will be larger for a shorter array due to the scaling property of the Fourier transform. The accuracy of the identified maximum will be affected by the width of the main lobe, and the possibility of separating multiple events will be affected as well.

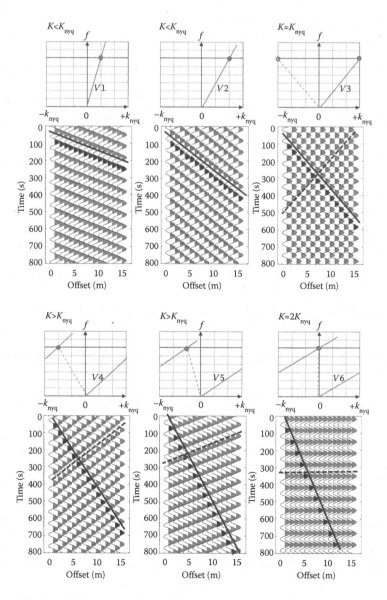

Figure 3.34 Spatial aliasing illustrated with a plane harmonic wave with decreasing veloc- ity. The decrease of the velocity produces an increase of the wavenumber, which in the third panel reaches the Nyquist wavenumber for the adopted spatial sampling. Spatial aliasing occurs, and apparent velocities are nega- tive in the *f*–*k* panels on the bottom. The continuous line in the seismic gather represents the actual phase velocity; the dotted line is the apparent phase velocity.

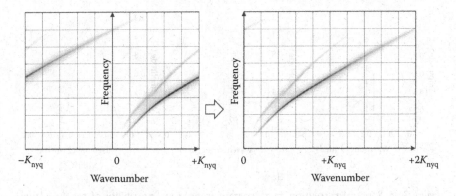

Figure 3.35 Unwrapping the negative quadrant of the f–k spectrum in off-end gathers to recover the correct representation of the propagation.

In theory, at a given frequency, the energy can propagate only at a discrete set of wavenumbers, one for each of the existing modes at the considered frequency. The wavenumber spectrum, at a constant frequency, should have spikes at the modal wavenumbers $k_1(f_1), k_2(f_1), ..., k_n(f)$ associated with the n existing modes. In reality, the wavenumber spectral resolution, derived from the spatial windowing of the acquisition (geophones spread length), allows only for smeared picks. This windowing effect may prevent an effective identification of modal curves. The ideal wavenumber spectrum and the observed spectrum for two different arrays are shown in Figure 3.36. The theoretical spectrum is shown in Figure 3.36a; the two events are resolved with an array of 96 m (Figure 3.36b) and are not resolved with an array of 12 m (Figure 3.36c).

As a result of the insufficient spectral resolution, multiple modes can be unresolvable, and the energy can appear as propagating at an intermediate velocity without corresponding to any of the real modes due to the mode superposition. What can be observed in the data depends on the true events and on the acquisition layout.

As an example, we show the picked energy maxima for a synthetic spectrum computed for the same layered model using two different arrays. The model has a velocity inversion, and higher modes dominate in the high-frequency range. The synthetic apparent dispersion curve for a short array (24 m) shows that modal superposition leads to a dispersion curve not corresponding to a modal curve because of insufficient spectral resolution. The absolute maximum of the spectrum is due to the superposition of the contributions of modes. The apparent dispersion curve shows a smooth passage to higher modes with increasing frequency (Figure 3.37b). When the array length is sufficient to separate modes (96 m in this example), the maximum follows the mode with the highest energy at each frequency (Figure 3.37a).

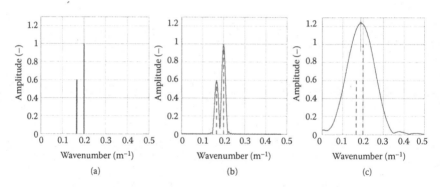

Figure 3.36 Spectral resolution at a constant frequency—possibility of resolving two modes with arrays of different lengths.

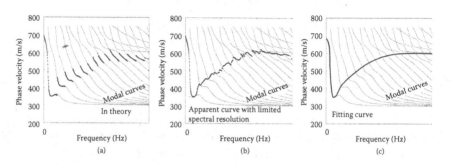

Figure 3.37 Example of modal superposition with two different arrays for the same synthetic case.

The example shows that the apparent velocity is affected by the modal superposition, depending on the array size. Similar results would be obtained with any of the processing techniques presented in Chapter 4 because the limitation is introduced by the acquisition and not by the processing.

Figure 3.38 represents the seismogram and picked maxima for an acquisition at a site composed of shallow, dry, overconsolidated clay over a clay sequence. The array length of 48 m used for the acquisition is not sufficient to separate the modes, and an apparent dispersion curve is observed in the range from 10 to 100 Hz.

Long arrays seem to be preferable because the longer the aperture of the spatial window (i.e., the array length), the higher the spectral resolution. On the contrary, with a short array there are less risks of lateral variations, insufficient *S/N* ratio, high-frequency attenuation, and spatial aliasing (for a given number of receivers).

The actual spectral resolution depends on the true array length. Yet, the computation of the spectrum requires evaluating the spectral density

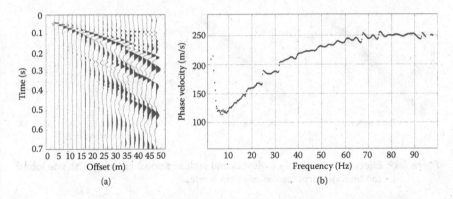

Figure 3.38 Example of modal superposition with real data. The picked curve from 10 to 100 Hz results from the superposition of multiple modes that are not resolved with the given acquisition parameters: (a) seismic gather; (b) apparent experimental dispersion curve.

on a number of frequency values much larger than the number of traces. This can be achieved in frequency–wavenumber analysis with the 2D Fourier transform via zero-padding the traces. The wavenumber spacing is crucial for an accurate picking of spectra but does not affect the actual spectral resolution and the size of the main lobe. Advanced techniques of array deconvolution, or synthetic aperture, can increase the spectral resolution.

3.4.2.4 Effects of side lobes

Spatial windowing introduces a fictitious spreading of energy in the f–k spectrum, and the side lobes might prevent higher modes from being identified and hiding the energy maxima associated with them (Linville and Laster 1966). The side lobes in the k domain might create apparent maxima that should not be confused with modes. This effect can be reduced using an appropriate tapering in the space domain, for example, a Hanning window. In Figure 3.39, the spectrum on the left is computed using a box window; the one on the right is computed using a Hamming window. The side lobes for a boxcar window can be computed as maxima of the sync function. The example on the right represents the set of maxima (including the first four side lobes on each side) of a 48 m array converted in velocity.

A generalized Gaussian window, without side lobes, helps in discriminating secondary maxima due to higher modes and those due to side lobes. Data are acquired without any special windowing, and this aspect refers more to processing (see Chapter 4).

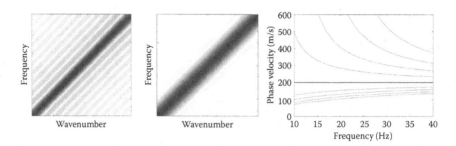

Figure 3.39 Effect of the side lobes without and with additional tapering. The side lobes can be misinterpreted as multiple modes.

3.4.3 Survey design

Considering the objectives of the surface wave test, the correct experimental design and the careful execution of the experiment are of primary importance for obtaining an optimal result. Many alternative approaches exist depending on the objective of the investigation and on the field of application. The spatial scale of surface wave testing can vary from millimeters to kilometers. Most of the acquisition parameters depend on the target depth and the desired resolution. In the following section, we discuss the acquisition geometries for active and passive tests at the intermediate scale of engineering applications for near-surface site characterization.

Designing a surface wave survey involves choosing the different elements of the experimental setup. The acquisition geometry is particularly important for its effects on the acquired data. The choice of the instruments (sources, receivers, and recording system) has a large impact as well. Often it imposes constraints because it is conditioned by the available instrumentation, by logistics, and by budget compatibility.

The term "acquisition geometry" usually indicates the space sampling of the wave field. It is characterized by the geometry of the receiver spread; the number and the position of the receivers define the total size of the array and the receiver spacing. In active tests, the position of the source, with minimum and maximum offset, is also a crucial parameter. The time sampling is defined by the sampling frequency and by the length of the records, or the time window.

In the next section, data acquisition on land is discussed. Surface waves at the seafloor in shallow water can be used with very similar approaches and methods. Some notes on the acquisition in this context will be given in Chapter 8.

3.4.3.1 Acquisition layout for active tests

An active surface wave test implies the use of a receiver array and of an active source. The most used scheme involves a linear array of evenly spaced receivers with in-line sources. The wave field is generated at the source location and is sampled along a radial line. The position and the direction, or azimuth, of the receiver line should be designed considering the local geology and trying to avoid known or likely lateral variations and dips, topographic changes, as well as surface or buried obstacles such as foundations or underground facilities. The acquired dataset will be a gather of traces with a constant spacing along the radius.

The acquisition parameters have to be designed depending on the aim of the test (i.e., the desired investigation depth and the resolution required for shallow layers). The parameters that have to be designed are the array length L and the receiver spacing ΔX, the source offset, and the time sampling parameters (Figure 3.40).

The *receiver spacing* is the spatial sampling frequency and affects the maximum wavenumber, which corresponds to the minimum wavelength. In turn, the shortest wavelength directly affects the possibility of resolving shallow layers, which can be important in some applications. As discussed in Section 3.3, the maximum wavenumber can be extended to $1/\Delta X$, which means that the minimum wavelength can be as low as the receiver spacing. In shallow applications, with spacing of a few meters, the limit often comes from the attenuation of the high frequencies and not from the sampling. Typically, values from 1 to 5 m are used for soil characterization.

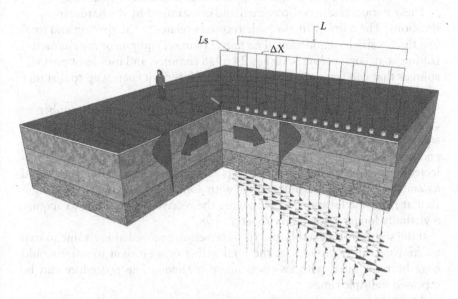

Figure 3.40 Schematic view of the active seismic acquisition—a linear array of receivers deployed radially from a surface source. The key geometrical parameters are indicated in the Figure.

The *array length* is important for many reasons. Provided that the signal is recorded with a sufficient SNR over the whole set of traces, a longer array provides a higher spectral resolution. Consequently, the mode separation possibility increases as does the accuracy of the identification of long wavelengths. If in theory there is no limitation to the maximum wavelength that can be observed with a given length, in practice the uncertainty increases as the ratio between the wavelength and the array length. It is important to stress that the effective length is limited by the SNR. If the source is not powerful enough, a long array can be counterproductive. Moreover, there is a trade-off between the improvement of the accuracy and the risk of lateral variations, which are more likely over a longer distance. Typical lengths are in the range from 20 to 100 m in shallow applications. The investigation depth is usually along the order of a half array length.

The *source offset* is the distance between the source and the closest receiver. The near-field effects create deviation from the linear wave hypothesis and can be considered as coherent noise. The distance affected by the near-field is a function of the wavelength, and empirical rules suggested in the literature indicate a half to one wavelength. Excluding near-offset traces will remove these effects but will also remove an offset range, which is valuable to estimate the high-frequency part of the wave propagation properties. The high frequencies might be too attenuated in the acquired traces and have too low a SNR to provide reliable information. Acquiring short offset data and removing the near offset from the low-frequency part are recommended, if necessary.

These parameters are also related and constrained by the hardware specifications. The number of available receivers relates to the spacing and total length of a single acquisition. The most common equipment for characterization at an engineering scale has 24 to 48 channels and uses light portable sources that can generate a wave field with sufficient energy up to distance of about 100 m.

The practical logistic limitations coming from the limited number of channels can be overcome using alternative schemes and geometries, but some processing approaches work only with evenly spaced arrays with end-off shots. Some acquisition procedures allow larger datasets to be collected by moving shots and/or receivers. Traces are subsequently sorted and assembled in a large seismic gather with some trace editing. Repeating a shot at the same location while moving the receivers allows one to acquire a virtually longer array (Gabriels et al. 1987).

If the source is repeatable and the receivers are moved along a line in two positions, as in Figure 3.37a, the final gather is equivalent to what would have been acquired with twice as many channels. The procedure can be repeated multiple times.

An alternative to varying the position of the receivers is the acquisition of a common receiver gather. A single receiver is deployed and left in place while the shot position is moved along a line at evenly spaced locations.

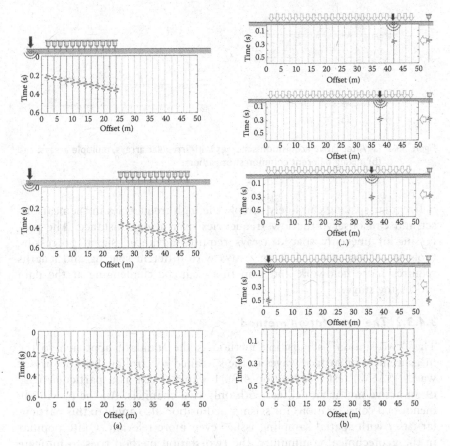

Figure 3.41 (a) Acquisition of a common shot gather merging two independent gathers, with the same shot position. (b) Acquisition of a common receiver gather with a single receiver.

The common receiver gather that is collected with this procedure (Figure 3.41b) for the reciprocity is equivalent to a common shot gather with the source at the receiver location. The repeatability of the source, and of the source phase in particular, is important for this type of acquisition.

Mixing multiple shot gathers, each with a plurality of receivers, could require more data processing. If the site does not present lateral variations, collecting multiple shot gathers with the same receiver array and sorting them with the offset could increase the aperture; this is the principle of the "walkaway test," as represented in Figure 3.42b. If the properties of the subsurface between the receiver array and the offset shot point are different from those under the receiver spread, the two gathers will not show a satisfactory phase continuity.

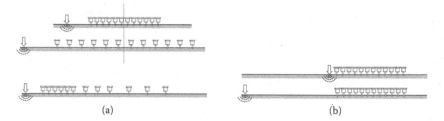

Figure 3.42 (a) Alternative acquisition schemes with irregular arrays, multiple arrays; and (b) mixing different common shot gathers.

Irregular spacing can finely sample the high frequencies in the near off-set and can observe the low frequencies over a long distance. The processing of unevenly spaced arrays requires advanced signal processing tools (see Chapter 4). Merging arrays with different spacing and lengths requires more fieldwork and sometimes can be challenging at the data processing stage.

3.4.3.2 The two-station method

The "two-station" acquisition method is the standard approach to the measurement of surface waves in the so-called spectral analysis of surface waves (SASW) test (see Section 4.3). It can be seen as a particular case of an active multichannel test with only two channels. All the previously mentioned considerations for seismic acquisition also apply to this particular case, with spatial sampling issues even more relevant. Quite popular in the geotechnical community, the two-station method tries to mitigate the strong limitations of spatial sampling with complex field operation. A single pair of receivers is used, and it has to be moved during the test to sample different wavelengths.

The distance between the source and the closest receiver is usually equal to the receiver spacing. For a given spacing, only a limited range of frequencies can be obtained due to spatial aliasing, attenuation, near-field effects, and so on. The configuration is changed, often changing the source as well, to construct the dispersion over a wide frequency range. Short receiver spacings and light sources are used for the high frequency (short wavelength), while wide receiver spacings and heavier sources are used for the low frequency (long wavelength). The set of values of the receiver spacing is designed considering the need of an overlap among the different ranges. Different schemes have been proposed to change the configuration. The common receiver midpoint scheme (Figure 3.43) is considered the most reliable for sampling the region with forward and reverse shots more homogeneously, which mitigates the uncertainties related to noise

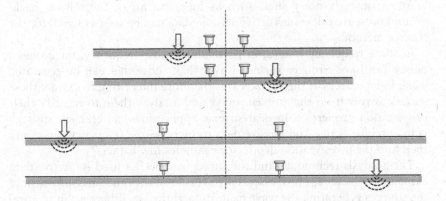

Figure 3.43 Two-station SASW acquisition with the common receiver midpoint scheme and forward and reverse shots.

and epistemic errors. A common source scheme in which only the receivers are moved is sometimes preferred for logistic considerations, especially when using heavy and difficult to deploy sources.

Reversing the source location compensates for phase distortions of the receivers and can help in identifying the effect of coherent noise. The effect of multiple modes, lateral variations, and other coherent noise types is more critical with two receivers than in multichannel shot gathers.

3.4.3.3 Acquisition of passive surface wave data

As discussed in the previous section, part of the ambient noise is a wave field generated by a number of sources of natural and human origin acting in the surroundings of the site (Bonnefoy-Claudet et al. 2006). In the low-frequency range, conventionally below 1 Hz, the noise sources are natural and mainly related to global geophysical events, and in particular to ocean waves. At a high frequency, conventionally above 1 Hz, the noise wave field is mainly generated by human activities (road traffic, industrial activities) even if atmospheric elements can largely contribute to the background vibrations and dominate the noise wave field in remote, quiet areas.

The ambient noises are often referred to as microtremors. The component of interest in shallow applications is the short period microtremor ($T < 1$ s, $f > 1$ Hz). The nature of this wave field makes its use particularly interesting in surface wave testing. Indeed the ambient noise wave field is generally dominated by surface wave components (Rayleigh and Love), with limited contributions from body wave propagation. The techniques based on the analysis of the ambient noise are called passive because of the absence of an active source.

At particularly noisy sites, such as in urban areas, the seismic background noise that degrades active seismic data can be seen as signal for the *passive* method.

In some frequency ranges, in particular at low frequency, the ambient noises can have stronger surface waves than those that can be generated using light sources at the surface. It is therefore interesting to extract these surface waves from the ambient noise and analyze them to identify their propagation properties. In near-surface applications, an efficient strategy is to combine active and passive data to increase the frequency range and therefore the investigation depth (see examples in Chapter 7).

The analysis techniques and the survey geometries used for active data cannot be directly applied, essentially because of the random component of the sources generating the wave field. In fact, the key difference in requirements affecting the geometry is related to the fact that the position of the source is not known and may change during the acquisition and for different frequency ranges. For a far source, the wave field is described by a value: the vector wavenumber k, with two components k_x and k_y. A linear array would detect the projection of the wavenumber along the array axis, obtaining an apparent velocity along the spread direction (see Section 4.6.3).

Hence, a 2D array is preferred, and the processing techniques have to be able to detect the wave parameters in two-dimensions. Details on the processing of passive data are presented in Chapter 4. Some considerations on the array are given in the following section. The performance of an array in enabling the estimation of the phase velocity depends on the array geometry and on the properties of the wave field. There is no universal agreement on the array design. The total array aperture, or diameter, is related to the maximum wavelength of interest; it is recommended to use an aperture from one-third of the maximum wavelength (Tokimatsu 1995) to a more conservative one wavelength (e.g., Asten and Henstridge 1984). In all directions, the station spacing should be designed considering the spatial aliasing— therefore being half of the minimum wavelength of interest.

The resolving power of an array depends on the maximum size and on the distribution of the receivers. The array performance can be evaluated using the theoretical array response function.

The array microtremor methods use a limited number of vertical receivers arranged in a 2D array. Common configurations are L-shaped and T-shaped, crosses, squares, triangles, and rectangular or hexagonal grids. The choice of the array also depends on the analysis method that is then used to process the data. A common choice for spatial autocorrelation (SPAC) processing is the triangle array with several nested equilateral triangles. This provides good results with a small number of geophones (Figure 3.44). The choice of the geometry is also constrained by the available equipment. Wireless receivers allow more complex and flexible geometries to be deployed.

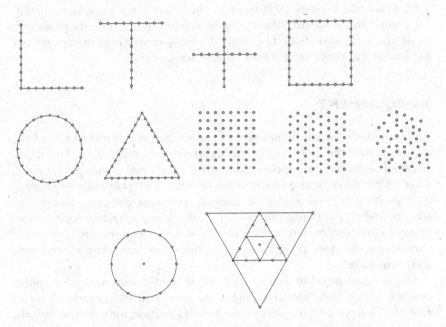

Figure 3.44 Examples of array configurations for passive surface wave tests.

The data acquisition is then fundamentally different from the active data acquisition. Ambient noise is recorded for a length of time sufficient to obtain a congruent number of data-segments. The acquisition parameters in time domain are set as a function of the desired frequency resolution and the maximum frequency of interest. The data segmentation is performed during the data processing, but often the maximum record length of the recording instrument will require recording multiple 30 s or 1 min gathers.

Interferometric techniques, with in-field data processing, can help in showing the improvement of the coherency of the events during the acquisition. The basic principles of passive surface wave interferometry are presented in Chapter 8.

The use of linear arrays for passive measurements has some appealing advantages from a logistic viewpoint. They allow using the same spread deployed for active tests, thus avoiding extra field operations, and allow the use of simple processing approaches. These are the main reasons for the popularity of the technique called refraction microtremor (ReMi), which indeed uses a linear array. It can be shown (see Section 4.6.3) that, in the case of a uniform and isotropic noise wave field, the response of the array can detect the real wavenumber.

However, the limitations induced by the linear array are severe; the linear array cannot identify the vector wavenumber but only its projection along the array direction. The method allows simple field operations but has limitations that can make the data of scarce value.

3.5 EQUIPMENT

The equipment for the acquisition of the surface waves has to record the wave field over the desired frequency range without distortions. The acquisition of seismic data requires three main functional elements: a source, a set of receivers, and an acquisition system. The positioning system is an important part of seismic acquisition. Knowing the exact location of sources and receivers is a crucial step in obtaining reliable velocity information. However, in near-surface surface wave testing, usually very simple surveying techniques are sufficient (e.g., measuring receiver positions with a tape measure).

The small acquisition devices for shallow applications are rather simple systems with a light source to input the energy in the ground. A set of receivers detects and transduces the ground motion into electric signals, which are transmitted as analog electric signals to a seismograph that digitizes and records the data. These are the key elements of the measurement chain, and their characteristics affect data quality.

Secondary components of the measurement chain, such as the data cable, trigger circuit, and control system, can also affect the data quality. They are important for the system's performance in terms of robustness and flexibility and for the efficiency of the field operations. A schematic view of the equipment for surface wave testing is represented in Figure 3.45. In the following section, we will discuss the main characteristics of sources, receivers, and seismographs.

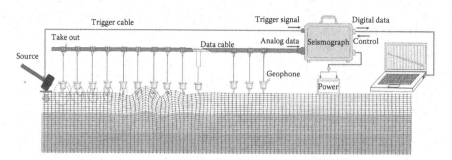

Figure 3.45 Schematic view of the field equipment. The essential elements are a source, receivers, and a digital acquisition device (usually a seismograph).

3.5.1 Sources

A seismic source is any device that releases energy into the Earth by generating seismic waves. In general, a seismic source should be able to generate high-amplitude signals without generating much coherent noise. For surface wave acquisition, the source should generate surface waves with enough energy to guarantee a sufficient SNR over the desired frequency range and offset range. In other seismic applications, a short compact wavelet is desirable. In surface wave acquisition, a wide frequency range is the objective. The challenge is often in having sufficient energy at the low-frequency end of the spectrum to increase investigation depth. The key parameters in the choice of the source are the energy and the frequency content. Logistical aspects can be very important—the site preparation, the cost, the source repeatability, the cycle time between shots, the environmental damage, the safety requirement, and the need for authorizations.

The comparison between sources is not straightforward because the source radiation depends upon environmental conditions at the surface and below the surface. A major factor to consider in surface wave data is the modal amplitude response. The amplitude spectrum of surface wave data largely depends on the subsurface. Even with a perfectly flat source spectrum in a lossless medium, the amplitude spectrum will show peaks and resonances and frequency ranges with high-amplitude attenuation.

Considering the data acquisition on land, sources can be classified as impulsive sources and vibrating sources. In shallow surveys, the most commonly used seismic sources are impulsive sources: hammers, sledgehammers, weight drops and accelerated weight drops, seismic guns, and explosives (Figure 3.46). Vibrating sources, standard in the exploration applications of the seismic methods, are used less in near-surface surveys. Vibrating sources include small controlled electromechanical vibrators, mini-vib, vibroseis, and Sosie™.

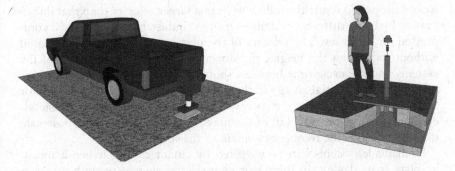

Figure 3.46 Weight drops, accelerated weight drops, and seismic guns are seismic sources used in surface wave data acquisition for near-surface characterization.

3.5.1.1 Impulsive sources

Impulsive sources input the energy into the ground with a short force application. The duration of the pulse directly affects the frequency content, as anticipated by Lamb (1904). Common impulsive sources can be chemical-based explosives (explosive, blasting caps, seismic guns, etc.) or mechanical (sledgehammer, weight drop, piezoelectric in holes).

Despite the number and variety of available seismic sources, probably the most used impact sources in shallow surveys are simple *sledgehammers* (weighing 1–15 kg) striking on metal or plastic plates. Indeed, a sledgehammer is easy to carry and use, and it has a low purchase and operational cost. It is easy to be triggered with an inertia switch on the hammer or with an electrical contact between hammer and metal plate. The triggering system is reliable and allows enhancement by in situ stacking (as has been shown earlier, the SNR increases with stacking with the square root of the number of shots).

The velocity at which a sledgehammer hits the plate can reach 15 m/s, depending on the hammer weight and on the force of the operator. Tests have shown that the impulsive dynamic force can exceed 20 kN. The weight of the hammer influences the frequency content of the generated pulse. A light hammer is preferable for the generation of high frequencies. The use of a metallic plate is recognized as increasing the frequency of the signal as well. On stiff surfaces, such as a rock outcropping or pavement, a very high frequency can be produced (more than 1000 Hz) with no need for striking plates. In most geological conditions and ambient noise levels, a sledgehammer can input sufficient energy for a 50–100 m array. Using a sledgehammer, there is virtually no site preparation or environmental damage.

Another popular source in shallow applications is the *weight drop* and the *accelerated weight drop*, also called thumper. It consists of a mass that is lifted using a winch or a piston and then released or accelerated to the ground. The mass can vary from a few kilograms to several tons. The height can vary as well, from less than a meter to tens of meters. Accelerated weight drops accelerate the falling mass to a larger velocity than that due to gravity by using different systems—springs, industrial elastic bands, compressed air, or gases. The velocity of the mass at impact can be increased without increasing the height, the dimension, and the portability of the systems and the cycle time between shots. The energy of a weight drop is related to the potential energy of the mass at the firing height; although for accelerated weight drops, the energy is increased by the work of the accelerating system during the fall of the mass. The use of a metallic baseplate usually increases the frequency content of the source.

Homemade systems can be triggered by circuit closing (when a metallic plate is used with an inertia or piezoelectric switch) or with a starter geophone. With homemade systems, the cycle time between shots can be slow, the movement in the field can be affected by logistic constraints, and the safety of the operation has to be considered.

In Figure 3.47, a single shot with a freefall weight drop of 130 kg from 3 m is compared with 20 stacked sledgehammer shots. Even if the overall SNR looks comparable, the spectral analysis reveals more energy at low frequencies in the weight-drop gather.

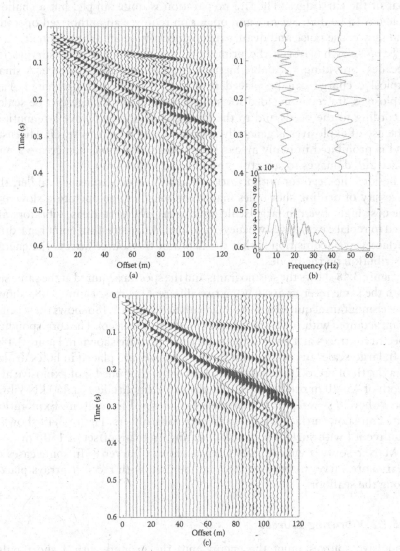

Figure 3.47 Comparison between (a) a vertically stacked sledgehammer acquisition and (c) a weight-drop acquisition. (b) Two traces at the same offset from the source are also compared in terms of spectral amplitude.

Other impulsive sources are *seismic guns*, ranging from modified rifles to iron water pipes with a coupler and nipple chamber on one end. Seismic guns typically input the energy by firing industrial cartridges into a shallow hole in the ground. The energy is greater than that of a sledgehammer, and the frequency content is similar. The use of seismic guns requires some preparation; after the purchase cost, the operation cost is essentially just that of the cartridges. The site preparation is quite simple, but a shallow hole has to be dug. Seismic gun operation requires an authorized operator and may cause noise and disturbance, especially in an urban context.

Explosives have been the primary source for land seismic surveys for decades, providing a scalable high-energy and broadband source. A small explosive charge can be placed in a shallow hole and detonated. Fast explosives are recommended for shorter pulses. The charge can be scaled according to the needs and to the local conditions and wave attenuation. The use of explosives is generally restricted and subject to specific licenses and is prohibited in many areas. The induced vibrations can create a nuisance and damages to structures and buried utility lines.

Besides the need for authorization and a licensed explosive handler, the necessity of drilling shot holes makes the speed of the operation slow and the cost high. Even in large-scale operations in desert areas, vibrators are used to replace explosives whenever possible. An interesting point regarding surface wave acquisition is that the depth of the source affects the energy distribution on the modes.

Figure 3.48 shows the seismograms and the spectra acquired at the same site with the same receiver array using two different sources. Figure 3.48a shows the seismogram acquired with a sledgehammer. Figure 3.48b shows the seismogram acquired with 100 g of explosives in a hole (1 m deep). The corresponding spectra for traces at a given offset from the source are shown in Figure 3.48c.

In large-scale surveys, several kg of explosives are placed in holes drilled to a depth of up to 20 m. Source studies show that one kg of explosive at a depth of 15–20 m produces roughly the same amplitude as a 450 kN vibrator with a 20 s sweep or a 20 ton weight dropped from 20 m. Exploration data can show surface waves up to very long offsets. Figure 3.49 shows a shot record with surface waves dominating up to an offset of 1500 m.

Marine sources will be briefly discussed in Chapter 8. In some cases of near-shore surveys, shots on land can be used with receiver arrays placed along the seafloor.

3.5.1.2 Vibrating sources

Impulsive sources input the energy into the ground with a short pulse of pressure lasting a few milliseconds. Therefore, the energy density in time (the source power) is high. An alternative approach is the use of longer controlled signals, using vibrating sources, with a low energy density.

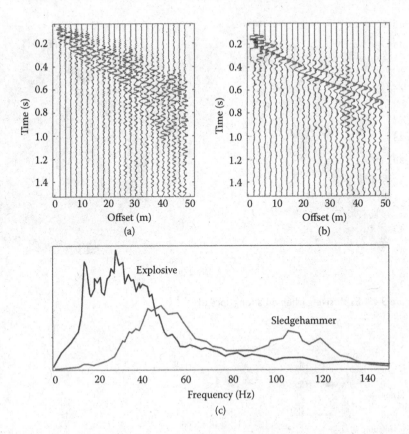

Figure 3.48 Seismogram comparison between a sledgehammer (a); an explosive in a shallow hole (b). The difference in frequency content is evident in the time signals and is confirmed by the corresponding spectra (c).

The typical example is the seismic vibrator, commonly used in the majority of land seismic reflection surveys. The energy is input using a controlled force signal lasting several seconds.

Nonimpulsive sources are called swept, controlled, or vibrating sources. They range from portable electromechanical shakers with a weight of less than 100 kg to large track-mounted hydraulic vibroseis weighting tens of tons. The principle of the vibrator is illustrated in Figure 3.50. A base plate is kept in contact with the ground by a hold down passive mass (the weight of the device or of the vehicle). An actuator imposes movement to a reaction mass, and the corresponding force is transmitted to the base plate. The schematic mechanical system is represented in Figure 3.50b.

The main mechanical parameters include the mass of the reaction mass M_r and of the base plate (M_b), the force of the actuator on the piston F_a,

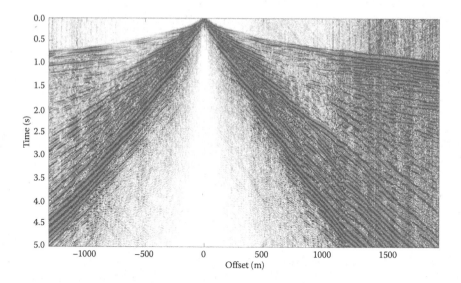

Figure 3.49 Explosive gather on a long spread.

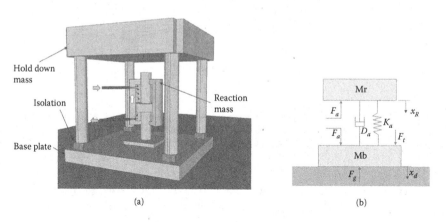

(a)   (b)

Figure 3.50 (a) Scheme of the vibrating source. (b) Representation of the mechanical system describing vibrating sources.

the force of the ground on the baseplate (and the opposite force of the baseplate on the ground), the force of the hold down mass F_t (often the frame of a vehicle) applied on the base plate through insulators, and the damping D_a and stiffness K_a of the actuator. The equation of motion for the two masses, considering the two masses as rigid, can be written as

$$-F_a = M_r \ddot{x}_r + D_a \left(\dot{x}_r - \dot{x}_b \right) + K_a (x_r - x_b) \tag{3.38}$$

$$F_a + F_t - F_g = M_b \ddot{x}_b + D_a(\dot{x}_r - \dot{x}_b) + K_a(x_r - x_b) \tag{3.39}$$

where x_r and x_b are the vertical displacements of the reaction mass and of the base plate, respectively. By summing Equations 3.38 and 3.39, it is possible to write

$$F_t - F_g = M_b \ddot{x}_b + M_r \ddot{x}_r \tag{3.40}$$

If the baseplate is isolated dynamically from the hold down mass, the dynamic force F_t is negligible and the force transmitted to the ground is equal to the weighted sum of inertia forces associated with the accelerations of the baseplate and of the reaction mass.

Small portable devices with a total weight of less than 100 kg can deliver forces up to 0.5 kN, while large-scale vibrators for seismic exploration have a peak force up to 400 kN. The maximum performance is actually a function of the frequency. Different factors limit the peak force in different frequency ranges. In the low-frequency range, there are the two factors limiting the energy that can be input into the ground: the peak-to-peak stroke of the reaction mass is the first limiting factor, at very low frequency, and the decoupling force. The boundary between the two depends on the magnitude of the reaction mass. A heavy vibrator with a high hold down weight, a large reaction mass, and a long stroke is best suited for high force at low frequency. In the high-frequency range, for hydraulic vibrators, the limit is usually related to the servo-valve bandwidth and the baseplate flexibility.

Quantification of the input source force is essential for some data processing techniques. For instance, the transfer function of a site can be obtained from the ratio in frequency domain between output (measured ground motion) and input (source force) (see Section 5.3). The source can be characterized using an accelerometer mounted on the frame.

Vibrating sources are used with different input signals. Long monochromatic signals guarantee a high SNR, but long operational times are required to sample the whole frequency range of interest. Sweeps and chirp signals are standard practice in reflection surveys. The advantage of the sweep for surface wave acquisition is that it covers the whole frequency range of interest in a single signal, and standard data processing can be applied.

3.5.1.3 Sweep signals

A sweep signal is a nonstationary function that can have the general form

$$A = A(t)\sin(2\pi f(t) + \varphi) \tag{3.41}$$

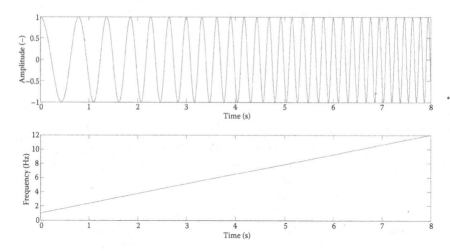

Figure 3.51 Example of a linear sweep and its time–frequency representation.

The frequency is therefore not constant but varies with time. Linear sweeps use a linear function for setting the frequency

$$f = c \cdot t + f_0 \qquad (3.42)$$

The frequency can increase with time (upsweep) or can decrease (downsweep). Upsweeps, such as the example in Figure 3.51, are preferred in reflection surveying because the correlation ghosts appear at an earlier (often negative) time and have less interference with the primary reflections.

With such a frequency function, the different frequencies are differently sampled (i.e., a different number of periods are recorded for each of them). To better sample the low frequencies, logarithmic or quadratic sweeps are preferable. The use of a sweep generates a signal with a relatively low amplitude compared to the impulsive sources. The correlation can be used to create a compact signal. The random noise is not correlated and is therefore filtered out.

The principle of the correlation is illustrated in Figure 3.52, where an input sweep is reflected by three reflectors (impedance contrasts in the subsoil). The uncorrelated time signal shows interference among the different shifted and scaled copies of the sweep. The correlation process acts as deconvolution, producing a series of copies of the sweep autocorrelation at the position of the spikes.

The recording time should be equal to the sweep length plus the listen time. The correlation produces a record that has full overlap of the sweep onto the recorded signal only for the duration of the listen time. Longer lags and negative times have partial overlap and a limited frequency content. The total energy is proportional to the duration of the sweep. Long sweeps

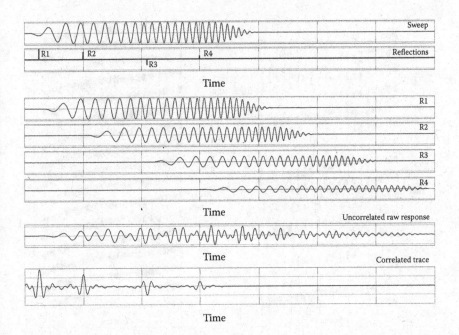

Figure 3.52 Principle of sweep correlation. The signal is the result of the convolution of the source signal with the Earth reflectivity series. It contains the super-position of shifted and scaled copies of the source signal. The correlation procedure decodes the recorded trace.

can be used for a higher SNR; the sweep rate in Hz/s affects the energy density in the different frequency ranges.

When acquiring data with controlled vibrating sources, the visual inspection of field records in time-offset domain can be challenging. The ambient noise over its full spectrum can create large amplitude vibrations, larger than those of the monochromatic signal. In the example shown in Figure 3.53, the raw gather is plotted on panel a and the filtered gather is plotted on panel b.

Alternatives to simple impulse and controlled sources are repetitive impulsive sources as the Sosie mini-Sosie, and Swept Impact Seismic Technique (SIST). These systems use impactors hitting the ground 5–15 times per second in a signal that can be 3–5 min long, with a pseudorandom series. A process similar to the correlation reconstructs a single impulse with high energy, compressing the series of pulses.

3.5.2 Receivers

The receivers are the first element of the recording chain. They transduce the ground motion into a measurable electric signal that is then transmitted, conditioned, and recorded by the acquisition device.

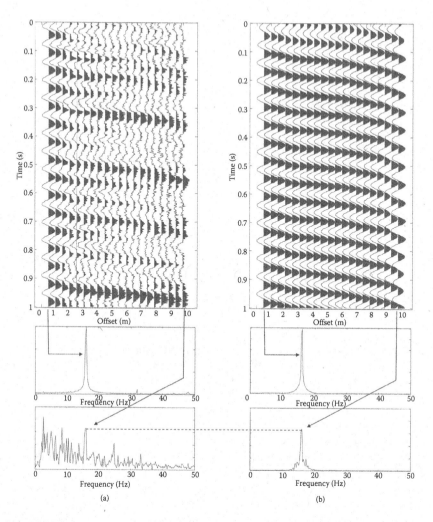

Figure 3.53 Harmonic multichannel record. (a) Raw data and (b) filtered data. In the far offset, the filtering enhances the coherency of the 16 Hz component.

The ground motion, on land, can be described in terms of particle displacement, velocity, or acceleration. The different sensing devices essentially can measure one or more components of the particles' velocity or acceleration, using a moving mass connected by a spring to the motion of the point to be measured. A soft spring allows the mass to stay in its initial position, providing a reference for measurement of displacement or velocity. A stiff spring causes the mass to move, with a residual displacement related to the acceleration. The mass motion, or the stress or strain, is measured using different physical principles.

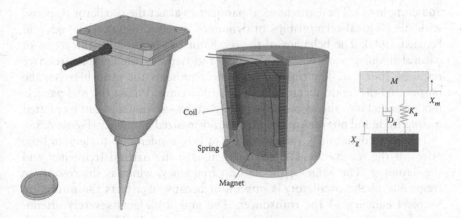

Figure 3.54 A geophone is an electrodynamic velocity transducer. Most geophones are of the moving coil type: a coil is suspended in a magnetic field and its oscillations generate a current.

In most shallow engineering surveys, the receivers are velocimeters called *geophones*. In some applications, especially with passive methods, the need to record a low frequency requires the use of very low-frequency geophones, called *seismometers*. In nondestructive testing, in pavement testing, and other applications where higher frequencies are needed, and often high-amplitude vibrations are generated, the use of *accelerometers* is preferred. Recently, the introduction of high-fidelity *microelectromechanical systems* (MEMS) digital accelerometers is also making accelerometers appealing for low-frequency applications (see Figure 3.54).

3.5.2.1 Geophones

Geophones are velocimeters (i.e., electrodynamic velocity transducers). Typically, they are of the moving coil type. A small coil is suspended by a spring in a magnetic field produced by a permanent magnet fastened to the casing. A similar system is obtained when the moving part is the magnet. The vibration of the soil causes a displacement of the geophone casing and magnet. Due to its inertia, the relative movement of the coil produces a small voltage proportional to the relative velocity. The axis of movement can be horizontal or vertical. When using a vertical geophone to transduce the vertical particle velocity, the tilt sensitivity has to be taken into account. Indeed the response is affected by the angle between the transducer and the vertical direction.

Ideally, the output of the geophone should be proportional to the input ground particle motion. In reality, a frequency-dependent transfer function, or response, is introduced by the geophone. It affects the transduced signal and the recorded seismic trace. The geophone can be considered as a single degree-of-freedom forced oscillator with mass m, a spring with stiffness k, and damping D. These mechanical parameters affect the oscillator response with the classical relationships of dynamics (see, for instance, Clough and Penzien 1993). The behavior of the oscillator can be described in terms of natural frequency and viscous damping ratio and by the complex receiver response function. The transfer response represents the ratio between the output and the input: it can be seen as a filter transforming the real particle motion signal into the recorded signal. The response function can be plotted as amplitude and phase as function of the normalized frequency (Figure 3.55).

Two main dynamic parameters have to be considered to understand the effect of the receiver on the recorded signals: the natural frequency and the damping. The value of the natural frequency, which is the resonance frequency of the oscillator, is important because it affects the minimum usable frequency of the transducer. The amplitude gets severely attenuated below the natural frequency. For example, a 40 Hz geophone will heavily attenuate the low frequencies used for surface wave measurements in shallow applications. For surface wave testing, geophones with a low natural frequency are typically used in order to obtain the desired

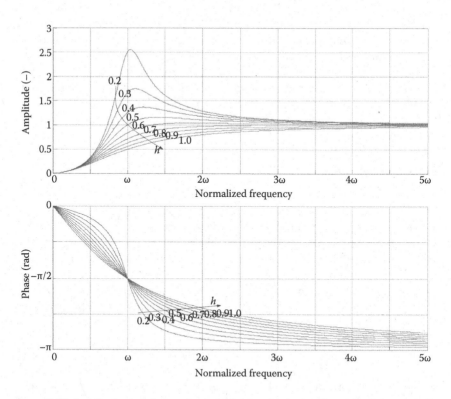

Figure 3.55 Amplitude and phase response of a geophone, as a function of the normalized frequency, for different values of damping.

investigation depth. Nevertheless, the trade-off between natural frequency and geophone roughness has to be considered. In order to obtain a low natural frequency, a large suspended mass has to be used. As a consequence, the instrument becomes heavier and less manageable in the field because the weight of the mass may easily damage the spring if the instrument is not carefully transported and deployed. Moreover, low-frequency geophones are more expensive. A compromise is represented by 4.5 or 2 Hz natural frequency geophones; 1 Hz geophones allow for deeper characterization but their cost is about one order of magnitude higher than 4.5 Hz geophones.

The effect of the damping on the response curve is important as well because it flattens the resonance peak at the natural frequency. It is recommended to tune the damping to get a flat response in the frequency band of interest. The damping of a geophone can easily be modified electrically by inserting a shunt resistor across the coil terminals. The current circulating in the coil passing through the shunt resistor produces a magnetic field opposing to the coil movement and damps it. The shunt resistor can be designed such that the damping factor is about equal to 0.6 to flatten the amplitude above the natural frequency as in Figure 3.55.

The sensitivity of the geophone can be expressed as the ratio of the output and the input in the flat-response frequency band. It is the output tension divided by the input particle velocity, and it is usually expressed in volt/mm/s.

In surface wave applications, the geophone response is critical for the amplitude response and the phase distortion.

In frequency domain, the recorded signal S_{rec} is the product of the complex spectrum of the true signal S_{true} by the receiver complex response function R

$$S_{rec}(f) = S_{true}(f) \cdot R(f) \tag{3.43}$$

Considering the amplitude A and the phase φ of the signal, the effect of the response of the receiver can be written as

$$A_{rec}(f) = A_{true}(f) \cdot A_R(f) \tag{3.44}$$

$$\varphi_{rec}(f) = \varphi_{true}(f) + \varphi_R(f) \tag{3.45}$$

The true amplitude is multiplied by the receiver response. This can attenuate the signal (and the noise) in certain frequency bands but does not influence the inferred phase velocity, which is essentially a function of the recorded phases. Geophones can also be used out of their flat band and below their natural frequency, obtaining attenuated signals. The recorded phase is affected by the receiver phase; the phase difference among multiple receivers, directly proportional to the wavenumber and affecting the velocity, is not affected by the receiver phase if the latter is constant on all traces. For this reason, using identical receivers is of fundamental importance to obtain good quality data for surface wave analysis.

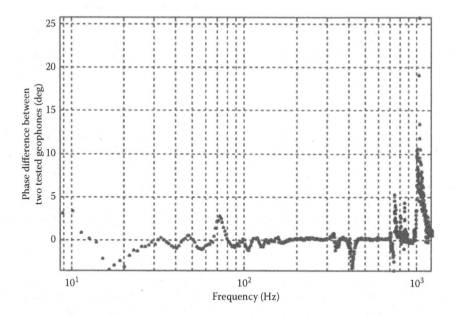

Figure 3.56 Real response of geophones measured on a shaking table: phase difference between two receivers.

The theoretical response curve describes the ideal response of the receiver. Real geophones have a different behavior at high frequency with spurious resonance noise due to undamped motion of the moving parts of the geophone along different directions. The real phase response is more relevant because the induced phase difference affects the estimated velocity. And the phase difference between two receivers can be large. Figure 3.56 shows the phase difference as a function of the frequency for two geophones tested on a shaking table (after Strobbia 2003). The phase distortion can become very large for weak excitation below half of the natural frequency.

A phase error of one degree can be relevant. The additional phase lag due to the receivers affects the phase velocity. For a two-receiver measurement, it can be computed as

$$V = \frac{2\pi}{\Delta\varphi}\Delta X \cdot f \qquad (3.46)$$

where f is the frequency, $\Delta\varphi$ is the phase difference, and ΔX is the distance. The uncertainty on the velocity is

$$\sigma_V = \frac{2\pi \cdot \Delta X \cdot f}{\Delta\varphi^2}\sigma_{\Delta\varphi} \qquad (3.47)$$

For example, a velocity of 600 m/s at 5 Hz and 5 m spacing would correspond to 15°. If the error is 5°, the obtained velocity is 900 m/s. Clearly, two-station measurements (see Sections 3.4.3.2 and 4.3) are the most sensitive to phase distortions, whereas the effects are mitigated and the estimate of phase velocity is robust in multichannel measurements.

The saturation can also be an important parameter. There is maximum input that clips the output for physical reasons. Indeed, the motion of the suspended mass is limited by the geometry of the casing. Saturation of the traces alters the frequency content and makes the experimental data useless.

3.5.2.2 Accelerometers and MEMS

In some applications, especially in nondestructive testing of pavement systems or structural elements, the high-frequency response is essential for estimating the surface wave components with very short wavelengths. Accelerometers can be more suitable for such applications. Conventional accelerometers are piezoelectric or piezoresistive. Piezoelectric accelerometers contain elements that are subjected to strain under acceleration. They rely on piezoceramic (such as lead zirkonate titanate) or single-crystal (such as quartz) piezoelectric elements to create a voltage. In general, they have an excellent high-frequency response. Piezoresistive accelerometers, often used in shock applications, are based on sensing flexure. They have a limited cost but typically a low sensitivity. Capacitive accelerometers can be used in servo mode and have a superior stability and linearity, but their cost is higher than the other accelerometers.

Recently MEMS sensors with the sensitivity, dynamic range, and low noise required for seismic acquisition have been developed. The term MEMS stands for microelectromechanical systems and indicates a combination of mechanical and electrical components built into very small devices, at the microscale. The first applications of MEMS were actually accelerometers. In principle, MEMS accelerometers could deliver a broader bandwidth and more accurate amplitude. One of the benefits is the possibility of recording the low frequency down to zero, hence also recording the gravity.

A typical MEMS accelerometer is composed of a moving proof mass attached through a mechanical suspension (e.g., a polysilicon spring) to a reference frame. The mass has radial fingers positioned between plates fixed to the frame. Acceleration causes deflection of the mass from its center position in one or more directions. The relative position of fingers and plates creates changes in the differential capacitance. The differential capacitance is measured electronically using modulation/demodulation techniques. A schematic representation of a MEMS accelerometer is reported in Figure 3.57.

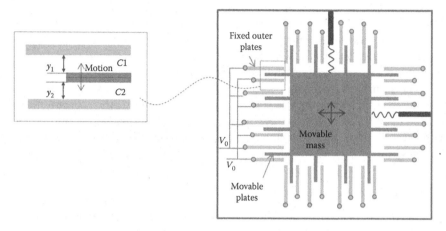

Figure 3.57 Scheme of a 2D MEMS accelerometer.

The mass of the moving part is as small as 1 μg, and the distance between the plates is of the order of 1 μm. Often, a single chip contains multicomponent MEMS accelerometers and measures the acceleration in the different directions. Besides the interest in multicomponent recording, it can also measure and correct the possible tilt of the sensor.

In acquisitions for deep exploration, the signals are digitized on the sensor or close to a group of sensors.

3.5.2.3 Receiver coupling and land streamers

The coupling of the receivers with the ground is a key element for obtaining good quality data. The receivers for land applications usually have steel spikes at the end of the casing. They are firmly planted in the ground or even planted and buried to minimize the local ambient noise. In some cases, base plates can be used, and the receiver's contact with the ground is assured only by its weight (for instance, on hard ground, where it is difficult to insert the spike).

An alternative to the use of receivers coupled individually to the ground is the land streamer (van der Veen et al. 2001; Vangkilde-Pedersen et al. 2006). The name comes from marine streamers, which are strings of hydrophones towed behind boats in marine seismic surveys. Land streamers consist of webbing supporting the data cable and the receivers, usually with plates for the coupling to the ground. Different designs, including wings (to avoid rotation), and different materials have been proposed. Compared to the marine acquisition, land data face more challenges related to the background noise, the coupling between receivers and ground, the cross-coupling between receivers, and the orientation of the receivers. Additional concerns can be the wear of the system and operational and logistic constraints. The coupling with the ground is less ideal than with planted or buried receivers, even if several comparisons show that similar data quality can be obtained in optimal conditions.

The streamer can be towed by a vehicle in a very efficient way in flat terrain or along roads. They can be very efficient in profiling, when the receiver array is moved by regular steps along a line, together with the source. Streamers for land applications can be short and have closely spaced geophones, offering a large increase in productivity.

3.5.2.4 Use of two-component receivers

The acquisition of multicomponent data can be necessary for some data-processing strategies. The polarization of the surface waves can be measured to estimate the contribution of different modes. It can also be used for joint or stand-alone inversion processes aimed at site characterization (see Chapter 8). In active surface wave acquisition, the motion components of interest are the vertical and the horizontal radial components. The polarization changes with the offset in real data. In Figure 3.58, the normalized particle velocity for nine two-component receivers and a harmonic source is plotted as a function of the offset.

With impulsive signals, the traces can show a more complicated hodogram, but often the ellipticity of the Rayleigh waves can be recognized. In Figure 3.59, the particle motion for a two-component acquisition with 20 evenly paced stations is plotted.

The particle motion can become complicated if multiple frequencies are present. Multiple frequencies, multiple modes, and the different excitability of vertical and horizontal components can create complex trajectories. The following example refers to a vibrating source generating a signal with a 7 Hz central frequency. The secondary peaks show different amplitudes on the vertical and horizontal components, and the combination of modes and frequencies creates complex but repeatable trajectories. In Figure 3.60, the hodograms represent cyclic signals following complex artistic paths.

3.5.2.5 Receivers for marine surveys

Geophones are usually sealed and can be used in moderately wet environments. When acquiring data in water, however, the use of hydrophones is needed. Hydrophones are normally piezoelectric devices, with a membrane sensing the pressure variations in the fluid. They have less coupling issues than the geophones and a wide frequency range in the response.

In seabed applications, sometimes geophones and hydrophones are combined in a cable (ocean bottom cable) or in nodes in which the digitalization is also performed (ocean bottom seismometers, ocean bottom nodes). In shallow applications, strings of hydrophones are the most common choice.

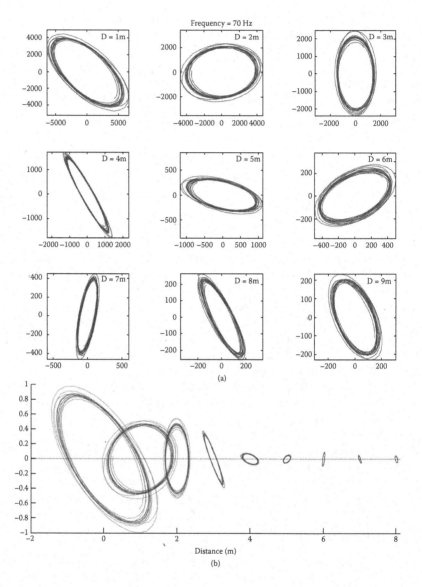

Figure 3.58 Particle velocity of the two components of in-line receivers': (a) normalized hodograms; (b) true amplitude hodograms as a function of source offset.

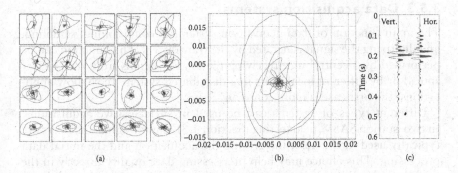

(a)　　　　　　　　(b)　　　　　　　　(c)

Figure 3.59 Hodograms of particle velocity for an impulsive source: (a) normalized hodograms; (b) example of hodogram at a given source offset; (c) vertical and horizontal velocity time histories.

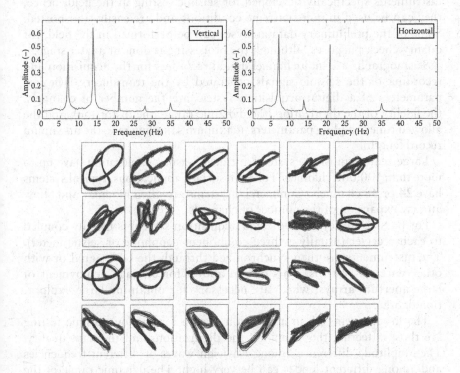

Figure 3.60 Amplitude spectrum of the horizontal and vertical component of the closest receiver (top). Normalised hodograms of the 24 recorder receivers (bottom).

3.5.3 Data acquisition systems

The main function of data acquisition systems is conditioning, sampling, and digitizing the signal generated by the transducers. This function can be performed by a central unit receiving the analog signals from the receiver spread or by different local units in a distributed system. In shallow applications, the first case is more common.

Different types of systems can be used in surface wave acquisitions. In two-station SASW field testing (Section 4.3), digital signal analyzers are typically used in the field, integrating the acquisition and the initial data processing. This choice may help in assessing data quality directly in the field, but usually signal analyzers are laboratory instruments, which are not designed for field operations. They require a power outlet and are not sturdy and waterproof. In multichannel acquisition, multichannel seismographs are the usual choice for field acquisition. These are multipurpose instruments specifically developed for seismic testing in the field; hence, they can be used in quite extreme conditions and are easily transported. Some of the preliminary data analyses can be performed in the field, for quality check purposes, although the processing is done at a later stage.

Seismographs are multichannel digital recorders for the acquisition and recording of the seismic signals generated by the transducers. The key parameters of a digital acquisition system are the number of channels, the fidelity of the digitization (the dynamic range, the distortion), and the allowed time sampling parameters (maximum sampling rate, the maximum record length).

Large-scale acquisition systems for deep seismic exploration have up to more than 100,000 channels. For near-surface applications, typical systems have 24 or 48 channels. Commercial systems are often modular and allow an easy expansion of the number of channels.

For passive surveys, a dedicated acquisition system is usually coupled to each receiver (usually a three-component geophone or seismometer). The instruments are then synchronized through the GPS signal or with other wireless systems. This option allows for an easy deployment of large aperture arrays, which are necessary for obtaining large exploration depths.

The key elements of an acquisition system for shallow seismic testing are those affecting the sampling, the digitization, and the data quality. The amplitude difference among Rayleigh waves at different frequencies and among different modes can be very high. The dynamic range of the seismograph affects the quality of data, the detectable frequency range, and the detectable modes. The dynamic range is a key characteristic of the analog-to-digital converter (A/D).

The A/D accuracy is measured in bits and describes the number of discrete values that can be produced within the input range. State-of-the-art instruments have 24-bit A/D converters, often based on sigma-delta oversampling, which corresponds to a 144 dB dynamic range. The resolution and quantization noise are discussed in Section 3.3.6.

The possible sample intervals are crucial in the choice of an acquisition system. They must allow the required sampling frequency and the maximum desired record length.

The minimum sampling interval can be as low as 10 microseconds on tens of channels. In surface wave acquisition for near-surface site characterization, usually a 1 or 2 ms sampling interval is sufficient because the analysis is performed in the frequency domain and the frequencies of interest are typically below 100 Hz (except for pavement testing where much higher frequencies are necessary). The maximum number of samples, combined with the largest sampling interval, gives the maximum recorded length, which is particularly important for passive measurements. The possibility of acquiring pretrigger data, based on circular buffering of the armed seismograph, enables the recording of data before the activation of the seismic source. This is necessary to evaluate the noise level at the moment of the acquisition and to enable synchronization of data in case of unreliable triggering. Moreover, a certain amount of pretrigger allows sampling windows different from the boxcar windows to be applied in order to mitigate leakage during the processing.

The analog filters are useful for some applications, but the built-in lowcut filter of some systems prevents them from acquiring low-frequency data (usually below 2 Hz). This feature is introduced in some systems to optimize the acquisition of seismic reflection survey, but it is unwanted for surface wave testing because it limits the investigation depth.

The bandwidth of the system indicates the frequency range where the system can record data reliably, and it obviously affects the measurable frequency range. Typical values range between 2 Hz and 20 kHz.

The system noise is measured in terms of system noise floor (crosstalk between channels). The system distortion usually is less than 10/million in the main seismic frequency range, and it is not critical for surface wave analysis.

Important parameters in coupling an acquisition system with receivers are the input signal range (in volts, as peak-to-peak maximum tension) and the available preamplifier gains. Indeed, the analog signals are usually amplified before their sampling and digitization. Automatic gains can affect the possibility of testing for surface waves at very short distances from the source and hence the possibility of obtaining very high resolutions close to the ground surface. This aspect may be critical for some applications, such as for pavement testing.

The trigger system and its accuracy are important properties for active data acquisition. In particular, high-accuracy triggering is necessary for stacking data in time domain. For surface wave analysis, stacking can be applied at a later stage of processing if single-shot data are recorded in the field. This strategy avoids the errors related to an asynchronous trigger. Accuracy of triggering is of paramount importance when multiple acquisitions with different source and/or receiver array positions are assembled to create a single large shot gather.

The possibility of recording an auxiliary channel is useful in controlled source acquisition. Indeed, monitoring of the input force is useful for the calculation of the experimental transfer function of the site (see Chapter 5).

The data transmission and data storage (internal or transmitted to the computer that is used to control the acquisition) can be important for the cycle time. The format of the saved data is usually one of the few standard formats for the seismic data (e.g., segy or seg2), which can then be converted in ASCII format for subsequent processing. Some systems use proprietary formats, although this choice is not recommended because it limits the possibility of sharing data.

Some software features can contribute to the efficiency of the field operations, for example, data acquisition and display filters (low-cut, high-cut, and notch) or software functions to test the system and the receiver spread. The possibility of visualizing the real-time output of the receivers or testing the receivers with a pulse test is useful for checking that all the receivers are working properly.

The system's robustness in various environmental conditions (operating temperature, waterproof rating, and so on) and the practical aspects of power consumption also have to be considered when selecting an acquisition system.

Chapter 4

Dispersion analysis

One of the main tasks in surface wave testing is to extract information about the geometric dispersion of surface waves from observations of the particle motion at two or more receiver locations. The information is usually presented in the form of dispersion curves that show the variation of surface wave phase or group velocity versus frequency or wavelength. In this chapter, we describe the most widely used methods for calculating phase velocity dispersion curves. In fact, group velocities are rarely used for near-surface applications.

The methods are presented with reference to the analysis of Rayleigh waves, which are commonly analyzed for site characterization, but they may be easily extended to the analysis of other dispersive waves, such as Love waves, Scholte waves, and guided waves (see Chapter 8).

The methods may be classified according to the procedure adopted:

- Direct assessments of the propagation parameters (e.g., wavelength in the steady-state Rayleigh method or phase delay in the two-station spectral analysis of surface waves [SASW])
- Regression methods, in which the propagation parameters are obtained by fitting the experimental data with the expected theoretical functions (multi-offset phase analysis [MOPA], spatial autocorrelation [SPAC], transfer function)
- Transform-based methods, in which the experimental data are transformed from the original space–time domain into a different domain in which the propagation parameters are easily identified as spectral maxima (e.g., frequency–wavenumber, frequency–slowness, frequency–velocity analysis)

In this chapter, most of the examples are generated from a reference real dataset. The active-source multichannel test has been performed in Florence (Italy), using 24 vertical geophones, with a natural frequency of 4.5 Hz, deployed on a linear array with 2 m spacing. The wave field was generated with a weight-drop system, with a mass of 130 kg in free fall from 3 m. The seismic source had an offset of 4 m from the receiver array. 12 single shot

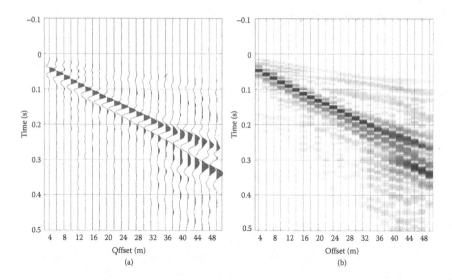

Figure 4.1 Reference dataset: stacked multichannel seismogram (shot gather) displayed as (a) wiggle traces and (b) density plot. The main surface wave event is the one arriving at the furthest offset at 0.3 s.

gathers have been acquired separately to allow for statistical assessment of signal-to-noise ratio (used in the SASW procedure, see Section 4.3) or for uncertainty evaluation (see Sections 4.4 and 4.8). The stacked shot gather is plotted in Figure 4.1.

4.1 PHASE AND GROUP VELOCITY

Nondispersive waves propagate with a single velocity, and their waveform propagates without changing shape. When there is dispersion, as for surface waves in heterogeneous media, the velocity depends on frequency, and the waveform changes shape as it propagates (see Chapter 2). Dispersive waves are characterized by two distinct velocities—the phase and group velocities (Rayleigh 1877). The phase velocity is the speed of propagation of a single phase of the waveform (e.g., a peak or trough), whereas the group velocity is the velocity of a packet or "group" of waves (Figure 4.2).

The phase and group velocity can be expressed mathematically as

$$V_{phase} = \frac{\omega}{k}$$

$$V_{group} = \frac{\partial \omega}{\partial k}$$

(4.1)

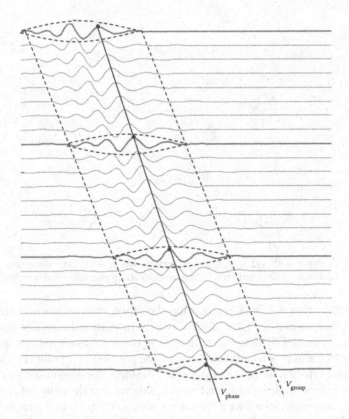

Figure 4.2 Phase velocity and group velocity.

The relationship between phase and group velocity is given by

$$V_{\text{group}} = V_{\text{phase}} + k \frac{dV_{\text{phase}}}{dk}$$

$$= V_{\text{phase}} \left(1 - k \frac{dV_{\text{phase}}}{d\omega} \right)^{-1}$$

(4.2)

From Equation 4.2, it is clear that when the derivatives of phase velocity are equal to zero (i.e., the material is nondispersive), the phase and group velocities are identical.

For multimode surface waves, each mode can be described by its frequency-dependent phase and group velocities (i.e., phase and group dispersion curves). Figure 4.3 shows an example for two modes.

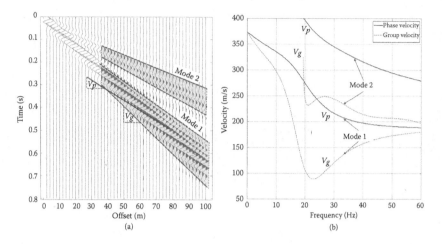

Figure 4.3 Example with synthetic data of surface waves with two dispersive modes: (a) synthetic shot gather; (b) phase velocity and group velocity dispersion curves.

From Equation 4.2, it is evident that the group velocity is lower than the phase velocity when the latter decreases with increasing frequency (as in the example of Figure 4.3). In this case, the system is defined as normally dispersive.

In the frequency–wavenumber domain, the position of a point corresponds to its phase velocity, while the tangent (i.e., the local slope) represents its group velocity. The example reported in Figure 4.4 shows that the points of the phase velocity dispersion curves with the same velocity (P1 to P4 in Figure 4.4a) are aligned along a straight line in the f–k domain (Figure 4.4b). On the contrary, the same value of group velocity for different modes (G1 to G3 in Figure 4.4a) corresponds to the same local derivative of the phase velocity dispersion curves in the f–k panel (Figure 4.4b). The f–k graph depicted in Figure 4.4b also shows constant phase velocity lines.

4.2 STEADY-STATE METHOD

The first surface wave method developed for near-surface site characterization was the steady-state Rayleigh method (SSRM), proposed by Jones (1958, 1962) and then adopted at the Waterways Experiment Station, in the United States (Ballard 1964).

Taking into account the limitation of available instrumentation at that time, the method was quite ingenious and allowed for an evaluation of the dispersion curve with a very simple experimental setup. Showing the potential of surface waves for site characterization, it paved the way for subsequent and more refined approaches that lead to the diffusion of surface waves in the engineering community.

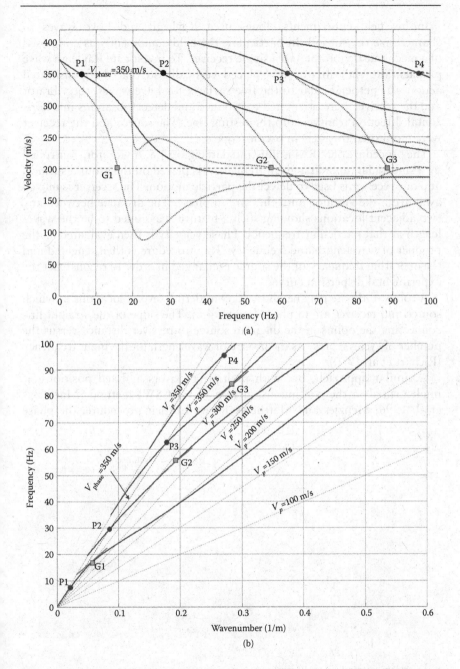

Figure 4.4 Example of surface waves with four dispersive modes. Phase velocity and group velocity dispersion curves are represented as (a) frequency versus velocity and (b) frequency versus wavenumber. Dotted straight lines in (b) represent constant phase velocity events.

In his field experiments, Jones used Rayleigh and Love waves to characterize the subsoil. In particular, the field equipment was composed of a mechanical vibrator and a single receiver. To investigate Rayleigh wave propagation, the vibrator was placed vertically so that the transmitted action was perpendicular to the free surface; for Love waves, the vibrator and the receiver were orientated to produce and detect vibrations in a horizontal direction transverse to the testing line. The source and the receiver were monitored through an oscilloscope.

The procedure of SSRM is illustrated in Figure 4.5. A vibrator, controlled by a sinusoidal input at a given frequency, is placed on the ground surface, and one receiver is used to detect the particle motion. The receiver is moved away from the vibrator until they are in phase. The distance between any two adjacent locations showing such a feature is assumed to be the wavelength at that particular frequency. Phase velocity is then evaluated as the product of wavelength and frequency. The procedure is then repeated and the operating frequency of the source is changed in order to reconstruct the experimental dispersion curve.

A robust estimate may be obtained considering several locations at which source and receiver are in phase (Figure 4.6). The slope of the straight line connecting the points in the diagram source-to-receiver distance versus the number of cycles represents the inverse of wavelength for the input frequency (Richart et al. 1970).

A similar approach, but with multiple receivers at fixed positions, is implemented in the continuous surface wave (CSW) method (Matthews et al. 1996; Menzies and Matthews 1996). In their procedure, the phase

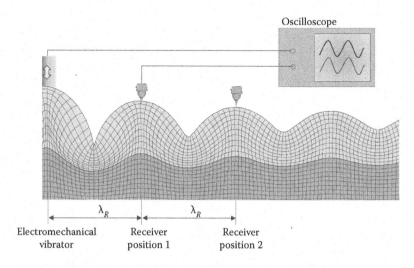

Figure 4.5 Steady-state Rayleigh method (SSRM), field procedure.

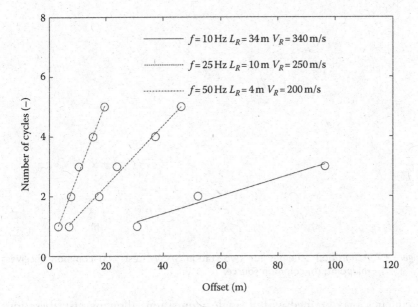

Figure 4.6 Determination of the average wavelength of Rayleigh waves by SSRM.

velocity is computed from the slope of the resulting phase angles plotted against source-to-receiver distance (see also Section 4.4).

4.3 SPECTRAL ANALYSIS OF SURFACE WAVES

Although the acronym SASW could in principle be used to designate any of the methods reported in the present chapter (because they all are based on the analysis of the propagation of surface waves in the frequency domain), it is conventionally used to address the following two-station procedure, which was proposed by researchers at the University of Texas, Austin, in the 1980s (Nazarian and Stokoe 1984; Stokoe et al. 1994).

SASW tests are performed with two receivers that are colinear with an active source (as shown in Figure 4.7). Typically, the receivers are placed such that the source offset (D) is equal to inter-receiver spacing (X) (Sànchez-Salinero 1987). The basic principle is that the velocity can be estimated as the ratio of the distance divided by time delay. For a single harmonic wave, the time delay is evaluated on a given phase of the signal. The method can be implemented using either harmonic sources or impulsive sources. The advantage of using harmonic sources is that the energy is concentrated at a given frequency, allowing high reliability in the determination. The advantage of using an impulsive source is that information over a wide range of

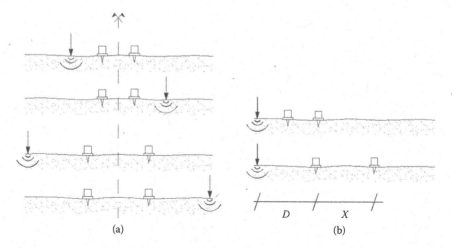

Figure 4.7 Acquisition schemes for two-station SASW tests: (a) common receivers midpoint; (b) common source.

frequencies is obtained with a single acquisition, allowing fast operations in the field.

The time domain signal recorded at each receiver is given by $s(x_m,t)\, m = 1,2$. Taking the temporal Fourier transform of each signal yields

$$S\left(x_m,\omega\right) = \left|S\left(x_m,\omega\right)\right| e^{i\left[\phi(\omega)+k(\omega)x_m\right]} \qquad (4.3)$$

where $\phi(\omega)$ is the arbitrary phase of the source. The cross-power spectrum between the two signals is

$$
\begin{aligned}
S_{12}\left(\omega\right) &= \overline{S}_1\left(\omega\right)S_2\left(\omega\right)\\
&= \left|S\left(x_1,\omega\right)\right| e^{-i\left[\phi(\omega)+k(\omega)x_1\right]} \left|S\left(x_2,\omega\right)\right| e^{i\left[\phi(\omega)+k(\omega)x_2\right]} \qquad (4.4)\\
&= \left|S\left(x_1,\omega\right)\right|\left|S\left(x_2,\omega\right)\right| e^{ik(\omega)(x_2-x_1)}
\end{aligned}
$$

where $\overline{S}$ denotes the complex conjugate of S. In practice, it is desirable to use ensemble averaging to reduce the variance of the measured cross-power spectra (Bendat and Piersol 2010)

$$\hat{S}_{12}\left(\omega\right) = \frac{1}{N}\sum_{k=1}^{N}\overline{S}_k\left(x_1,\omega\right)S_k\left(x_2,\omega\right) \qquad (4.5)$$

where N is the number of spectra that are averaged. The effects of uncorrelated noise decrease with a factor equal to the inverse of the square root of N.

We may extract the phase from the cross-power spectrum

$$\theta_{12}(\omega) = \arg\left(\hat{S}_{12}(\omega)\right)$$

$$= k(\omega)(x_2 - x_1)$$

(4.6)

Thus, we may calculate the phase velocity dispersion curve by rearranging Equation 4.6 as follows

$$V_R(\omega) = \frac{\omega(x_2 - x_1)}{\theta_{12}(\omega)}$$

(4.7)

Figure 4.8a shows an example of a phase spectrum with a pair of receivers separated by 8 m extracted from the reference dataset presented at the beginning of the chapter. This example is typical of phase spectra measured using a transient source. The large uncertainty in the phase spectrum at high frequencies is likely due to the inability of the source to generate sufficient energy and to geometric attenuation of the waves. The plot is presented as a *wrapped phase* with values of $\theta_{12}(\omega)$ ranging from π to $-\pi$ (i.e., modulo 2π). To apply Equation 4.7, the user must unwrap the phase for the frequency range of interest by adding integer multiples of 2π in the following manner[01]

$$\theta_{unwrap}(\omega) = \theta_{wrap}(\omega) \pm 2n\pi, \quad n = 0,1,2\ldots$$

(4.8)

Choosing the proper values of n can be a source of uncertainty in SASW tests. This step is often conducted using some automated algorithms (e.g., Poggiagliolmi et al. 1982), but many problems can arise due to the influence of noise that can prevent the correct identification of multiple integer cycles and can produce fictitious jumps in the wrapped phase (see, e.g., Figure 4.9). Operator judgment may solve this problem, but it still remains a subjective procedure (Al-Hunaidi 1992).

Other spectral quantities can help in defining the frequency range in which the information is reliable. The autopower spectra at the two receivers (Figure 4.8c and d) provide an indication of energy distribution. Where their values are high, frequency ranges are likely to be those in which the generated signal is stronger; hence, the latter prevails over uncorrelated noise. This is the reason why light sources (e.g., a small hammer) are used for small receiver spacing because these sources generate much more energy at a higher frequency. Conversely, heavy sources (e.g., a massive weight

---

[01] Negative values of θ are often used to represent a phase due to propagation delay. It may be necessary to multiply Equation 4.8 by –1.

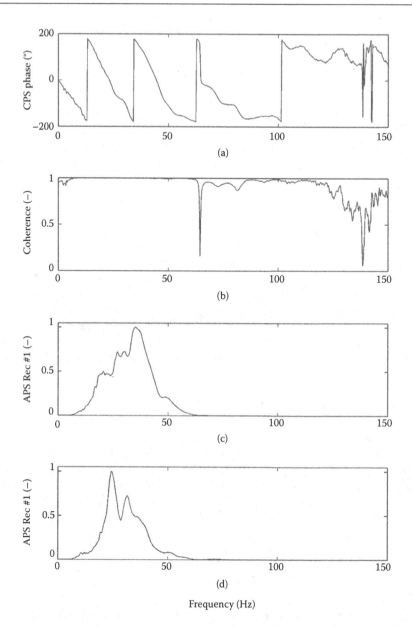

Frequency (Hz)

Figure 4.8 Example of spectral quantities for the interpretation of a two-receiver SASW test (source: 130 kg weight drop; inter-receiver distance: 8 m): (a) cross-power spectrum (wrapped); (b) coherence function; (c) normalized autopower spectrum (Receiver 1); (d) normalized autopower spectrum (Receiver 2).

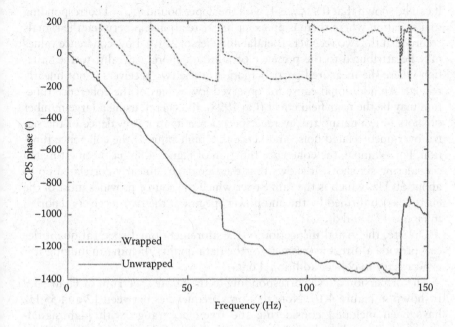

Figure 4.9 Phase unwrapping for the cross-power spectrum of Figure 4.8a. At about 140 Hz, the anomaly represents a typical example of failure of the unwrapping procedure in identifying phase jumps.

drop system) generate strong signals with low-frequency content; hence, they are used for wide spacings. For the example of Figure 4.8, the auto-power spectra indicate that the energy is concentrated in the frequency range between about 10 and 60 Hz. The assessment of signal quality is made using the "coherence function," namely a spectral quantity obtained comparing different registrations that is a measure of the degree by which input and output signals are linearly correlated. A value close to unity is an index of good correlation; hence, the recorded signals can be considered genuine and unaffected by ambient noise. The coherence function between two receivers, γ_{12}, is defined as

$$\gamma_{12}^2(\omega) = \frac{\hat{S}_{12}(\omega) \cdot \overline{\hat{S}_{12}(\omega)}}{S_{11}(\omega) \cdot S_{22}(\omega)} \tag{4.9}$$

where the upper bar denotes the complex conjugate. The ordinary coherence function provides a measure of how the measured particle motion at the first receiver is related to the particle motion recorded at the second receiver.

It can be shown that $0 \leq \gamma_{n_1 n_2}^2 \leq 1$, with the upper bound $\gamma_{n_1 n_2} = 1$ corresponding to a situation where there is an exact linear relationship between the signals recorded at the two receivers (Bendat and Piersol 2010). Low coherence values may be attributed to the presence of noise or, more generally, to the situation where the measured particle motion at the two receivers is not linearly related. An additional cause for observed low values of the coherence function may be the near-field effect (Lai 1998). In general, using a large number of shots in evaluating the average spectra results in a prevalence of the signal over uncorrelated noise and hence in a high value of the coherence function. For example, the coherence function of Figure 4.8b has been estimated considering six shots. It shows that the signals are linearly correlated up to about 60 Hz, which is the range over which the source provides most of the energy (as confirmed by the autopower spectra at the two receivers reported in Figure 4.8c and d).

On site, the visual inspection of the aforementioned spectral quantities can provide a direct assessment of the data quality, helping in judging if it is necessary to collect additional data.

The dispersion curve corresponding to the phase spectrum of Figure 4.9 is shown in Figure 4.10. Note that only frequencies between 12 and 55 Hz have been included considering the frequency range with high signal-to-noise ratio and the limitations related to near-field effects, which will be discussed hereafter.

Information for a limited frequency range is typically obtained with a given receiver spacing. This is due to many factors such as wave attenuation, spatial aliasing, near-field effects (see the discussion at the end of

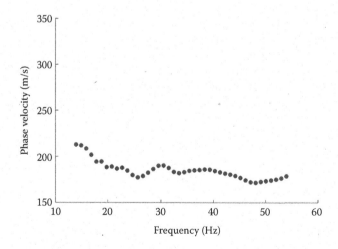

Figure 4.10 Experimental dispersion curve estimated from the phase of the cross-power spectrum reported in Figure 4.9 (receiver spacing: 8 m).

this section). If data at other frequencies are required, the test may be repeated using different receiver spacing according to the measurement schemes of Figure 4.7. The common receivers midpoint scheme is preferable because of its symmetry. Moreover, the source may also be moved to the opposite side of the array to determine the dispersion curve in the "reverse" direction in an effort to average any differences in phase velocity due to lateral inhomogeneity at the site and/or internal phase distortions in receiver responses. The common source array is easier to be implemented in the field, especially if heavy sources are used.

In usual practice, the receiver spacing is doubled in the new setup, implementing a geometrical progression. Dispersion curves from individual receiver spacings are combined to form a composite dispersion curve for the site.

The SASW technique has several important limitations that arise from the use of only two receivers. First, spatial aliasing complicates interpretation of the test results. To avoid spatial aliasing, the wavenumber must be less than the Nyquist wavenumber. In the context of the SASW test

$$k(\omega) \leq k_{\text{Nyquist}}$$

$$\frac{\theta_{12}(\omega)}{x_2 - x_1} \leq \frac{\pi}{x_2 - x_1} \tag{4.10}$$

$$\theta_{12}(\omega) \leq \pi$$

Thus, the "jumps" in the phase spectrum and, more specifically, the need to choose n in Equation 4.8 are the result of spatial aliasing. As noted earlier, this aspect of SASW test interpretation often results in uncertainty. Although "automatic" phase unwrapping algorithms are available, they are frequently fooled by poor quality data, and manual interpretation of the phase spectrum is usually required.

Second, the wavenumber resolution is generally poor for receiver spacings normally used in SASW tests. The main lobe in the aperture smoothing function (see Chapter 3) is wide and spans the entire range of wavenumbers defined by $-k_{\text{Nyquist}} \leq k \leq k_{\text{Nyquist}}$. The most important practical implication of the poor wavenumber resolution is that it is impossible to resolve multiple surface wave modes in the SASW test. The smoothing produced by convolving the aperture smoothing function with multiple modes propagating at different wavenumbers yields only a single apparent wavenumber. Indeed, the procedure based on the phase of the cross-power spectrum (Equation 4.7) yields a single value of the time delay and hence of phase velocity for any testing setup, with no possibility to discriminate different modes of propagation.

Another main issue is related to near-field effects (i.e., deviations from the usual assumption that the propagation is dominated by plane Rayleigh waves). Indeed, the influence of body wave components may affect the motion detected at the first receiver, and the consequent distortion leads to an underestimation of the actual phase velocity.

Initially, it was proposed to discard all data for which the distance between the source and the first receiver was less than one-third of the obtained wavelength (Heisey et al. 1982). However, some numerical analysis showed that this criterion underestimates the extension of the near-field (Sànchez-Salinero 1987).

Usually, a near-field extension of a half wavelength is assumed for a normally dispersive soil (i.e., where stiffness is increasing with depth), whereas about two wavelengths is a more prudential estimate for a strongly inverse dispersive soil (i.e., where a soft layer is present below or is trapped between stiffer ones). These indications are based on numerical simulation of the complete wave field generated by a point source for different layer configurations (Tokimatsu 1995). Nevertheless, there still is an open discussion about the influence of direct and reflected/refracted body waves on the recorded signals.

It is important to remark that because the Rayleigh waves are dispersive in layered media, the distance at which the condition of far-field can be assumed is not a constant but depends on the frequency analyzed. For example, considering a wrapped cross-power spectrum phase (Figure 4.8a), discarding data affected from a near-field that is a half wavelength long is equivalent to cutting information given by an initial section of 180° (i.e., the portion between 0 and the first jump).

Finally, because of attenuation, data relative to inter-receiver distance higher than about three wavelengths are usually strongly affected by noise (Stokoe et al. 1988); therefore, it is preferable to discard them.

In summary, assuming that no strong contrasts of stiffness are present, for each receivers configuration, the following restrictions are applied to select reliable dispersion curve data points

$$\frac{D}{3} < \lambda < 2D \tag{4.11}$$

in which λ is the estimated wavelength and D is the inter-geophone distance that is taken equal to the distance source-first geophone. A visual interpretation of these filtering criteria is given in Figure 4.11, showing the whole set of data that can be obtained by a given experimental setup and the subset that is considered acceptable.

As seen here, the construction of the experimental dispersion curve is strongly affected by the operator's experience because a selection of significant and corrected data is actually required. Moreover, because this is time-consuming and involves many manipulations of the data, it is usually conducted in the office after data collection in the field.

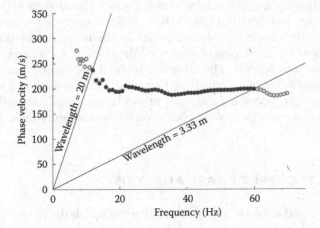

Figure 4.11 Example of application of filtering criteria to the dispersion curve corresponding to one source–receiver configuration with 10 m spacing; only the data points between the two limiting wavelengths are considered acceptable (full black dots).

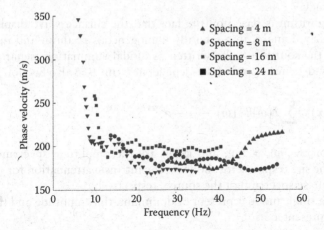

Figure 4.12 Example of experimental dispersion curve estimated with four testing setup in SASW test.

The whole ensemble of data collected using a series of geophone configurations (Figure 4.12) has to be assembled to create a single global dispersion curve covering a wide enough range of frequency. This is necessary because the automated inversion algorithms (see Chapter 6) have to compare the numerical dispersion curve to the experimental one by computing a norm of the distance between the two. Usually for any subset in frequency (or wavelength), the average value of phase velocity is assigned to the central frequency (or wavelength) of the subset. Still, there are a few problems— one is that the choice of reducing points in the frequency or wavelength domain must be made consistently with the domain successively adopted for the inversion process. The other one is about the significance of adopting the mean value of a population that hardly can be seen as a statistical distribution (this because the number of overlapping information in a given frequency range is arbitrary, depending on testing configurations, quality of data, etc.).

4.4 MULTI-OFFSET PHASE ANALYSIS

MOPA is a surface wave analysis technique in which the phase versus offset of particle motion is processed to estimate the phase velocity (Strobbia and Foti 2006). It can be considered an extension of the SASW method for multichannel arrays. The procedure also allows for the identification of lateral variations and a robust estimation of the velocity. Moreover, traces can be weighted as a function of their uncertainty, thus improving the accuracy.

The algorithm is based on the fact that the surface wave displacement in frequency domain, in a laterally homogeneous medium and in far offset from the source, can be written as modal summation, separating the frequency-dependent and offset-dependent terms (see also Section 2.4.2)

$$s(\omega,x) = \sum_m I(\omega) R_m(\omega) \frac{e^{-\alpha_m(\omega)x}}{\sqrt{x}} e^{i(\omega t - k_m(\omega)x + \varphi_0(\omega))} \tag{4.12}$$

where $I(\omega)$, $R_m(\omega)$, $\alpha_m(\omega)$, and $\varphi_0(\omega)$ are the amplitude spectrum of the source, the site response for mode m, the intrinsic attenuation for mode m, and the phase spectrum of the source, respectively.

When a single mode is present or dominant, the amplitude and the phase can be represented as

$$A(\omega,x) = I(\omega) R(\omega) \frac{e^{-\alpha(\omega)x}}{\sqrt{x}} \tag{4.13}$$

and

$$\varphi(\omega,x) = -k(\omega) \cdot x + \varphi_0(\omega) \tag{4.14}$$

Hence, for a single mode and a single frequency, the amplitude and the phase are an exponential and a linear function of offset from the source, respectively (Figure 4.13).

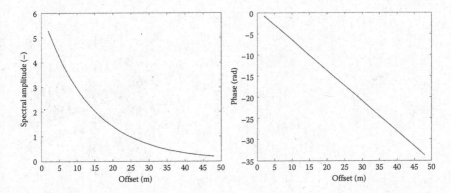

Figure 4.13 Theoretical behavior of the phase and the amplitude for one frequency as a function of the offset.

The extraction and inspection of the phase in experimental data is the key step of MOPA. It allows the estimation of the phase velocity as well as the identification of lateral variations and of near-field effects.

Figure 4.14 shows the reference shot gather together with the corresponding seismograms after narrow-band filtering at 9 and 19 Hz, respectively. The filtered seismograms show graphically the velocity of the propagating energy at the considered frequency. This experimental evidence can be implemented in an efficient processing procedure.

The time shift of a harmonic signal is related to its phase

$$s(t) = \sin(\omega t + \varphi) = \sin(\omega(t - \Delta t)) \tag{4.15}$$

gives

$$\varphi = -\omega \cdot \Delta t = -2\pi f \cdot \Delta t \tag{4.16}$$

In practice, it is possible to estimate the time shift of the harmonic (stationary and infinite) function from the phase at a given frequency. The phase can be extracted using the Fourier transform. Figure 4.15a and b report the amplitude and the phase, respectively, for the 19 Hz frequency of the reference dataset (Figure 4.15c).

In Figure 4.15b, the phase is reported in a modulo-2π representation and needs to be unwrapped to describe the linear propagation along the array. The simplest unwrapping procedure changes absolute jumps greater than π radians to their complements to 2π. Figure 4.16 shows the wrapped phase and its corresponding unwrapped phase.

The unwrapped phase is directly related to the time delay for the considered frequency. In the example of Figure 4.17, considering the 46 m

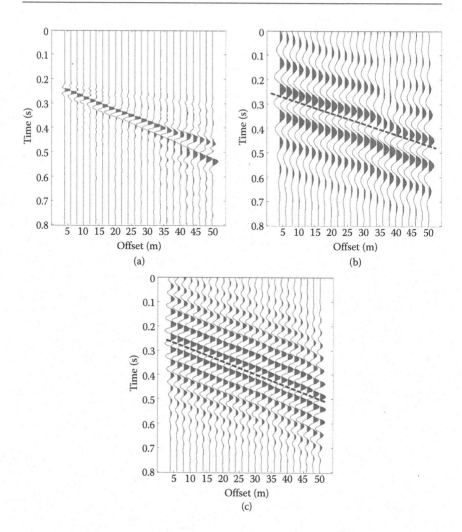

Figure 4.14 (a) Raw seismogram (b) band-pass filtered around 9 Hz, and (c) around 19 Hz. The slope in time-offset represents the phase velocity at the considered frequency.

interval between the first receiver and the last receiver, the phase difference of about 28 radians gives a time delay of 0.236 s and a phase velocity of 195 m/s.

If multiple shots with the same receiver spread are acquired, the effect of the incoherent noise on the data can be assessed with a rigorous statistical approach. For example, Figure 4.18a shows the traces recorded at the 10th geophone for all the available shots in the reference dataset. In Figure 4.18b, the individual phase values at the frequency of 19 Hz are

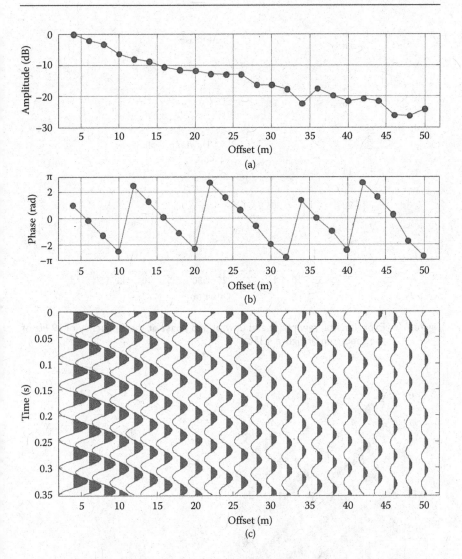

Figure 4.15 (a) Amplitude and (b) phase of the 19 Hz harmonic as a function of offset for the reference dataset. The corresponding harmonic traces are represented in (c).

represented with the dots and the mean is indicated with the dashed line. The shaded area indicates the interval between the mean and one standard deviation. The error bar in Figure 4.18c summarizes the distribution of the phase at 19 Hz at the considered receiver No. 10.

With a multichannel array, a phase distribution can be obtained for each receiver station, for each considered frequency. Considering three

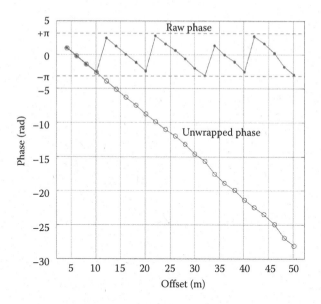

Figure 4.16 Experimental raw phase and unwrapped phase at the frequency of 19 Hz of the spectrum of Figure 4.15b.

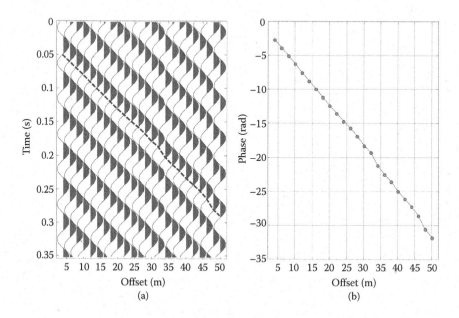

Figure 4.17 (a) Normalized harmonic components at 19 Hz with the annotated velocity and (b) phase versus offset graph for the considered frequency.

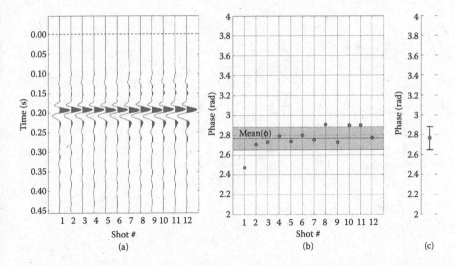

Figure 4.18 Statistical analysis of the phase: (a) traces recorded by the 10th receiver of the array; (b) extracted phase for each shot and the average value (the shaded area represents the ±σ interval); (c) average phase and the corresponding standard deviation plotted as error bar.

frequency values, the phase distribution is plotted in Figure 4.19. Statistical data can be used to obtain the best estimate of the wavenumber at each frequency and its associated standard deviation by using the following procedure.

The linear model describing the phase in a laterally homogeneous medium in the far-field of the source is described by Equation 4.14, which can be written for each receiver as

$$\varphi_i = k \cdot x_i + \varphi_0 \tag{4.17}$$

where the phase φ_i at the offset x_i depends on the unknown wavenumber k. Considering an array with N receivers, we get

$$\begin{cases} \varphi_1 = k \cdot x_1 + \varphi_0 \\ \varphi_2 = k \cdot x_2 + \varphi_0 \\ \quad\vdots \\ \varphi_N = k \cdot x_N + \varphi_0 \end{cases} \tag{4.18}$$

that in matrix form can be expressed as

$$\mathbf{\Phi} = G \cdot M \tag{4.19}$$

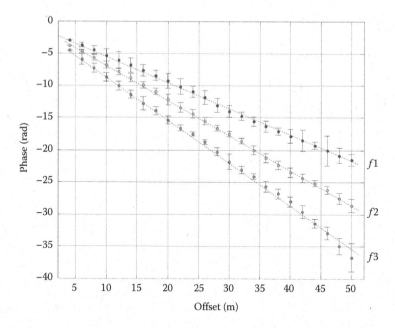

Figure 4.19 Phase distribution versus offset for three different frequencies accounting for the standard deviation of the experimental data.

where $\mathbf{\Phi} = [\varphi_1, \varphi_2, \ldots, \varphi_N]^T$ is the vector of the experimental phases, $M = [k, \varphi_0]^T$ is the vector of the unknown model parameters, and $G = \begin{bmatrix} x_1 & 1 \\ x_2 & 1 \\ \vdots & \vdots \\ x_N & 1 \end{bmatrix}$ is the data kernel matrix, depending on the geometry (Menke 1989).

The model parameters can be estimated in the least-squares sense by the pseudoinverse G^{-g}

$$M = G^{-g} \cdot \mathbf{\Phi} \tag{4.20}$$

with $G^{-g} = (G^T \cdot G)^{-1} \cdot G^T$.

The experimental values of the phase can be weighted according to the corresponding standard deviation in order to estimate the uncertainty of the estimated wavenumber. The weighted least-squares estimation can be obtained by replacing the pseudoinverse with the corresponding weighted version G_W^{-g}, written as (Tarantola 2005)

$$G_W^{-g} = \left(G^T \cdot W^T \cdot W \cdot G \right)^{-1} \cdot G^T \cdot W^T \cdot W \tag{4.21}$$

where W is the diagonal matrix containing the weights (i.e., the standard deviation of the phase).

With the assumption of normally distributed and independent data, the linear relationship between the experimental phases and the model parameters

$$M = G_W^{-g} \cdot \Phi \tag{4.22}$$

implies that

$$\sigma_M^2 = G2_W^{-g} \cdot \sigma_\Phi^2 \tag{4.23}$$

where σ_M^2 and σ_Φ^2 are the covariance matrices of the model parameters and of the data, respectively, and $G2_W^{-g}$ is the matrix containing the squares of the elements of G_W^{-g} (Santamarina and Fratta 2010).

Adopting this procedure, the mean value of the wavenumber and the associated standard deviation can be estimated at each frequency of interest. The wavenumber is then used to compute the phase velocity V, the standard deviation of which can be estimated with a first-order approximation

$$V = \frac{2\pi \cdot f}{k} \tag{4.24}$$

$$\sigma_V = \frac{2\pi \cdot f}{k^2} \sigma_K \tag{4.25}$$

Figure 4.20 shows the mean and standard deviation of the unwrapped phase for the reference dataset.

Figure 4.21 shows the experimental dispersion curve with the associated standard deviation. However, the standard deviation does not describe the shape of the probability density function of the phase velocity.

This procedure also allows an assessment of the linearity of the phase in the experimental dataset. Indeed, as discussed earlier, the phase is a linear function of the offset only in a laterally homogeneous medium and in the far-field of the source. These hypotheses are not always verified. Indeed, real data may be recorded at sites where lateral variations are present, and for a given array and source configuration, some of the receivers may fall in the near-field, for some frequencies.

For example, near-field effects can be detected in the reference dataset. Figure 4.22 shows amplitude and phase as a function of offset, for the frequency of 5.9 Hz. The corresponding wavelength is about 33 m, and despite the low standard deviations, the phase of the first six traces appears anomalous with respect to the trend.

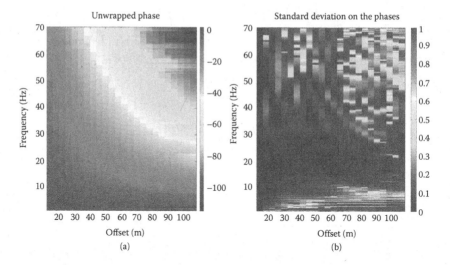

Figure 4.20 (a) Mean and (b) standard deviation of the phase as a function of offset and frequency, for the reference dataset.

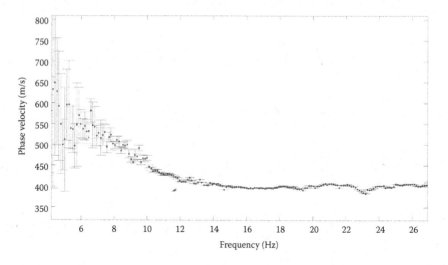

Figure 4.21 Dispersion curve obtained with MOPA. The dispersion curve is obtained from the best estimate, and the experimental uncertainties are statistically estimated.

It is clear that, at an offset of about 16 m, the slope of the phase versus offset changes. It can also be observed that, approximately at the same offset, the attenuation also has a large variation. This anomalous behavior is likely to be related to the near-field effect. Indeed, observing the behavior at different frequencies, it can be found that the distance at which

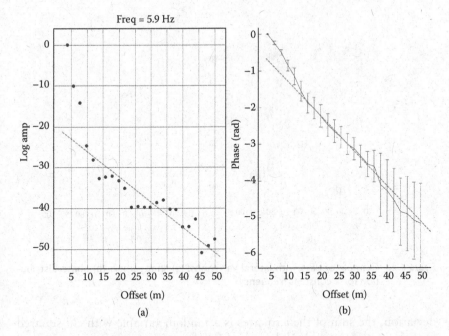

Figure 4.22 (a) Amplitude after geometric spreading correction and (b) unwrapped phase versus offset at the frequency of 5.9 Hz, for the reference dataset.

the linearity is lost changes with the wavelength. When lateral variations are present at the site, the change of slope tends to be located at a specific position along the array.

The identified near-field region can be removed from the data; at every frequency, only the traces below a near-field criterion, i.e. the traces with an acceptable linearity, are considered.

Similarly, the deviation from the theoretical linear behavior can indicate the presence of lateral variations. An example of severe lateral variation is provided in Figure 4.23. The shot gather is plotted in Figure 4.23a and the phase versus offset in Figure 4.23b.

A simple visual inspection of the data and of the phase plots can help with identifying the lateral variations. However, a more rigorous procedure can be implemented; given the estimated phase uncertainties, the linearity of the phase can be statistically tested. The principle is that if the uncertainty on the different phase points is low, the regression misfit cannot be too large; a large misfit implies that the model is not linear.

If each of the N individual prediction errors is assumed to be a Gaussian independent random variable x, each with zero mean and unit standard

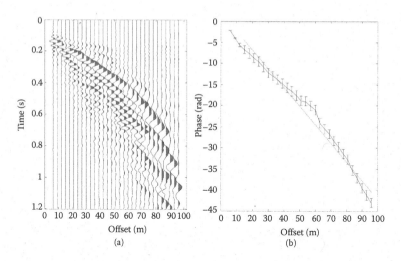

Figure 4.23 Example of data with lateral variations: (a) shot gather; (b) phase distribution for a constant frequency.

deviation, the sum of their squares is a random variable with chi-squared probability density with N degrees of freedom

$$\chi^2 = \sum_{i=1}^{N} x_i^2 \tag{4.26}$$

Therefore, the total misfit in the least-squares sense has a chi-squared distribution. If the different errors have different standard deviations, the overall misfit can be predicted by weighting with the covariance matrix C and including the vector of the means μ, to give a random variable with the same probability density

$$\chi^2 = (x - \mu)^T C(x - \mu) \tag{4.27}$$

Given the best fitting linear trend, the corresponding total misfit is compared with the misfit distribution predicted from the phase uncertainties. A 95% confidence criterion can be adopted in the chi-squared test of the final misfit, to identify automatically the presence of nonlinearity.

If lateral variations are identified in the data, a single dispersion curve cannot be reliably extracted. A possible approach for identifying automatically sharp lateral variations and for handling cases such as the one in Figure 4.23 has been proposed by Vignoli and Cassiani (2009). The phase distribution is analyzed to identify the consistent phase knee-points, and the data are split into subsets where the linearity hypotheses are verified. A continuous analysis of the local phase gradient to extract the local wavenumber with an extension of the MOPA procedure has been proposed by Vignoli et al. (2011).

This technique may also be easily adapted to simultaneously determine the dispersion and attenuation curves by using the magnitude and phase of the cross-power spectrum to determine a complex-valued wavenumber at each frequency (see Section 5.3).

4.5 SPATIAL AUTOCORRELATION

The SPAC method was developed by Aki (1957, 1965) to calculate dispersion curves using microtremor data. The original implementation of the method used a circular receiver array with an additional receiver located at the center of the circle. We will take the more general approach of the extended spatial autocorrelation (ESAC) method (Ohori et al. 2002), which does not impose such restrictions on the geometry of the array.

4.5.1 Single source

The basis of the method is the observation that the spatial autocorrelation function may be obtained via an inverse spatial Fourier transform of the wavenumber

$$R(\chi) = E\left[s(\mathbf{x},t)s(\mathbf{x}+\chi,t)\right]$$

$$= \frac{1}{2\pi}\int_{-\infty}^{\infty} S(\mathbf{k})\,e^{-i\mathbf{k}\cdot\chi}d\mathbf{k} \tag{4.28}$$

where χ is the spatial lag and $E[..]$ denotes the expected value of a quantity. Considering a unit-amplitude plane wave described by

$$s(\mathbf{x},t) = e^{i(\omega_0 t - \mathbf{k}_0 \cdot \mathbf{x})} \tag{4.29}$$

the wavenumber spectrum of the signal is

$$S(\mathbf{k}) = \delta^2(\mathbf{k} - \mathbf{k}_0) \tag{4.30}$$

For the sake of compactness, frequency dependency is omitted reporting the expressions for a given frequency ω_0.

Substituting Equation 4.30 into Equation 4.28 yields

$$R(\chi) = \frac{1}{2\pi}\int_{-\infty}^{\infty} \delta^2\left(\mathbf{k}-\mathbf{k}_0\right)e^{-i\mathbf{k}\cdot\chi}d\mathbf{k}$$

$$= \frac{1}{2\pi}e^{-i\mathbf{k}_0\cdot\chi} \tag{4.31}$$

$$= \frac{1}{2\pi}\left[\cos(\mathbf{k}_0\cdot\chi) - i\sin(\mathbf{k}_0\cdot\chi)\right]$$

Because the autocorrelation function is an even function of the spatial lag, we can limit our interest to the real part of the solution. Thus

$$R(\chi) = \frac{1}{2\pi}\cos(\mathbf{k}_0 \cdot \chi) \qquad (4.32)$$

Finally, the autocorrelation function may be normalized to obtain the auto-correlation coefficient:

$$\rho(\chi) = \frac{R(\chi)}{R(\chi = 0)} \qquad (4.33)$$

$$= \cos(\mathbf{k}_0 \cdot \chi)$$

Equation 4.35 may be used to estimate the wavenumber (and thus phase velocity) from observations of the spatial autocorrelation coefficient of the wave field. The most efficient means of calculating experimental spatial autocorrelation coefficients is via the normalized cross-power spectra between signals at two receiver positions

$$\rho(\chi, \omega) = \frac{\Re\left(\hat{S}_{ij}(\omega)\right)}{\sqrt{\hat{S}_{ii}(\omega)\hat{S}_{jj}(\omega)}} \qquad (4.34)$$

where $\hat{S}_{ij}(\omega)$ is the ensemble average cross-power spectrum between receivers i and j, and $\hat{S}_{ii}(\omega)$ and $\hat{S}_{jj}(\omega)$ are the autopower spectra at the two receivers.

Figure 4.24 shows examples of the use of the spatial autocorrelation method to determine the dispersion curve from an active surface wave test using a linear array of receivers. In this case, the wavenumber and spatial lag may be represented by scalar quantities. The resulting dispersion curve is shown in Figure 4.25.

A major concern in SPAC analysis of active-source data is that the signal for the evaluation of the experimental autocorrelation coefficients at the reference receiver may be affected by near-field effect, which is not accounted for in the formulation. Moreover, the method provides the estimate of a single phase velocity; hence, it does not allow for the separation of modes of propagation, providing an apparent phase velocity irrespective of the wavenumber resolution.

4.5.2 Isotropic wave field

It may be of interest to consider an *isotropic wave field*, which is comprised of random waves propagating in all possible directions with equal probability. This ideal situation may be reasonably assumed to represent a passive

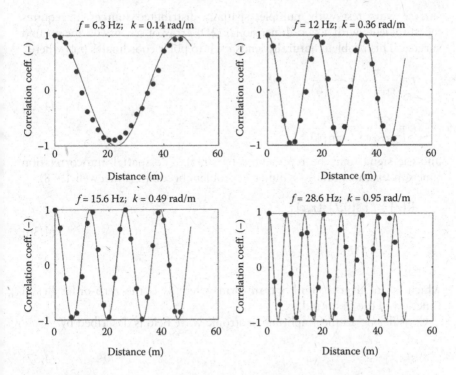

Figure 4.24 Example of fitting obtained at different frequencies with SPAC analysis on active-source multistation data for the reference dataset.

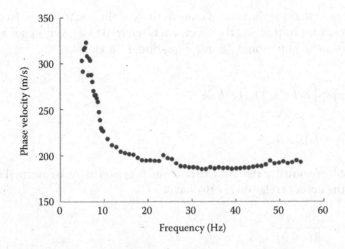

Figure 4.25 Example of experimental dispersion curve estimated with SPAC procedure on active-source multistation data.

surface wave test with multiple, spatially distributed sources. It requires measurements with a two-dimensional (2D) array of sensors on the ground surface. This problem naturally lends itself to polar coordinates (r,ϕ) where

$$r = \frac{x}{\cos(\phi)} = \frac{y}{\sin(\phi)}$$

$$k = \frac{k_x}{\cos(\phi)} = \frac{k_y}{\sin(\phi)} \tag{4.35}$$

and the signal may be represented by $s(r, t)$. The spatial autocorrelation function for this radially symmetric problem becomes (Bracewell 1978)

$$R(r) = E\left[s(0,t)s(r,t)\right]$$

$$= \int_0^\infty S(k) J_0(kr) k \, dk \tag{4.36}$$

which is an inverse Hankel transform where J_0 is the zero-order Bessel function of the first kind.

Considering a unit-amplitude, isotropic wave field is described by

$$s(r,t) = e^{i(\omega_0 t - k_0 r)}, \tag{4.37}$$

its wavenumber spectrum is given by

$$S(k) = \delta(k - k_0) \tag{4.38}$$

(Again, note that for the sake of compactness of the equation, the frequency dependency is omitted and the spectrum is referred to a given frequency ω_0). Substituting Equation 4.38 into Equation 4.36 yields

$$R(r) = \int_0^\infty \delta(k - k_0) J_0(kr) k \, dk$$

$$= J_0(k_0 r) k_0 \tag{4.39}$$

As noted previously, the autocorrelation function may be normalized to obtain the autocorrelation coefficient

$$\rho(r) = \frac{R(r)}{R(r = 0)}$$

$$= J_0(k_0 r) \tag{4.40}$$

To comply with the assumption of radial symmetry, it is necessary to calculate the experimental autocorrelation coefficient using one receiver as a reference (i.e., the receiver at the "origin" of the array). Thus

$$\rho(r,\omega) = \frac{\Re\left(\hat{S}_{0j}(\omega)\right)}{\sqrt{\hat{S}_{00}(\omega)\hat{S}_{jj}(\omega)}} \tag{4.41}$$

where the subscript 0 denotes the reference receiver.

In his original implementation of the method, Aki (1957, 1965) obtained Equation 4.40 by taking the azimuthal average of the spatial autocorrelation function, although this required the use of a circular array (i.e., constant r) as noted previously. The current implementation has no such requirement. Equations 4.33 and 4.40 may be used to determine the wavenumber by minimizing the least-squares error between the theoretical and experimental autocorrelation coefficients as a function of x and r, respectively. A limitation of the SPAC method is that it may be used only for unidirectional (Equation 4.33) or isotropic (Equation 4.40) wave fields (Horike 1985). For active tests, this poses no problem because the user creates unidirectional waves via a controlled source. For passive tests, however, the wave field consists, in general, of multiple sources and may not conform to either of these conditions.

4.6 TRANSFORM-BASED METHODS

The wave field transformations are basic tools in the seismic data processing (Yilmaz 1987) because they allow the separation and identification of different seismic events. The use of global transforms of the dataset for the analysis of the dispersion has been discussed by Nolet and Panza (1976). By means of the wave field transformations, the experimental data are transformed into images of the dispersion curve of each mode. The experimental data are transformed from the original space–time domain into a different domain (ω–p or f–k) in which the propagation parameters are easily identified as spectral maxima. The formal equivalence of different transformations can be proven considering the mathematical properties of the different transforms (Santamarina and Fratta 2010), and there is practically no difference in the obtained dispersion curves (Foti 2000).

In theory, transform-based methods allow the identification of several distinct Rayleigh modes, which could be highly beneficial for the inversion process. In the transformed domain, modes can often be separated even when their presence is not visually detectable in the untransformed data. Nevertheless, the poor spatial resolution (see Chapter 3) often prevents such a possibility, especially in near-surface applications. Indeed, depending on the aperture of the array, it may be not possible to clearly identify different modes, and the result of the dispersion analysis is an apparent dispersion curve influenced by several modes of propagation (see Section 3.3).

4.6.1 Frequency–wavenumber domain

The original wave field, detected in the space–time domain, can be decomposed into its components at constant frequency and wavenumber. In this way, the seismic-gather is converted into an image of the energy density as a function of the frequency and of the wavenumber. This transform is often used in seismic processing because it allows for events having different frequencies, wavenumbers, and apparent velocities to be separated and filtered out. Different wave phenomena are separated and can be muted and suppressed (*f–k* dip filtering). In particular, in seismic processing, the ground-roll (i.e., surface wave components in the wave field) is viewed as high-energy coherent noise, and its removal is an important step.

The use of *f–k* transform for the processing of surface waves has been proposed by Nolet and Panza (1976). Once the modal wavenumbers have been estimated for each frequency, they can be used to evaluate the dispersion curve recalling that phase velocity is given by the ratio between frequency and wavenumber. An example of application is shown in the following: a synthetic seismogram in time-offset domain (Figure 4.26a) with a single non-dispersive event and a velocity of 200 m/s is transformed into *f–k* domain; the resulting image (Figure 4.26b) represents the energy density as a function of the frequency and of the wavenumber. A single velocity corresponds to a straight line from the origin; the line slope increases with the velocity.

Using a large number of signals (256), Gabriels et al. (1987) were able to identify six experimental Rayleigh modes for a site, which were then used for the inversion process. The possibility of using modal dispersion curves is a great advantage with respect to methods giving only a single dispersion curve (e.g., the two-station method and multistation data processed with MOPA or SPAC methods) because having more information means a better constrained inversion. Nevertheless, it has to be considered that in standard practice the number of receivers for engineering applications is typically small, and reduced spatial sampling strongly affects the resolution of surface wave tests. Receiver spacing influences aliasing in the wavenumber domain, so that if high-frequency components are to be sought, spacing must be small. On the contrary, the total length of the receiver array influences the resolution in the wavenumber domain (see Chapter 3).

The straightforward path to obtain an *f–k* spectrum is the application of the Fourier transform in both time and space on field data (see Section 3.3.7). Typically, the number of samples in time domain and the length of the recording are sufficiently high to obtain a good resolution in the frequency domain, whereas the number of samples in the space domain is limited,

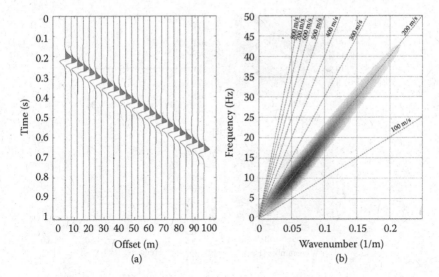

Figure 4.26 Example of synthetic data with a single nondispersive event: (a) synthetic shot gather; (b) frequency–wavenumber spectrum showing a constant phase velocity with frequency.

so that zero padding is applied in space to achieve a reasonable resolution in the wavenumber domain (Figure 4.27a). The space windowing of data is also required to avoid side lobes, which generate local maxima in the spectrum that could be erroneously picked as multiple modes. An example of spectrum computed without tapering is shown in Figure 4.27b. The direct picking of a spectrum with poor spectral resolution, computed without zero padding, is shown in Figure 4.27c.

An example of dispersion analysis in the frequency–wavenumber domain is reported in Figure 4.28, considering the reference dataset. The computation of the spectrum is performed with Hanning tapering and zero-padding in space to a total of 2048 traces, then the spectrum has been computed with a conventional fast Fourier transform algorithm.

The main limitation of using the 2D Fourier transform to directly obtain the *f–k* spectrum is that traces have to be uniformly spaced. The *f–k* spectrum for nonequispaced arrays can be evaluated using other techniques of spectral estimation.

The frequency–wavenumber spectrum may be estimated via a process called *frequency-domain beamforming* (Johnson and Dudgeon 1993). The method is conceptually similar to the one based on the τ–p transform (Section 4.6.2).

Let $s(x_m,t)$ be the signal observed at the mth receiver of an array of M receivers. The column vector

$$S(\omega) = [S(x_1,\omega), ..., S(x_M,\omega)]^T \qquad (4.42)$$

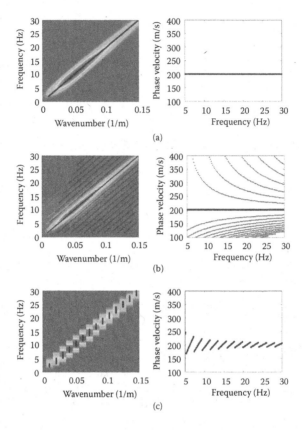

Figure 4.27 Example of *f–k* spectra of the synthetic seismogram (shown in Figure 4.26) and picked dispersion curves. (a) The spectrum (left) and the curve (right) computed with Hanning windowing and zero padding. (b) The spectrum (left) computed without tapering and the curves (right) corresponding to local maxima. (c) The spectrum (left) computed without zero padding and the picked curve (right).

contains the temporal Fourier transform of the signal for each receiver, and

$$e(k) = \left[e^{-ik \cdot x_1}, \ldots, e^{-ik \cdot x_M} \right]^T \quad (4.43)$$

is a *steering vector* (T denotes the transpose of the vector). The output of the conventional frequency-domain beamformer is given by (Johnson and Dudgeon 1993)

$$Z(\mathbf{k}, \omega) = \sum_{m=1}^{M} w_m S(\mathbf{x}_m, \omega) e^{i\mathbf{k} \cdot \mathbf{x}_m} \quad (4.44)$$

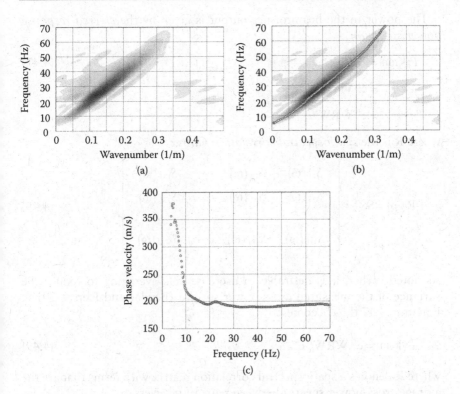

Figure 4.28 Example of dispersion analysis in the frequency domain using the 2D Fourier transform: (a) *f–k* spectrum; (b) picking of maxima in the *f–k* spectrum; (c) experimental dispersion curve.

where

$$\mathbf{W} = \begin{bmatrix} w_1 & 0 & 0 \\ 0 & \ddots & 0 \\ 0 & 0 & w_M \end{bmatrix} \qquad (4.45)$$

is a diagonal matrix containing *shading weights* for each receiver. It is convenient to express Equation 4.44 in quadratic form as follows

$$Z(\mathbf{k},\omega) = e^{H}\mathbf{W}\mathbf{S} \qquad (4.46)$$

where *H* denotes the Hermitian transpose of a vector.

The power in the beamformer output is given by the *steered response power spectrum*

$$P(\mathbf{k},\omega) = ZZ^H \qquad (4.47)$$

$$= e^H WSS^H W^H e$$

$$= e^H WRW^H e$$

where R is the *spatiospectral correlation matrix*

$$\mathbf{R}(\omega) = \mathbf{S}\mathbf{S}^H = \begin{bmatrix} S_{11}(\omega) & S_{12}(\omega) & \cdots & S_{1M}(\omega) \\ S_{21}(\omega) & S_{22}(\omega) & \cdots & S_{2M}(\omega) \\ \vdots & \vdots & \ddots & \vdots \\ S_{M1}(\omega) & S_{M2}(\omega) & \cdots & S_{MM}(\omega) \end{bmatrix} \qquad (4.48)$$

As noted earlier, it is desirable to use ensemble averaging to reduce the variance of the measured cross-power spectra (Bendat and Piersol 2010). Equation 4.47 thus becomes

$$P(\mathbf{k},\omega) = e^H W\hat{\mathbf{R}}W^H e \qquad (4.49)$$

where $\hat{\mathbf{R}}$ denotes a spatiospectral correlation matrix with terms that are the average cross-power spectra between pairs of receivers.

The frequency beamforming technique can be applied to receiver arrays of any shape and is often used to analyze passive data collected with 2D geometries on the ground surface (see Section 7.2.1).

For active-source data with linear arrays, the formulation can be simplified because the source position and the direction of propagation are known; hence, the vector wavenumber $\mathbf{k}$ and the receiver position $\mathbf{x}$ can be simplified to scalar values k_x and x_m. The optimal shading weights are defined according to an approximation of the geometric spreading of Rayleigh waves (Zywicki 1999)

$$w_m = \sqrt{x_m} \qquad (4.50)$$

The processing technique proposed by Park et al. (1999) for the multichannel analysis of surface waves (MASW) method is equivalent to a conventional frequency-domain beamforming with shading weights equal to

$$w_m = \frac{1}{\left| S(x_m,\omega) \right|} \qquad (4.51)$$

so that the normalized amplitude of each receiver is 1.0.

A variety of more advanced signal processing techniques are available that may generally be called adaptive array processing methods (Johnson and Dudgeon 1993). Examples include minimum variance, linear prediction, Pisarenko harmonic decomposition, eigenvector, and multiple signal classification methods. These methods generally provide a higher resolution than the conventional beamforming technique. A detailed presentation of these methods is beyond the scope of this book; interested readers are referred to Johnson and Dudgeon (1993) for a discussion of these advanced methods and to Zywicki (1999) for examples of the application of these methods to engineering surface wave analyses.

4.6.2 Frequency–slowness analysis (MASW)

The use of the frequency–slowness (ω–p) transform for the analysis of dispersive waves was proposed by McMechan and Yedlin (1981). It is based on the slant stack (τ–p) transform or linear Radon transform, one of the basic tools of seismic data processing. The τ–p transform allows the decomposition of a wave field into its plane wave linear components. In this case, the Radon transform is defined as

$$\bar{u}(\tau,p) = \int_{-\infty}^{\infty} u(\tau + px, x)dx \tag{4.52}$$

The transform gives the energy density as a function of τ (the time delay at zero offset) and p (the ray parameter, or slowness). The linear Radon or τ–p transform of a record in time-offset domain stacks the wave field along a straight line of slope p for each value of τ (Figure 4.29a). A straight line in the time-offset domain is described by a constant τ and p; a linear event in the x–t domain with a slope p and intercept τ would map into a single point in the τ–p plane.

On real discrete data, Equation 4.52 is converted to the discrete slant stack transform in which the integral is substituted by a sum

$$\bar{u}(\tau_l, p_k) = \sum_{j=1}^{N} u(\tau_l + p_k x_j, x_j) \tag{4.53}$$

where $\tau_l = l\Delta\tau$ and $p_k = k\Delta p$. For the computation, the minimum and maximum value of p (p_{min} and p_{max}) and the step Δp must be set a priori; p_{min} and p_{max} are easily estimated on the basis of the inspection of the

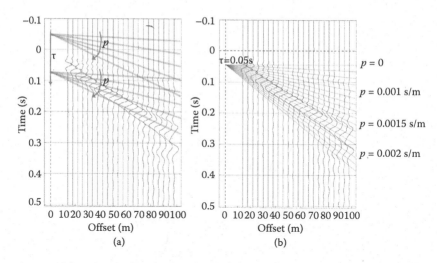

Figure 4.29 (a) Schematic representation of the principle of the τ–p transform, and (b) the data in time offset are summed along straight lines, with intercept τ and slope p.

experimental seismogram; Δp depends on the frequency content of the signal. To avoid aliasing, the following sampling condition has to be satisfied

$$\Delta p \le \frac{1}{2\pi f_{\max} x_{\max}} \qquad (4.54)$$

where $f_{\max}$ and $x_{\max}$ are the maximum frequency and the length of the transformed data in the x direction, respectively.

It is important to observe that the τ–p transform can be easily computed using the 2D Fourier transform, that is, the f–k spectrum (Buttkus 2000)

$$U(f,k) = \int\limits_{-\infty}^{\infty} \int\limits_{-\infty}^{\infty} u(t,x)\, e^{-i2\pi(ft-kx)}\, dx\, dt \qquad (4.55)$$

and its inverse

$$u(t,x) = \int\limits_{-\infty}^{\infty} \int\limits_{-\infty}^{\infty} U(f,k) e^{i2\pi(ft-kx)}\, dk\, df \qquad (4.56)$$

If $U(f,k)$ is computed along a straight line $k = fp$ for the fixed slowness value p, then

$$U(f,fp) = \int\limits_{-\infty}^{\infty} \int\limits_{-\infty}^{\infty} u(t,x) e^{-i2\pi f(t-px)}\, dx\, dt \qquad (4.57)$$

Substituting $\tau = t - px$, one gets

$$U(f,fp) = \int\limits_{-\infty}^{\infty} \int\limits_{-\infty}^{\infty} u(\tau + px, x)e^{-i2\pi ft} \, dx \, d\tau \qquad (4.58)$$

And remembering that

$$\bar{u}(\tau, p) = \int\limits_{-\infty}^{+\infty} u(\tau + px, x) dx \qquad (4.59)$$

we have

$$U(f, fp) = \int\limits_{-\infty}^{\infty} \bar{u}(\tau, p)e^{-i2\pi f\tau} \, d\tau \qquad (4.60)$$

which has the inverse Fourier transform

$$\bar{u}(\tau, p) = \int\limits_{-\infty}^{\infty} U(f, fp)e^{-i2\pi f\tau} \, df \qquad (4.61)$$

showing finally that the $t-p$ transform can be determined by first transforming the wave field $u(x,t)$ into the $f-k$ domain and then calculating the one-dimensional (1D) inverse Fourier transform along a straight line $k = fp$ for each p value. The details of the algorithm for direct and inverse τp transform can be found in Buttkus (2000).

Figure 4.30 reports a synthetic seismogram consisting of two linear events propagating with different velocities (285 and 133 m/s) and with different initial times. The two events also have different frequencies (16 and 32 Hz, respectively). The two slownesses (0.0035 and 0.0075 s/m) are clearly mapped in the corresponding $\tau-p$ transform (Figure 4.30b).

If the $\tau-p$ image is Fourier transformed along the time direction, a frequency–slowness image can be produced (Figure 4.31a). The conversion into frequency–phase velocity (Figure 4.31b) is straightforward.

The $\tau-v$ and the frequency–phase velocity transform of the reference dataset are reported in Figure 4.32.

In comparison to the analysis in the frequency–wavenumber domain, the localization of peaks in the frequency–slowness domain defined by this technique is directly informative about the shape of the dispersion curve. If the two methods are also equivalent in principle, the one based on the slant stack transform gives a clearer and faster image of the dispersion.

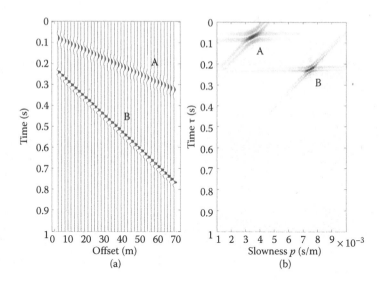

Figure 4.30 Example of τ–p transform of a simple synthetic with two events: (A) velocity = 285 m/s, t_0 = 0.08 s, f = 16 Hz; (B) velocity = 133 m/s, t_0 = 0.22 s, f = 32 Hz): (a) Time-offset data and (b) τ–p transform.

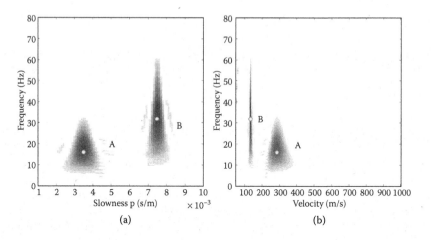

Figure 4.31 (a) Slowness–frequency and (b) velocity–frequency spectra computed from the τ–p transform of Figure 4.30.

4.6.3 Refraction microtremor method

In passive surface wave surveys, the ambient noise (microtremor) is usually recorded with 2D arrays of sensors (Aki 1957; Lacoss et al. 1969; Ohori et al. 2002). The 2D receiver distribution allows the identification of surface wave propagation direction generated by uncontrolled

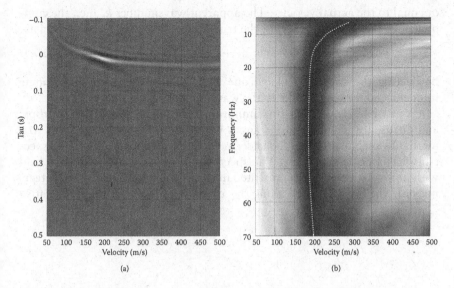

Figure 4.32 Examples of (a) τ–v and (b) frequency–velocity panels for the reference dataset.

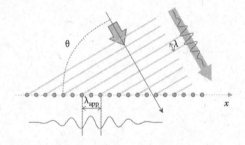

Figure 4.33 When detecting a plane wave with a linear array, an apparent wavelength longer than the true wavelength is observed. Therefore, there is an apparent velocity higher than the true velocity. The apparent in-line wavenumber is smaller than the true wavenumber.

and unknown sources. This approach requires specific acquisition and processing techniques. Louie (2001) proposed an alternative approach to conducting passive measurements using linear arrays and the equipment available for seismic refraction investigations; this approach is known as the refraction microtremor (ReMi) technique.

The obvious limitation is that, in the presence of a single source, the measured apparent velocity depends on the (unknown) angle ϑ between the array direction and the source azimuth (Figure 4.33). For a single plane wave recorded by an array in the direction x, the apparent velocity is equal to the true velocity divided by $\cos\vartheta$; thus, it is always greater than or equal to the actual velocity. The apparent wavenumber k_x (i.e., the component in the array direction of the true wavenumber k) can be written as

$$k_x = k \cdot \cos\vartheta \qquad (4.62)$$

Neglecting the spectral leakage due to the finite array length, a single frequency appears as a spike in the wavenumber spectrum.

It is possible to overcome this limitation if the microtremor is an isotropic and diffused wave field, with sources homogenously distributed at all azimuths. In such a situation, considering a single frequency, the k_x–k_y spectrum has a polar symmetry. The spectrum of an ideal diffused wave field with unit wavenumber is represented in Figure 4.34a. A linear array detects not only the in-line energy along the x axis with the positive and negative

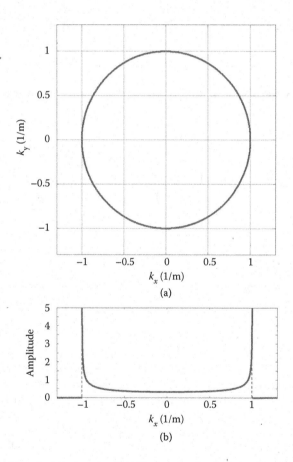

(a)

(b)

Figure 4.34 (a) The ideal spectrum of a diffused wave field with unit wavenumber in the k_x–k_y domain. (b) The corresponding spectrum observed by a linear array—the spectral maxima are on the true wavenumber.

actual wavenumbers, but it also detects the energy traveling with lower apparent wavenumbers. The true positive and negative wavenumbers are stationary points in the projection onto the array direction, and they result in spectral maxima (Figure 4.34b). Considering an array in the x direction, the spectrum of k_x can be written as (Strobbia and Cassiani 2011)

$$P(E) = \frac{1}{\sqrt{1 - \cos^2 \vartheta}} \frac{1}{\pi} \tag{4.63}$$

The spectrum represented in Figure 4.34b does not change with the direction of the array due to the polar symmetry of the wave field.

The finite length of the recording array introduces some complexity in the ReMi spectrum that needs to be considered. Figure 4.35 shows the ReMi spectra for a given frequency for plane waves with uniform azimuthal directions and for two different array lengths (96 and 48 m). Note also that, for any finite length array, the maximum of the 1D spectrum will be at a wavenumber smaller than the true value.

The resulting ReMi spectrum in the $f-k$ domain is plotted in Figure 4.36a for a simple layered medium, assuming a perfectly isotropic diffused wave field. Half of the wave field will have a positive propagation direction and half a negative propagation direction, with respect to the orientation of the x axis. For a homogeneous wave field, the spectrum is symmetric. This property can be used for a check on the assumption of uniform noise distribution in the experimental dataset, which is a fundamental hypothesis for the ReMi method. The symmetry of the spectrum is a necessary condition but not a sufficient condition to prove that the noise wave field is isotropic. A symmetric crossline distribution of sources can produce a symmetric spectrum with a very large overestimation of the actual phase velocity.

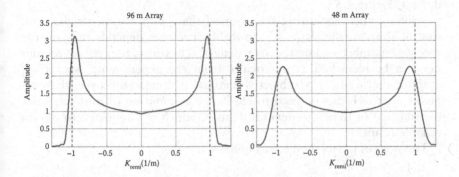

Figure 4.35 The theoretical ReMi wavenumber spectra for two different lengths of the linear array, for a perfectly isotropic noise wave field. The true unit wavenumber is indicated with vertical dashed lines. The maxima are located at values of k smaller than the true wavenumber of the wave field.

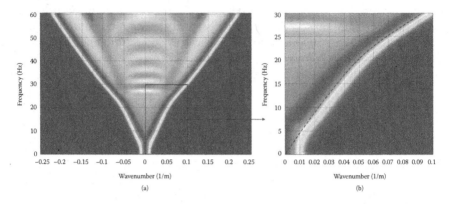

Figure 4.36 (a) Synthetic ReMi spectrum for a perfectly isotropic noise. The spectrum is symmetric; (b) The close-up shows the relationship between the spectral maximum and the true wavenumber (indicated as a black dashed line).

The identification of the true wavenumber from the ReMi spectrum is not straightforward. The true wavenumber is plotted as a dotted line in Figure 4.36b; at each frequency, the maximum of energy is close but not coincident to the true wavenumber. In the original ReMi method (Louie 2001), one way of estimating the wavenumber is by manual picking in a zone between the maximum and the point where the spectral amplitude decreases abruptly. The manual picking is subjective and could introduce large errors in the estimated wavenumber and, consequently, on the experimental phase velocity. To show the difference between the ReMi spectrum and the active data spectrum, with an in-line source, we compare the spectra and the theoretical position of the true wavenumber for a synthetic dataset (Figure 4.37).

Strobbia and Cassiani (2011) showed that, with a uniform distribution of sources, the shape of the ReMi spectrum depends only on the true wavenumber and on the array length. Therefore, the true wavenumber can be identified automatically from the experimental spectrum, without the need for a subjective manual picking. Moreover, the proposed approach can be used to check the assumption of uniform noise distribution in the experimental dataset, which is a requisite to obtain reliable results.

The steps of the method are as follows:

1. The data are segmented into normalized, overlapping subwindows.
2. For each subwindow, the elementary spectrum (either $f–k$ or $f–p$ or $f–v$) is computed, normalized, and stacked; positive and negative quadrants are plotted.

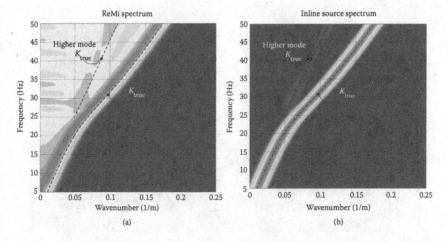

Figure 4.37 Comparison between (a) the ReMi spectrum for a perfectly isotropic noise and (b) the spectrum with an in-line source. The position of the true wavenumber, for the fundamental and first higher modes, is plotted as dashed lines on both spectra. The maximum of the ReMi spectrum is not on the true wavenumber.

3. The frequency range of interest is picked. The spectrum is normalized and split into two parts (negative and positive wavenumbers), which can be independently inverted, fitting the data with the theoretical shape of the spectrum.
4. The two inverted wavenumbers are compared to check the assumption of uniform noise distribution.

An example of ReMi *f–k* spectrum, computed from a 15 min noise record on a 24 channel array, is plotted in Figure 4.38.

The inspection of the spectrum qualitatively shows an acceptable symmetry and suggests the presence of higher modes as secondary local maxima. An example of the fitting of a section of the spectrum is presented in Figure 4.39, considering the frequency of 30 Hz.

The results of the inversion procedure for the whole frequency range are shown in Figure 4.40. Figure 4.40a shows the wavenumbers for the positive and negative quadrants, and Figure 4.40b shows the associated misfit; Figure 4.40c shows the two dispersion curves, together with the curve from an active acquisition obtained on the same array. For this case, the agreement between active and passive data is remarkably good.

Another example is reported in Figure 4.41. In this case, the spectrum is markedly asymmetric, with a sharp variation of the wavenumber at low frequencies. An asymmetric spectrum can indicate the presence of a dominant noise source in one direction.

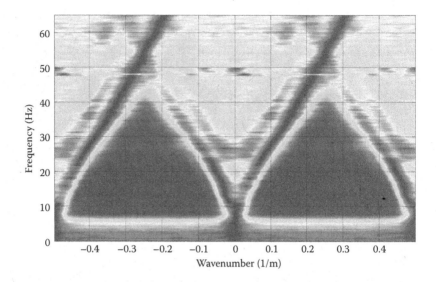

Figure 4.38 Experimental ReMi spectrum for a site where the basic hypothesis of uniform distribution of the seismic sources appears to be verified because of the symmetry of the spectrum.

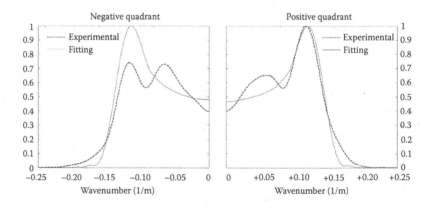

Figure 4.39 The two quadrants of the ReMi f–k spectrum (Figure 4.38) are inverted independently to estimate the wavenumber. The fitting of the slice at 30 Hz is shown here.

A very interesting case occurs when the source is *nearly* in-line. Indeed, sometimes the array is deployed purposefully and aligned with some known source of ambient noise in an attempt to mimic active tests. In this case, the spectral shape is similar to those of a single source and the spectral maximum is expected on the true wavenumber. The ReMi spectrum should be picked as an active spectrum in such condition.

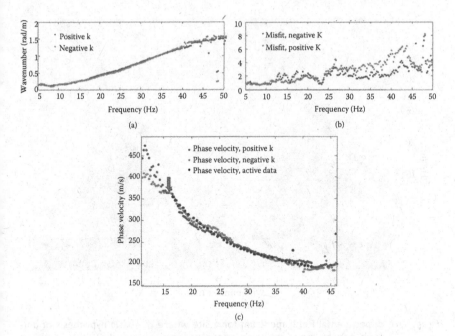

Figure 4.40 Results from the application of the automated ReMi procedure on the data-set of Figure 4.38: (a) estimated wavenumbers for the positive and the negative quadrants; (b) misfit between the experimental spectrum and the theoretical function; (c) experimental dispersion curve from the ReMi procedure and from an active-source test at the same site.

4.7 GROUP VELOCITY ANALYSIS

Analysis of group velocity is used for seismological applications related to the characterization of the Earth's crust on the basis of high-period teleseismic signals. This technique is also applied to microtremors for geological basin characterization, but their limited resolution prevents a diffuse use for near-surface characterization.

The multiple filter method proposed by Dziewonski et al. (1969) for the analysis of earthquake signals is based on the use of band-pass frequency filters. Applying narrow band-pass filters with different center frequencies, different wave groups are separated in the signal (Figure 4.42). These different packets of oscillation can be associated with different modes of propagation, and the peak of each envelope can be used to evaluate the corresponding wave group time delay and hence its group velocity (Figure 4.42b).

The main problem concerning the application of this technique to active-source measurements is related to interference between the modes, which occurs frequently when different modes have similar velocities or when the

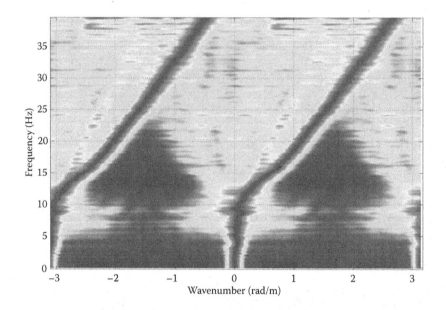

Figure 4.41 Experimental ReMi spectrum for a site where the basic hypothesis of uniform distribution of the seismic sources appears not to be verified because of the lack of symmetry in the spectrum.

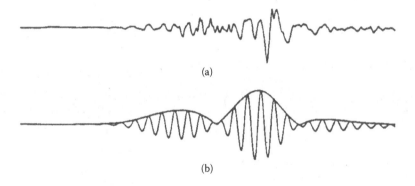

(a)

(b)

Figure 4.42 Example of narrow band-pass filtering: (a) original earthquake signal; (b) filtered signal. (From Dziewonski, A. et al., *Bull Seismol Soc Am*, 59, 427–444, 1969.)

source is too close to the receivers. Such problems can prevent the method from being effective in resolving the different dispersion curves. Gabriels et al. (1987) reported such difficulties in their attempt to analyze the data collected in a multistation session using the multiple filter technique.

The analysis of group velocity can also be used to separate the contribution of Rayleigh modes before the evaluation of phase velocity (Al-Hunaidi 1994;

Karray and Lefebre 2009). The separation can be obtained with the time-variable filter technique, originally proposed to filter out noise (Pilant and Knopoff 1964) and then applied to the analysis of multimode signals for seismological and geophysical applications (Landisman et al. 1969) to cut off the energy that is not associated with a selected wave group.

4.8 ERRORS AND UNCERTAINTIES IN DISPERSION ANALYSES

Uncertainties in dispersion analysis arise from uncertainties in the experimental data and uncertainties introduced during the processing of the dataset.

Data uncertainties are due mainly to noise in the recorded signals and to geometrical uncertainties related to the location and tilting of the receivers. The influence of several sources of uncertainty has been studied by O'Neill (2003) using numerical simulations. He reports minimal influence from geophone tilt and coupling, whereas positional errors, static shift, and additive Gaussian noise introduce larger uncertainties in the experimental dispersion.

Uncertainties in recorded signals are associated with coherent noise and uncorrelated noise. The latter is externally generated noise (environmental noise) and can be studied via the statistical distribution of the recorded signals, if many repetitions of the test in a given configuration are available. Coherent noise is due to events generated by the seismic source (i.e., near-field effects). An approach for evaluating uncertainties in phase analysis has been provided in Section 4.4.

The estimation of the uncertainty is not trivial when the experimental dispersion curve is determined using transform-based methods. Indeed, difficulties arise in properly quantifying how the data error is propagated through a series of complex data processing steps—from the acquisition of the signals in the time-offset domain (t,x) to the calculation of the dispersion curve.

The uncertainty associated with the experimental dispersion curve in multistation surface wave testing is more easily determined by a direct measurement of the statistical distribution of primitive and derived surface wave data. Even though this approach requires considerably more time and effort to be implemented, it does not involve simplifying assumptions (e.g., small variance of the raw data), and it is exempt from major technical difficulties.

Figure 4.43a reports the ensemble of 12 independent dispersion curves obtained with frequency–wavenumber analysis of each available shot gather for the reference dataset. The experimental dispersion curve can be subdivided into two regions: a high-frequency region with very low values of the coefficient of variation and a low-frequency region with higher values

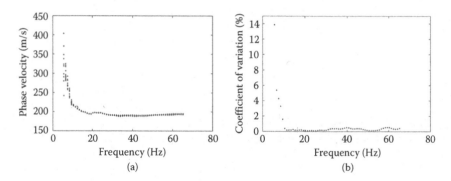

Figure 4.43 Assessment of the uncertainty associated with the experimental dispersion curve: (a) experimental dispersion curves obtained from the frequency–wavenumber analysis of each shot for the same experimental setup; (b) coefficient of variation for the phase velocity at each frequency as obtained from the ensemble of data points in (a).

of the coefficient of variation (Figure 4.43b). These trends are in agreement with other studies on different datasets (Tuomi and Hiltunen 1996; Marosi and Hiltunen 2004a; Lai et al. 2005). Lai et al. (2005) also showed that the assumption of Gaussian distribution is reasonable for the population of phase velocity at each frequency.

The aforementioned uncertainties are limited to repeatability of the test in the same testing setup and with dispersion analysis performed using the same computer code. Larger uncertainties are to be expected because the measurement and processing techniques are varied. In this respect, several comparative analyses at benchmark testing sites provide an insight into the variability that can be expected, even if it is not simple to observe well-defined trends and consistent differences among different processing techniques (Cornou et al. 2007; Boore and Asten 2008; Tran and Hiltunen 2011; Kim et al. 2013). Uncertainties in surface wave analysis are further discussed in Section 6.5.2.

Chapter 5

Attenuation analysis

The spatial attenuation of surface waves as they propagate away from the source is associated with the geometric spreading and with the intrinsic energy dissipation caused by material damping. The estimation of the intrinsic attenuation component can be used for the characterization of the dissipative properties of the medium, typically assuming a viscoelastic constitutive model. This chapter deals with the methods used to estimate the surface wave attenuation parameters to be used in a subsequent inversion process aimed at estimating the small-strain damping ratio profile (see Chapter 6).

Attenuation analysis is generally performed on multichannel measurements acquired using a linear array of receivers and an active source. The procedure requires accurate measurements of the amplitude of the surface wave particle motion. It is essential that the effects of the noise are considered and that the amplitude perturbations are minimized. For example, the verticality and physical coupling of each receiver should be checked carefully. Moreover, an accurate calibration of the receivers is needed to guarantee a uniform response by the array. If methods based on the source force are used, the receiver sensitivity is required for the conversion of the electrical signal to physical units, and it has to be specifically evaluated in the calibration process.

5.1 ATTENUATION OF SURFACE WAVES

A complete discussion of surface wave propagation in viscoelastic media is presented in Chapter 2. In this section, we briefly review the main parameters related to the attenuation of surface waves. The frequency-dependent complex wavenumber k_n^* defines the propagation of dispersive surface wave modes in linear viscoelastic media. The phase velocity and the attenuation constant for the nth mode are related to the complex wavenumber by

$$k_n^* = Re\left(k_n^*\right) + Im\left(k_n^*\right) = \left(k_n - i\alpha_n\right) = \left(\frac{\omega}{v_n} - i\alpha_n\right) \tag{5.1}$$

Table 5.1 Properties of layered system used to generate the examples of Figures 5.1 and 5.2

Layer	Thickness (m)	Vp (m/s)	Vs (m/s)	Dp (–)	Ds (–)	ρ (Mg/m³)
1	5	400	200	0.04	0.04	1.9
2	5	500	250	0.03	0.03	1.9
3	5	600	300	0.025	0.025	1.9
Half-space	∞	800	400	0.02	0.02	1.9

The real part of the wavenumber is the physical wavenumber, a function of the real physical phase velocity v_n; the imaginary part of the complex wavenumber is the attenuation constant. The frequency-dependent attenuation constant α_n is the coefficient of the exponential function describing the spatial decay of the spectral amplitude for a single mode in the far-field

$$A_n(\omega, r) = A_{n,0}(\omega) \cdot \frac{1}{\sqrt{r}} \cdot e^{-\alpha_n(\omega) \cdot r} \tag{5.2}$$

where r is the source to receiver distance. To show the relationship between complex wavenumber, velocity, and damping, let us consider the simple layered medium of Table 5.1.

The solution of the Rayleigh eigenvalue problem for the medium of Table 5.1 is obtained following the procedure detailed in Chapter 2. The real and imaginary parts of the complex wavenumber are plotted in Figure 5.1.

Relating the complex wavenumber to the material properties is not a straightforward process. Indeed, even in a medium with homogeneous velocity and intrinsic absorption, both wavenumber and attenuation coefficient are frequency dependent. A representation showing the relationship with the material properties can be obtained using two different parameters, derived from the real and imaginary part of the wavenumber. The first one is the modal phase velocity, which is $v_n = \dfrac{\omega}{Re(k_n)}$. It has a more intuitive and direct relationship with the layer velocities.

The second parameter is the Rayleigh modal *phase damping ratio* (Misbah and Strobbia 2014), which is defined using an analogy with the body wave damping ratio in a viscoelastic medium, as

$$D(\omega) = \frac{Im\left(k^{*2}\right)}{2 \cdot Re\left(k^{*2}\right)} \tag{5.3}$$

The Rayleigh phase damping ratio is a modal property, related to the damping ratio of the subsurface layers. The same information in Figure 5.1 is represented in terms of phase velocity and phase damping ratio in Figure 5.2.

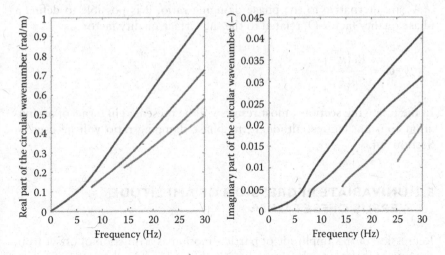

Figure 5.1 Real and imaginary parts of the complex wavenumber of the first four modes for the model of Table 2.1.

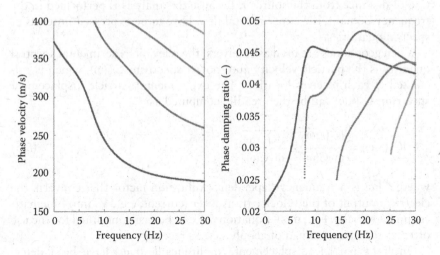

Figure 5.2 Phase velocity and phase damping ratio curves of the first four modes for the model of Table 2.1.

The variation of damping as a function of frequency is related to the geometric dispersion.

The two sets of parameters (the real and imaginary parts of the complex wavenumber, or phase velocity and phase damping) are equivalent. Nevertheless, the phase damping curves allow a more intuitive assessment and evaluation of attenuation data.

As an alternative to the phase damping ratio, it is possible to define a phase quality factor Q, related to the subsurface quality factor, as

$$Q(\omega) = \frac{Re\left(k^{*2}\right)}{Im\left(k^{*2}\right)} \qquad (5.4)$$

In the following sections, most results will be presented in terms of attenuation constant curves, although the phase damping ratio will be used in Section 5.4.

5.2 UNIVARIATE REGRESSION OF AMPLITUDE VERSUS OFFSET DATA

Regression of the amplitude of particle motion as a function of offset from the source is the straightforward method for the evaluation of attenuation coefficients. With no loss of generality, in the following, reference is made to vertical particle motion on the ground surface $u_2(r,t)$, which is a function of distance from the source r. Because the analysis is performed in the frequency domain, the experimental data have to be expressed in terms of spectral quantities.

When geophones are used as receivers, the relevant experimental spectral quantity is the particle velocity auto power spectrum $G_{rr}(\omega)$, which is calculated at each receiver location. The experimental particle displacement spectrum $U_2(r,\omega)$ can be then readily computed as·

$$|U_2(r,\omega)| = \frac{\left|\dot{U}_2(r,\omega)\right|}{\omega \cdot C(\omega)} = \frac{\sqrt{G_{rr}(\omega)}}{\omega \cdot C(\omega)} \qquad (5.5)$$

where $C(\omega)$ is a frequency-dependent calibration factor that converts the electrical output of the velocity transducer (current, e.g., V) into kinematical units (velocity, e.g., m/s). Equation 5.5 can be easily modified for the use of accelerometers, instead of geophones, as receivers.

Once the particle displacement amplitudes $|U_2(r,\omega)|$ have been determined, the Rayleigh attenuation coefficients $\alpha_R(\omega)$ can be computed from a nonlinear regression based on the equation that expresses particle motion as a function of source offset for a harmonic excitation (see Section 2.4.2)

$$|U_2(r,\omega)| = F \cdot Y(r,\omega) \cdot e^{-\alpha_R(\omega) \cdot r} \qquad (5.6)$$

where $Y(r,\omega)$ is the Rayleigh geometric spreading function describing the geometric attenuation of multimode Rayleigh waves in vertically heterogeneous media, and F is the magnitude of the vertical harmonic source applied at the free surface. The geometric spreading function can be calculated from the solution of the Rayleigh forward problem, if the shear wave velocity profile of the site is available (Rix et al. 2000). The latter can be determined from the inversion of the experimental dispersion curve or assumed from other available data. The experimental attenuation coefficients $\alpha_R(\omega)$ determined with the regression process are apparent attenuation coefficients that may represent the combination of several modes of Rayleigh wave propagation.

Examples of regression of particle displacement spectra are provided in Figure 5.3. As the frequency increases, the effects of mode superposition on geometric attenuation produce marked oscillations in the theoretical trend of displacements as a function of offset. This aspect is particularly pronounced in the examples of Figure 5.3 because the site is inversely

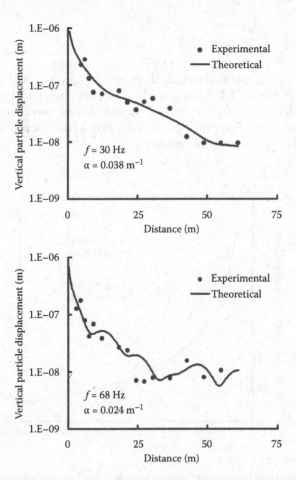

Figure 5.3 Examples of regression of vertical particle displacement amplitude versus offset. (From Rix, G. J. et al., *J Geotech Geoenviron Eng*, 126, 472–480, 2000.)

dispersive with a stiff crust over softer materials (Rix et al. 2000). Moreover, the oscillations in the geometric spreading function multiply for increasing frequency. For this reason, the accuracy in the evaluation of the attenuation coefficients is strongly dependent on the accuracy of the available shear wave velocity profile, especially for the high-frequency values.

A simplified approach can be devised assuming that the propagation of surface waves is dominated by the fundamental mode of Rayleigh wave propagation. Under this assumption, the geometric attenuation is described by the factor associated with surface wave propagation in homogeneous media, which is proportional to the inverse of the square root of the offset from the source. The equation of motion of surface wave propagation (Equation 5.6) simplifies into

$$\left| U_2\left(r,\omega\right) \right| = F \cdot \frac{e^{-\alpha_R(\omega)\cdot r}}{\sqrt{r}} \tag{5.7}$$

An example of regression of particle displacement spectra with Equation 5.7 is reported in Figure 5.4. Clearly in this case, the theoretical particle displacements follow a much simplified pattern. Errors associated with this simplification grow whenever higher modes play a relevant role in the propagation and in the increase of frequency.

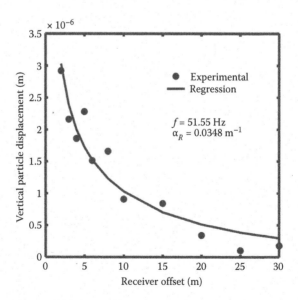

Figure 5.4 Examples of regression of vertical particle displacement amplitude versus offset with simplified expression for the geometric attenuation. (From Rix G. J. et al., *Geotech Test J*, 24, 350–358, 2001.)

In order to avoid uncertainties associated with the coupling of the source to the ground, the regression process also can be performed considering the amplitude of the input force F as an unknown variable. This also allows a straightforward extension of the technique to experimental data collected from sources other than controlled harmonic ones (e.g., impulsive sources), which are cheaper and more time efficient.

This procedure assumes that the particle motion on the ground surface is only caused by surface wave propagation, neglecting body wave contributions; hence, the position of receivers should be selected in order to minimize near-field effects, which are not taken into account in the analysis.

5.3 TRANSFER FUNCTION TECHNIQUE AND COMPLEX WAVENUMBERS

Dispersion and attenuation of surface waves are two aspects of the same phenomenon; hence, a robust and elegant approach can be devised to obtain the two pieces of information simultaneously. In Section 4.4, the regression of phase versus offset was introduced as a method to evaluate dispersion of surface waves, whereas in Section 5.2, the regression of particle amplitudes has been used to analyze surface wave attenuation. The two approaches can be unified by introducing the concept of displacement transfer function, which allows for the experimental dispersion and attenuation curves to be evaluated simultaneously using a single set of measurements. The procedure is then complemented by the coupled inversion that will be introduced in Section 6.4.3 to get a robust and consistent framework for fully coupled analysis of surface wave data leading to the simultaneous estimate of shear wave velocity and damping ratio profiles (see also the examples in Section 7.3).

In a linear system, which in this case is a linear viscoelastic medium, the ratio between the output and the input signals in the frequency domain is called the frequency response function, or transfer function, of the system (Oppenheim and Willsky 1997). In the case of surface wave testing, the input signal is the harmonic force $F \cdot e^{i\omega t}$ applied by the source (e.g., a vertically oscillating shaker) and the output signal is the vertical displacement on the ground surface $U_2(r,\omega)$ measured at a distance r from the source.

For far-field measurements, the vertical displacement $u_2(r,\omega)$ in a linear viscoelastic vertically heterogeneous medium by a harmonic source $F \cdot e^{i\omega t}$ located at the ground surface can be written as (see Section 2.4.2)

$$u_2(r,\omega) = F \cdot Y(r,\omega) \cdot e^{i[\omega t - \Psi(r,\omega)]} \tag{5.8}$$

where $\Psi(r,\omega)$ is the complex-valued phase angle and $Y(r,\omega)$ is the geometrical spreading function. Hence, the displacement transfer function $T(r,\omega)$ between the harmonic source and the receiver is given by

$$T(r,\omega) = \frac{u_2(r,\omega)}{F \cdot e^{i\omega t}} = Y(r,\omega) \cdot e^{-i \cdot \Psi(r,\omega)} \tag{5.9}$$

Assuming $\Psi(r,\omega) \approx k^*(\omega) \cdot r$, where $k^*(\omega) = \left[\dfrac{\omega}{V_R(\omega)} + i\alpha_R(\omega) \right]$ is the complex wavenumber, the implicit dependence of the complex-valued phase angle on the source-to-receiver distance is eliminated. Equation 5.9 thus becomes

$$T(r,\omega) = \frac{u_2(r,\omega)}{F \cdot e^{i\omega t}} = Y(r,\omega) \cdot e^{-i \cdot k^*(\omega) \cdot r} \tag{5.10}$$

Because $T(r,\omega)$ can be directly measured on site if the input source is adequately monitored, Equation 5.10 can be used as a basis of a nonlinear regression analysis for determining the complex wavenumber $k^*(\omega)$. The optimization procedure can be implemented with a classical least-squares technique; however, because $k^*(\omega)$, $Y(r,\omega)$, and $T(r,\omega)$ are complex-valued quantities, the actual implementation of the algorithm requires the definition of norm valid in a pre-Hilbert space (Parker 1994).

The assumption $\Psi(r,\omega) \approx k^*(\omega) \cdot r$ is equivalent to considering the phase angle $\Psi(r,\omega)$ as the result of a single mode of propagation. As a consequence, the method determines apparent values of Rayleigh phase velocities and attenuation coefficients, which can be influenced by modal superposition if there is not a single dominant mode (see Section 2.4.2).

The procedure requires knowledge of the geometric spreading function $Y(r,\omega)$, which can be calculated if a shear wave velocity model of the site is available. An iterative procedure is therefore necessary. At first, the experimental dispersion and attenuation curves are obtained assuming the geometric spreading function $Y(r,\omega)$ to be proportional to $1/\sqrt{r}$, such as for an homogeneous medium. These curves are inverted to obtain approximate profiles of shear wave velocity and material damping ratio, which are then used to calculate an improved estimate of $Y(r,\omega)$. The subsequent iteration uses the updated $Y(r,\omega)$ to determine more accurate dispersion and attenuation curves. The procedure is repeated until convergence.

Figure 5.5 shows an example of nonlinear regression to determine the complex-valued wavenumber $k^*(\omega)$, hence $V_R(\omega)$ and $\alpha_R(\omega)$. For the sake of a clear graphical representation, the results are reported in terms of phase and amplitude diagrams, but the regression is performed directly on the

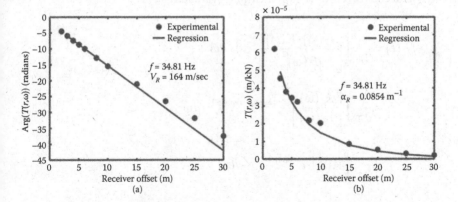

Figure 5.5 Example of regression of displacement transfer function versus offset for coupled estimation of phase velocity and attenuation coefficient: (a) phase versus offset plot; (b) amplitude versus offset plot. (From Rix G. J. et al., *Geotech Test J*, 24, 350–358, 2001.)

complex-valued transfer function. The analysis is repeated at other frequencies to retrieve dispersion and attenuation curves.

Displacement transfer function measurements require an accurate monitoring of the input force. This can be implemented with a load cell for impulsive forces (e.g., sledgehammers) or measuring the acceleration of the armature mass for a controlled source (e.g., a shaker or a Vibroseis) acting in swept-sine mode. Coupling between the sources and the ground represents a significant source of uncertainty. These difficulties can be circumvented by reformulating the transfer function method in terms of deconvolution of the seismic traces.

Deconvolution of a signal $f_2(t)$ with a signal $f_1(t)$ is represented in the frequency domain as the ratio between the Fourier transform of the two signals, $F_2(\omega)$ and $F_1(\omega)$, respectively

$$F_{21}(\omega) = \frac{F_2(\omega)}{F_1(\omega)} = \frac{F_2(\omega) \cdot \overline{F_1(\omega)}}{\left|F_1(\omega)\right|^2} \qquad (5.11)$$

The spectrum F_{21} contains information about the interstation phase delay and attenuation and represents the wave propagation between two stations (Dziewonki and Hales 1972). The phase information is entirely in the numerator on the right side of Equation 5.11, which corresponds to the cross-correlation of the two signals, $f_2(t)$ and $f_1(t)$, used in the two-station spectral analysis of surface waves (SASW) test for the determination of the phase velocity (see Section 4.3).

The deconvolved time signal can be evaluated as

$$
f_{21}(t) = \frac{1}{2\pi} \int_{-\infty}^{+\infty} \left[\frac{F_2(\omega) \cdot \overline{F_1(\omega)}}{|F_1(\omega)|^2} \right] \cdot e^{i\omega t} \cdot d\omega
$$

$$
= \frac{1}{2\pi} \int_{-\infty}^{+\infty} \left[\frac{A_2(\omega)}{A_1(\omega)} \right] \cdot e^{i\left[\omega t - \phi_2(\omega) + \phi_1(\omega)\right]} \cdot d\omega \qquad (5.12)
$$

$$
= \frac{1}{2\pi} \int_{-\infty}^{+\infty} A(\omega) \cdot e^{-i\phi(\omega)} \cdot e^{i\omega t} \cdot d\omega
$$

The function f_{21} represents a signal generated by a δ-impulse source acting at the position of the first receiver and detected at the second receiver (Dziewonki and Hales 1972); therefore, F_{21} is equivalent to the transfer function of the system.

Considering a set of multistation measurements of particle velocity along a straight line on the ground surface, the experimental transfer function $\tilde{F}(r,\omega)$ can be estimated via deconvolution of the whole ensemble of signals

$$
\tilde{F}(r,\omega) = F_{1i}(\omega) = \frac{F_i(\omega)}{F_1(\omega)} \qquad (5.13)
$$

where $F_i(\omega)$ is the Fourier transform of the ith signal detected at distance r from the source; $F_1(\omega)$ is the Fourier transform of the signal detected by the closest receiver; and $F_{1i}(\omega)$ represents the ith deconvolved signal.

The experimental transfer function can be used in a regression process to estimate the dispersion and attenuation curves of surface waves. Considering Equation 5.8, if the response of the receiver placed at $r = r_1$ is used as the reference trace, the theoretical transfer function can be written as

$$
\tilde{F}(r,\omega) = \frac{u_2(r,\omega)}{u_2(r_1,\omega)} = \frac{Y(r,\omega) \cdot e^{-i\Psi(r,\omega)}}{Y(r_1,\omega) \cdot e^{-i\Psi(r_1,\omega)}} \qquad (5.14)
$$

Assuming $\Psi(r,\omega) = k^*(\omega) \cdot r$, the implicit dependence of the complex-valued phase angle on the source-to-receiver distance is eliminated and Equation 5.14 becomes

$$
\tilde{F}(r,\omega) = \frac{Y(r,\omega)}{Y(r_1,\omega)} \cdot e^{-i \cdot k^*(\omega) \cdot (r - r_1)} \qquad (5.15)
$$

where $k^*(\omega)$ is the complex wavenumber (see Equation 5.1).

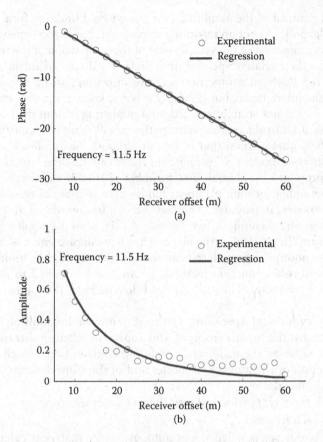

Figure 5.6 Example of regression of deconvolution transfer function versus off-set for coupled estimation of phase velocity and attenuation coefficient: (a) phase versus offset plot; (b) amplitude versus offset plot. (From Foti, S., *Geotechnique*, 53, 455–461, 2003.)

The theoretical transfer function of Equation 5.15 is used in a nonlinear regression analysis to estimate the complex wave number $k^*(\omega)$ from the experimental values of the transfer function, obtained by applying Equation 5.13 to experimental data. An example of a comparison between the experimental data and the best fitting theoretical transfer function is reported in Figure 5.6.

5.4 MULTICHANNEL MULTIMODE COMPLEX WAVENUMBER ESTIMATION

The effect of the modal superposition may be severe on the velocity estimation (see Chapter 3) and even more on the attenuation measurement. Indeed, the interference among multiple modes produces amplitude oscillations.

The local gradient of the amplitude, for a wave field resulting from the sum of multiple modes, is not an intrinsic site property, but it also depends on the offset. It is a function of the velocity and of the attenuation of the individual modes and their relative spectral amplitude. As discussed in the previous sections, the Rayleigh geometric spreading function can be introduced to consider the interference, but generally it is not known a priori. Deriving it from the data is not straightforward, and predicting it from the shear wave velocity model introduces uncertainties that are difficult to quantify.

A multichannel method that is able to separate the amplitude contributions of multiple modes and estimate the modal attenuation curves has been recently proposed by Misbah and Strobbia (2014). It estimates the complex wavenumber for multiple linear modes over an array of receivers using multiple sources, if available. The method can be considered an extension of beamforming techniques (see Section 4.6.1), with a complex propagation constant. In particular, the method has been implemented as an extension of the multiple signal classification (MUSIC) method (Schmidt 1986). As for the transfer function method, it can also be applied to nonevenly spaced receiver arrays. The analysis workflow includes the following steps:

1. The cylindrical spreading effect is removed by multiplying each trace by the square root of the source-to-receiver distance. This step removes the geometric spreading contribution, which is not a function of the number of modes and of their interference, from the amplitude decay.
2. Each trace is transformed into the frequency domain.
3. For each frequency
   a. The frequency-dependent autocovariance matrix is calculated as
   $$S[k] = \overline{\left(T[k] - \mu_{T[k]}\right)}\left(T[k] - \mu_{T[k]}\right).$$
   b. The eigenvalues $\lambda[k]$ and eigenvectors $V[k]$ of the auto covariance matrix $S[k]$ are evaluated.
   c. The noise eigenvectors $V_n[k]$ are extracted from the full set of eigenvectors $V[k]$.
   d. The dot product of the steering vector and each of the noise eigenvectors is calculated to obtain the pseudospectrum. If different shot locations are available, the traces from all shots are analyzed together, estimating the properties of the same set of events, recorded as different signals coming from different sources.
4. The complex wavenumbers for the different modes are evaluated as the values associated with the maxima of the pseudospectrum.

The example presented below refers to a site with a rather simple stratigraphy of fluvial sediments, with interbedded sandy and clayey layers. The raw seismic data are plotted in Figure 5.7.

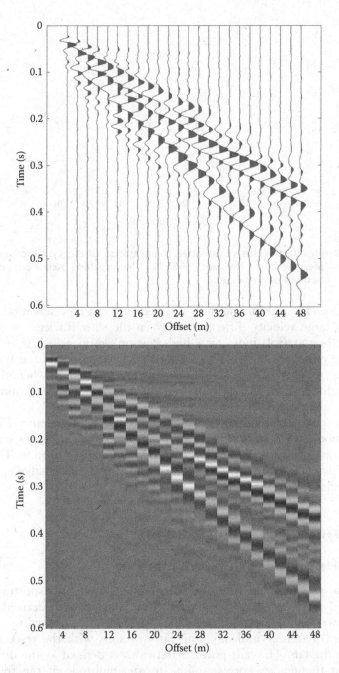

Figure 5.7 Real data seismogram acquired with 24 vertical geophones, spaced 2 m, and with a sledgehammer on plate as source. The data are plotted as wiggle trace (top) and variable density (bottom).

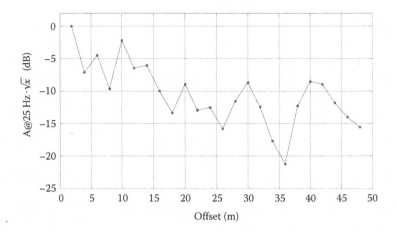

Figure 5.8 Amplitude versus offset, after compensation of the cylindrical geometric spreading, at the frequency of 25 Hz. The amplitude oscillations are due to the modal superposition.

The presence of multiple modes is clear in both representations of the data. A large velocity difference between the slow Rayleigh waves and the fast first arrivals indicates a high Poisson's ratio, which is associated with the presence of a shallow water table. The analysis of the amplitude versus offset shows large oscillations visible in Figure 5.8, where the spectral amplitude at 25 Hz is plotted versus the offset after correction for the cylindrical geometric spreading.

In this case, the amplitude has oscillations due to interference. The number of modes is not so obvious from the inspection of the gather, but it can be assessed looking at the *f–k* spectrum of the gather (Figure 5.9). There are at least three modes in the frequency range 0–50 Hz. The analysis gives the results shown in Figure 5.10.

5.5 OTHER SIMPLIFIED APPROACHES

5.5.1 Half-power bandwidth method

Attenuation curves can be extracted directly from the *f–k* spectra, which are used for the dispersion analysis using an approach derived by the half-power bandwidth (Badsar et al. 2010), usually adopted to evaluate structural damping and damping ratio of soils with the resonant column in the lab. The half-power bandwidth is defined as the difference between frequencies corresponding to an amplitude of the frequency response of a system equal to $1/\sqrt{2}$ the maximum amplitude (Clough and Penzien 1993).

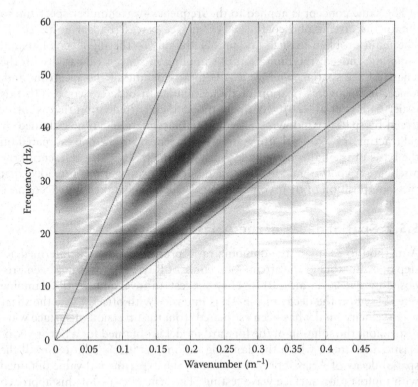

Figure 5.9 Plot of *f–k* spectrum of the gather of Figure 5.7. Multiple modes are present, and at least three modes are visible in the range 20–50 Hz.

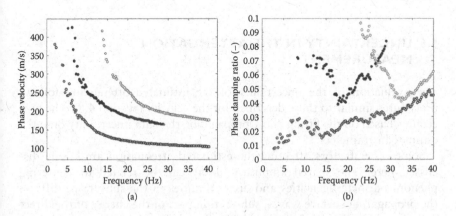

Figure 5.10 (a) Modal phase velocity and (b) modal phase damping ratio for the data of Figure 5.7. Three modes have been identified, and the damping ratio of the fundamental mode ranges between 1% and 4%.

The same concept is applied to the frequency–wavenumber spectrum to obtain the attenuation coefficient of surface waves (Badsar et al. 2010). The main problem in such an approach is due to the influence of spatial sampling; indeed, the limitation of sampling in space associated with the length of the spatial window results in a widening of the spectrum and, hence, in an overestimation of the damping. In order to cope with this issue, Badsar et al. (2010) evaluate the artificial attenuation that is introduced by spatial windowing. True spatial attenuation can be obtained by subtracting the artificial attenuation coefficient from the one estimated on the basis of the half-power bandwidth. The method is able to identify the attenuation coefficient for different modes of propagation, provided that the superposition of their contributions is adequately taken into account.

5.5.2 Spatial decay of the Arias intensity

A method to estimate the damping ratio profile without the intermediate step of estimating of the attenuation curve of Rayleigh waves has been proposed by Badsar et al. (2011). They suggest an inversion for the damping ratio based on the decay of the Arias intensity with offset. Once the shear wave velocity model has been estimated from the inversion of surface wave dispersion, the solution of the forward model is obtained for a linear visco-elastic medium in which the damping ratio is iteratively adjusted until the spatial decay of Arias intensity is close to the experimental value obtained with multistation surface wave testing. The crucial issue for this approach is that the solution of the forward problem requires an accurate elastic model in terms of shear wave velocity and Poisson ratio profiles. Hence, the S- and P-wave velocity profiles need to be evaluated independently and with high accuracy.

5.6 UNCERTAINTY IN THE ATTENUATION MEASUREMENT

The evaluation of the uncertainty of the estimated attenuation follows procedures similar to those described for the velocity (Section 4.8). The procedures based on the fitting allow propagating the data uncertainty onto the estimated parameters.

Some other factors affecting the estimated attenuation are briefly discussed here. The spatial amplitude decay can be related to scattering phenomena: heterogeneities and discontinuities in the subsurface diffract the propagating surface waves, subtracting part of the energy of the direct main wave front which is analyzed. Small-scale features such as inclusions, cavities, cracks, boundaries, and topographic irregularities can reflect a large portion of the energy. For example, the transmission and reflection

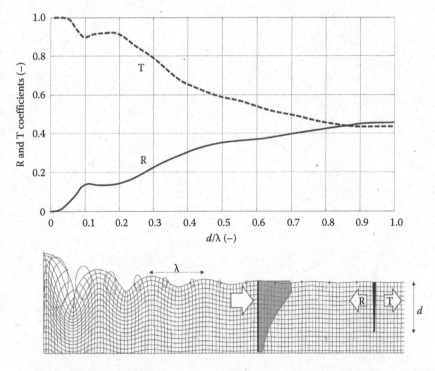

Figure 5.11 Reflection and transmission coefficients for a surface-breaking crack in a homogeneous medium in two-dimensional. (From Rodríguez-Castellanos, A. et al. *Geofís Int* 46(4), 241–248, 2007.)

coefficients for a surface-breaking vertical crack in a homogeneous medium are plotted in two-dimensional (2D) in Figure 5.11. As the ratio of the length of the crack with respect to the wavelength increases, the quantity of reflected energy increases. Such phenomena are not accounted for in the attenuation analysis and may produce a marked overestimation of the intrinsic attenuation and, hence, of the damping ratio of the subsoil.

Chapter 6

Inversion

The aim of this chapter is to illustrate the basic theory for the solution of the inverse problem associated with the propagation of surface waves. This is the final step of data interpretation when surface wave testing is used for site characterization. It involves a mathematical operation, called *inversion*, by which the experimental dispersion and/or attenuation curve is processed to obtain the unknown profile of shear wave velocity and/or shear damping ratio at the site.

The theory of inversion is a subject of fundamental importance in geophysics as well as in applied science and engineering because it is concerned with the inference of parameters characterizing a physical system or even a time-dependent process from a set of experimental measurements. Sophisticated imaging techniques used in medicine such as computerized X-ray tomography (CT), positron emission tomography (PET), and magnetic resonance imaging (MRI) are based on exploiting powerful algorithms for the solution of complicated nonlinear inverse problems. Also, the capabilities of nondestructive testing methods that are widespread in several areas of engineering are founded on the solution of inverse problems. Furthermore, inverse problems are a subject of great interest to mathematicians who developed a formal theory based on functional analysis and integral equations. Several monographs are dedicated to this subject (e.g., Menke 1989; Groetsch 1993; Parker 1994; Engl et al. 1996; Aster et al. 2005; Tarantola 2005; Kirsch 2011) with content and flavor that may be quite different depending on the background and the adopted perspective of the author(s).

The main focus of this chapter is the application of discrete inverse theory to surface wave measurements for near-surface site characterization. Background material on nomenclature, classification, and solution strategies of linear and nonlinear inverse problems is provided before showing applications specifically devoted to surface wave inversion. The chapter is subdivided into five main sections.

Section 6.1 illustrates some basic features and conceptual issues related to the theory of inverse problems as applied to geophysics and then

273

contextualizes them with reference to the parameter identification problem associated with surface waves. The inherent ill-posedness of inverse problems is discussed in conjunction with the strategies that may be adopted for its mitigation.

Section 6.2 reviews forward modeling associated with the propagation of surface waves and highlights its importance in the solution of the corresponding inverse problem. One-dimensional (1D), layered versus continuous models are discussed jointly with the discretization scheme adopted to represent the spatial variability of model parameters.

Section 6.3 illustrates a few algorithms commonly used to solve geophysical inverse problems starting with the steady-state Rayleigh method (SSRM), an empirical procedure adopted in the early days of surface wave testing. The SSRM was the precursor of modern spectral analyses of surface waves (SASW) and multichannel analyses of surface waves (MASW) testing.

Section 6.4 is the core of the chapter, and it is devoted to illustrating some of the details of the standard theory for the solution of linear and nonlinear inverse problems associated with surface waves that are relevant for the inversion of the experimental attenuation and dispersion curve, respectively. After reviewing a few definitions on the goodness of data fitting, the elegant and efficient formalism of the Moore–Penrose generalized inverse matrix is introduced to systematically solve the linear inverse problem. This allows us to address the solution of all categories of linear inverse problems, namely the under-, over-, and mixed-determined problems. Mitigation of ill-posedness is then discussed through the introduction of zeroth and higher-order Tikhonov regularization methods. Treatment and solution of nonlinear inverse problems are illustrated in conjunction with the uncoupled and joint inversion of the Rayleigh dispersion curve. Several classes of inversion algorithms are subsequently presented with an in-depth discussion of Occam's algorithm. The section ends with a brief discussion of the importance of a priori information in surface wave inversion.

Section 6.5 is concerned with the assessment of the influence that experimental errors in surface wave measurement have on the inversion process. The sources of uncertainty in determining the experimental dispersion and attenuation curves are examined, followed by an application of the first-order second-moment (FOSM) reliability method to determine how this uncertainty of the experimental data is projected, via the inversion algorithm, into uncertainty of model parameters, which are the shear wave velocity and shear damping ratio profiles at a site. The section concludes with a brief discussion on the importance that awareness of the inherent trade-off between model resolution and uncertainty has when solving an inverse problem, specifically that associated with surface waves, and is followed by an outline on the use of the Bayesian approach as an alternative method (with respect to the classical frequentist approach) to solve inverse problems.

6.1 CONCEPTUAL ISSUES

6.1.1 Forward and inverse problems in geophysics

Given the set of medium parameters $\{\lambda = \lambda(x_2), \mu = \mu(x_2), \rho = \rho(x_2)\}$ defining the physical and mechanical properties of a site and their variability with depth (see Figure 2.19), the problem of determining the dispersion and attenuation curves $V_R(\omega)$ and $\alpha_R(\omega)$ associated with that site is often referred to as the Rayleigh *direct* or *forward* problem. Conversely, if $V_R(\omega)$ and $\alpha_R(\omega)$ are given, then the problem of determining the unknown medium parameters $\{\lambda = \lambda(x_2), \mu = \mu(x_2), \rho = \rho(x_2)\}$ defines the Rayleigh *backward* or *inverse* problem.

If the aforementioned Rayleigh forward problem is interpreted as a *deterministic–mechanistic process*, the dispersion and attenuation curves can be viewed as special types of "response functions" of a physical system to a given "excitation." The system is represented by the soil deposit, whereas the excitation may be small perturbations of the initial equilibrium conditions of the system (free-vibration problem) or a source with a certain geometry and time variation (e.g., vertical, time-harmonic point or line force, Dirac-type impulse). In this interpretation, direct problems are concerned with determining the effects (i.e., the response function) induced on a physical system by certain causes (i.e., the excitation). In inverse problems, the roles of causes and effects are reversed, and the objective is to determine the causes that generated the observed effects (Engl 1993; Groetsch 1993).

If the deterministic–mechanistic process is viewed as a formal mapping $G: G(m) = d$, the forward problem corresponds to the problem of computing d from knowing G and m. In this mapping, G is a *mathematical operator* representing the model of the deterministic process, m is the excitation, and d is the response. Numerical simulations are nothing but approximate solutions of forward mathematical problems.[01] As for the inverse problem, there are two options: the first is the *causative problem* (inverse problem of category 1), which corresponds to determining m given G and d, and the second is the *model-identification problem* (inverse problem of category 2), where the objective is to compute G from knowing m and d. Figure 6.1 shows these alternatives graphically.

The inverse problem associated with the propagation of surface waves belongs to the inverse problem of category 2. *Any* problem concerned with geophysical prospecting and seismological studies of the Earth's structure where the objective is to determine some type of geological information about

---

[01] In the theory of linear systems, the forward problem is described by the *convolution* operator between an *input* signal, $m(t)$, and the *impulse response* function $G(\delta(t))$, which is the response of the system to a Dirac delta excitation $\delta(t)$. The symbol d(t) denotes the *output* signal. In the frequency domain, the impulse response function $G(\delta(t))$ becomes the *transfer* function of the system.

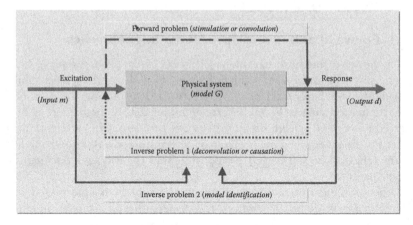

Figure 6.1 Solution of forward and inverse problems associated with a physical system. The problem is described through a deterministic process G formalized by the mapping G: G(m) = d, where G is the mathematical operator, m is the excitation, and d is the response. (Modified from Lai, C. G., *Surface Waves in Geomechanics: Direct and Inverse Modelling for Soil and Rocks,* Springer-Verlag, New York, 2005.)

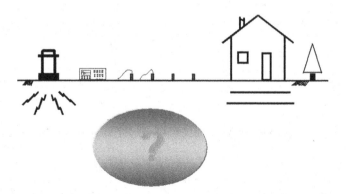

Figure 6.2 Essence of the geophysical inverse problem. Find one or more unknown geometrical (e.g., shape and size of a buried anomaly or the spatial variability of the contacts among different geological units) and/or mechanical (e.g., shear and constrained moduli) fields in the interior of the ground surface from measurements at its boundary and/or along boreholes (cross-well tomography).

the Earth's interior from measurements at its surface (Figure 6.2) belongs to the same category. Category 2 also includes the geodetic inverse problem of determining the shape of the Earth from gravimetric measurements.

Instead, the problem of retrieving information on source parameters of an earthquake from the analysis of recorded seismograms data belongs to inverse problems of category 1, which is also called the category of *deconvolution problems*. Determining the input signal of a seismometer from the knowledge of the output signal and the seismometer response is another example of a deconvolution problem.

Model-identification is often denoted interchangeably as a *parameter-identification* or *parameter-estimation problem* because the model G of a physical system or a process[02] is usually defined a priori from the laws of physics. Consequently, model-identification actually becomes parameter-identification because the original objective has been converted to a problem of determining a discrete set of model parameters.[03]

In recent years, several important inverse problems arising in modern science and engineering have spurred a renewed interest for inverse problem theory, the developments and ramifications of which, even in the restricted field of geophysical prospecting, have been spectacular. With regard to surface waves, solution of the associated parameter identification problem lies at the very core of the success achieved by sophisticated techniques used in nondestructive testing and in geophysical–geotechnical site characterization, which are in essence powerful solvers of inverse problems.

6.1.2 Ill-posedness of inverse problems

Inverse problems are generally known to be *ill-posed* or *unstable*, which is perhaps their most relevant feature if compared with the *well-posedness* of the corresponding forward problems (Engl 1993; Kirsch 2011). According to Hadamard (1923), a mathematical problem is said to be *well-posed* or *stable* if it satisfies the following three conditions:[04]

1. For all admissible data, a solution exists (*existence*).
2. For all admissible data, the solution is unique (*uniqueness*).
3. The solution depends continuously on the data (*stability*).

If any of these three conditions fails to hold, Hadamard called the problem *ill-posed*.[05] In reality, the *Hadamard's postulates* of well-posedness apply

---

[02] Typically, the process is represented by a system of partial differential equations, integral equations, and/or integro-differential equations and their associated initial and/or boundary conditions. Sometimes G is an algebraic operator that may be reduced to a matrix.

[03] In general, these are the constant or variable coefficients of the aforementioned differential/integral equations.

[04] A precise definition of *well-posedness* should also specify the functional space in which the solution is supposed to exist and the restrictions that a given set of data must satisfy to be considered admissible.

[05] At the origin of the Hadamard's definition of *well-posedness* is the idea that a mathematical model of a physical problem *must always* be well-posed. It was later recognized, however, that there are physical phenomena that do not necessarily satisfy this requirement (e.g., quantum physics).

to both *forward* and *inverse* problems. However, it is only in recent years that the importance of *instability* in the solution of forward problems has been widely recognized in studies of nonlinear dynamics and chaos theory[06] where small perturbations in the initial data produce unpredictable changes in the solution (Parker 1994).

In inverse problems, the two conditions that are most often violated are *uniqueness* and *stability*.[07] Of particular relevance in the solution of parameter identification problems is violation of uniqueness that is the existence for a given problem and set of data of more than one solution. In the deterministic–mechanistic interpretation of a forward problem given in Section 6.1.1 (Figure 6.1), this would correspond to a physical process where distinct causes may yield the same effect, a situation that is not so uncommon in physics and engineering. For instance, given a distribution of mass inside the Earth, one can uniquely predict the gravity field around the planet (forward problem). However, there are several different distributions of mass in the Earth that give exactly the same gravity field. Thus, the problem of inferring the mass distribution from the observation of the gravity field (inverse problem) has multiple solutions (Tarantola 2005).

An example drawn from structural engineering is that of a loaded, elastic beam. Although for a given loading configuration a unique deflected shape of the beam can be predicted (forward problem), the same deflected shape may be obtained from different loading patterns. Hence, the problem of determining the load configuration corresponding to a deflected shape (inverse problem) is *ill-posed* because the solution is not unique.

For the surface wave inverse problem, nonuniqeness implies that a given experimental dispersion (or attenuation) curve may correspond to more than one shear wave velocity (or material damping ratio) profile of the soil deposit.

From a mathematical point of view, nonuniqeness in the solution of an inverse problem is caused by a lack of sufficient information to constrain the solution. Alternatively, the available information is not completely independent, at least in certain regions of the domain of definition of the solution[08] (solution space).

---

[06] A classical example is the *butterfly paradigm* where a butterfly flapping its wings in New York generates a hurricane in San Francisco! Even the messing up of in an electronic, typewritten document produced by the introduction of an extra blank character is an example of *instability* of a forward problem; a small perturbation (the extra blank character) may have "catastrophic" consequences in the document formatting with page jumps, spoiling of figures, associated captions, and so on.

[07] This happens particularly in nonlinear problems such as the inverse problem associated with the propagation of surface waves, which will be discussed later in this chapter.

[08] A quotation attributed to Lanczos (1961) states that "A lack of information in the solution of a problem cannot be remedied by any mathematical trickery."

Two strategies can be used to enforce uniqueness in the solution of an inverse problem. The first is to add *a priori information* about the solution of the problem. For the parameter identification problem of surface waves, this might be, for instance, independent knowledge of model parameters in one or more layers (obtained, for example, from certain geotechnical tests) or the layer thickness obtained from borehole logs. Mass density and Poisson ratio are model parameters usually assumed to be known in advance because of the weak sensitivity of the dispersion (and attenuation) curve with respect to variations of these parameters (Nazarian 1984).

Another strategy to enforce uniqueness is to constrain the solution to satisfy certain requirements such as *smoothness* and *bounds*. In some cases, adding a priori information to the solution may be regarded as a constraint. An obvious example is that of requiring the model parameter to vary within a prescribed range or to be nonnegative (e.g., material damping ratio). There are, however, constraints of a different nature because they enforce features of *global* behavior to the solution.

In discussing strategies to enforce uniqueness, it should be remarked that the available methods are relatively simple for ideal, error-free measurements; however, the situation is more complicated for data containing bias and random errors, as will be discussed in Section 6.5.

Violation of the *stability* condition in the Hadamard's definition of well-posedness is also an important concern in the solution of forward and inverse problems. For instance, the solution of the *Fredholm integral equation of the first-kind* is very sensitive to small perturbations of the initial data caused by noise and unavoidable measurement errors, which may result in very large changes in the solution (Groetsch 1993). It is important to emphasize that the instability of these problems is a feature inherent to their nature and have nothing to do with a particular type of numerical algorithm used to solve them. For *linear* inverse problems with discrete and continuous linear operators, a *stability analysis* is usually carried out through the method of *singular-value decomposition*[09] (Strang 1988) as will be shown in detail in Section 6.4.

Very unstable parameter identification problems can be solved using mathematical techniques known as *regularization methods* that approximate the ill-posed problem with a parameter-dependent family of neighboring well-posed problems (Tikhonov and Arsenin 1977). Because some of these regularization methods admit a variational formulation (e.g., Tikhonov regularization) where the objective is the minimization of appropriate functionals, they can also be applied successfully for the solution of nonlinear inverse problems. Sections 6.4.2 and 6.4.3 will discuss Tikhonov regularization methods in some detail.

---

[09] Using this method, the *smallest singular value* controls the amplification of the perturbation errors, whereas the rate of decay of the singular values arranged in order of decreasing magnitude is used as a measure to quantify the *degree of instability* of a given inverse problem.

6.1.3 Inversion strategies: Local versus global methods

Solving the parameter identification problem associated with surface wave motion using the dispersion data as response functions is equivalent to solving an *inverse eigenvalue problem* or an *inverse spectral problem*[10] (Kirsch 2011). In fact, the objective is to determine certain coefficients of the two sets of first-order, linear ordinary differential equations (see Equations 2.70 and 2.71) from the knowledge of their eigenvalues ω/V_R (or ω/V_L).

In practice, a parameter identification problem is usually solved by converting it into a *parameter optimization* problem, the solution of which is then found from the stationary condition of an unconstrained or constrained functional (Parker 1994). In general, the techniques used to solve nonlinear optimization problems such as the inversion of an experimental dispersion curve in surface wave testing can be broadly divided into *global-search* (GS) and *local-search* (LS) methods. The functional of a nonlinear optimization problem will have, in fact, several stationary points in the solution space, and a question arises about finding the global extremum. LS procedures are iterative schemes that, starting from an initial guess of the solution, generate a sequence of improved approximations converging under suitable conditions to the solution. Most LS methods are calculus-based techniques that linearize a nonlinear functional at each iteration until a stationary point is reached. These techniques require the functional to be sufficiently smooth so that its *Fréchet*[11] derivatives (with respect to model parameters) exist and are continuous. Furthermore, even if all the smoothness requirements for the functional are satisfied, the sequence of approximations of the solution is guaranteed to converge only if the initial guess is sufficiently close to the solution. However, the most important limitation of LS procedures is that even when they succeed in finding a stationary point, there are no simple means to determine whether it is a local or a global stationary point in the solution space.

This dilemma is addressed by GS procedures, which are optimization techniques where the search for a global stationary point is conducted through an exploration of the entire solution space. This can be done either systematically by defining a grid or randomly as in Monte Carlo simulations. Other GS methods include simulated annealing and genetic inversion. Usually GS techniques are computationally more expensive than LS methods; however, they are more robust and reliable in finding the global extremum in the solution space. A further discussion on the differences between LS and GS techniques is reported in Section 6.4.3.

---

[10] The study of *inverse eigenvalue problems* is a standard topic in inverse theory. A classical paper on the subject is Kac (1966).

[11] The Fréchet derivative is defined on Banach spaces, and it is used to define the derivative of a functional needed for the solution of an optimization problem using the calculus of variations.

With regard to the parameter identification problem associated with surface waves, ordinarily the strategies adopted for the inversion of surface wave data belong to the category of LS procedures using modal or apparent dispersion and attenuation curves as response functions. The dispersion curves are usually computed from measurements of surface wave phase velocity, although group velocities have also been used. Typically, the dispersion curves are calculated with respect to the fundamental mode of propagation, which implicitly assumes that this mode governs the experimental dispersion curve measured at the site. The approach is substantially correct only if the fundamental mode is effectively dominating the measured seismograms such as what happens in *normally dispersive* soil deposits where the mechanical impedance increases regularly with depth (see Chapter 2). However, there are circumstances in which the experimental dispersion curve also reflects the contribution of higher modes. In this case, what is really measured in the field is not a "modal" but rather an "apparent" dispersion curve or a combination of both (see Section 2.4). The shape of these experimental curves is strongly influenced by the superposition of various modes of propagation. It should be pointed out, however, that the nature of the measured dispersion curve depends not only on the characteristics of the subsurface, but it is also affected by the configuration of the receivers adopted during acquisition (see Chapter 3). Typical examples where the distinction between modal and apparent dispersion curves is relevant include *inversely dispersive* soil deposits, which are media where the mechanical impedance varies irregularly and/or abruptly with depth (e.g., soil profiles characterized by the presence of a top stiff layer). In these situations, the solution of the inverse problem must properly take into account modal superposition. O'Neill (2004) shows an example of the use of such a type of inversion algorithm.

Dispersion and attenuation curves each represent one possible type of response function; however, there are other choices, in principle. In the frequency domain, these include displacement amplitudes and phase spectra. In the time domain, they might be the seismograms. The ability to successfully invert the response functions measured at the free surface of the ground to determine reliable estimates of model parameters in the interior depends to a significant degree on the response function selected to describe the medium response to a dynamic excitation. Factors governing the selection of appropriate response functions include the ability to experimentally measure the field variables, the capability to solve the corresponding inverse problem,[12] and the information content and sensitivity associated with the selected response function with respect to the desired model parameters.

---

[12] This also depends on whether an analytical or a numerical procedure is used to compute the *Jacobian* of the response function with respect to model parameters.

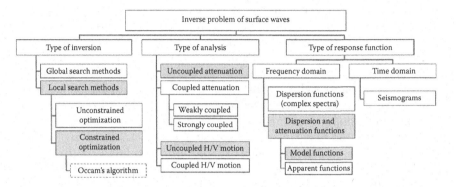

Figure 6.3 Algorithms for the solution of the parameter identification problem associated with surface waves. The shaded boxes indicate the most commonly used
methods in geophysical–geotechnical prospecting to date. (Modified from
Lai, C. G., Simultaneous inversion of Rayleigh phase velocity and attenuation for near-surface site characterization, PhD Thesis, Georgia Institute of
Technology, 1998.)

Figure 6.3 shows a sketch illustrating the combination of some of the possible algorithms that could be conceived for the solution of the parameter
identification problem associated with surface waves.

Coupled and uncoupled H/V motion denotes the kinematics of surface
wave motion considered in the inversion analysis. Usually, only the vertical
component of Rayleigh particle motion is taken into account. Love waves
or horizontal components of Rayleigh waves are rarely used even if it would
be desirable to increase the information content of measured surface wave
data (Tokimatsu 1995; Strobbia 2003).

6.2 FORWARD MODELING

In geophysics and other applied sciences, the ability to solve a parameter-
identification problem often relies on the capacity to actually solve the corresponding forward problem. This is particularly true in the case of nonlinear
problems such as the inversion of Rayleigh or Love dispersion curves. Solution
of the forward problem in turn implies a definition of a mathematical model
of the physical system under study (Figure 6.1). In surface wave modeling for
near-surface site characterization, the issues to be addressed in defining the
mathematical model are, at a minimum, the following:

- Geometrical modeling of the subsurface
- Numerical modeling of the subsurface
- Constitutive modeling of geomaterials

Geometrical modeling of the subsurface is concerned with the idealization related to the spatial variability of the physical properties of the medium. If the soil deposit is assumed laterally homogeneous and the physical properties are considered to vary only in the direction of gravity, then the geometrical model may be unidimensional. Two-dimensional (2D) or three-dimensional (3D) models are mathematical idealizations in which the properties are also allowed to vary in the horizontal direction. Unidimensional modeling is the standard in surface wave testing, although 2D profiling has already been proposed in advanced applications.

Numerical modeling of the subsurface refers to the discretization scheme adopted for the representation of the spatial variability of the field variables. In unidimensional models, a vertically inhomogeneous medium is usually idealized by a stack of homogeneous layers, representing different geological units, overlying a homogeneous half-space. This is a *discontinuous* discretization scheme in contrast to a subsurface model in which the material properties vary continuously with depth (Rix and Lai 2014)[13] (Figure 6.4).

In surface wave testing, the choices for geometrical and numerical modeling have important implications in the ability to successfully invert dispersion and attenuation data. Selection of a model that is not consistent with the real subsurface conditions may lead to grossly erroneous results. For instance, the adoption of a unidimensional geometrical model at a site characterized by strong lateral variability in physical and mechanical properties would make the interpretation of surface wave measurements severely biased.

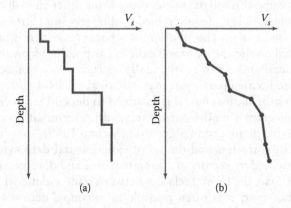

Figure 6.4 Subsurface numerical modeling: (a) layered (i.e., discontinuous) model and (b) continuous model.

---

[13] In the practical implementation of this numerical scheme, the medium is discretized by a stack of homogenous layers of arbitrary small thickness and the physical and mechanical properties within the individual layers such as V_S and V_P are typically assumed to vary *linearly*.

Similarly, the use of a continuous model (Figure 6.4) at a testing site characterized by strong contrasts in mechanical impedance among different layers would be inappropriate. On the contrary, in a homogenous soil deposit, the shear wave velocity at a site is expected to increase monotonically with depth because of the influence of overburden pressure. In such situations, surface wave measurements should be more appropriately interpreted adopting a subsurface model in which the material properties vary continuously with depth.

Constitutive modeling of geomaterials is related to the assumptions adopted to describe the mechanical response of soils and rocks subjected to low amplitude, dynamic excitations. As discussed in Chapter 2, linear elasticity and linear viscoelasticity have proved to be effective in describing phenomena of wave propagation in geomaterials at low strain levels. As such, they are the two most commonly adopted constitutive models.

Poroelasticity and Biot theory, which explicitly recognize the multicomponent nature of saturated geomaterials, are rarely used because the solution of the forward problem is considerably more complicated. However, special care should be taken when using one-constituent theories to interpret surface wave measurements at sites characterized by the presence of the water table. In fact, wave propagation phenomena associated with geophysical testing induce an undrained response in saturated materials due to the rate of application of the seismic excitation compared with their hydraulic conductivity. Correspondingly, in soil constitutive modeling, the values assumed for the Poisson ratio should correctly reflect the real position of the water table at the testing site because this soil parameter is markedly different in dry and water-saturated geomaterials under undrained loading (Foti and Strobbia 2002).

As mentioned earlier, the solution of a parameter identification problem is founded on the ability to solve the corresponding forward problem, which practically means to synthetically reproduce a set of experimental data associated with a particular mathematical model. In *discrete* inverse theory, the mathematical model is assumed to depend on a certain, finite number of unknown model parameters, the determination of which is the objective of the inversion algorithm (Menke 1989).

The model parameters and the set of experimental data may be conveniently represented by vectors of adequate size **m** and **d**, respectively. In the most general case, the formal relation between data and model parameters is implicit; however, it is often possible to uncouple data **d** from model parameters **m** and write the forward problem as follows

$$\mathbf{G(m)} = \mathbf{d} \tag{6.1}$$

where **G** is a nonlinear, vector-valued function operator reflecting the type of relationship existing in a particular mathematical model between input data and model parameters. For the forward problem associated with surface waves, the model may be a linear elastic, multilayered medium. In this case, the model parameters are the velocity of propagation of S-waves of the individual soil layers.[14] The set of experimental data is represented by a discretized dispersion curve describing the frequency dependence of Rayleigh (or Love) phase (or group) velocity (modal or apparent) over a finite frequency range. Equation 6.1 can then be rewritten by making the meaning of vectors **m** and **d** explicit, as follows

$$G(V_S) = V_{R/L} \qquad (6.2)$$

where $V_S = [(V_S)_1, (V_S)_2, \dots (V_S)_i \dots (V_S)_{nl}]$ is the vector comprising the shear wave velocities of the individual layers and $V_{R/L} = [(V_{R/L})_1, (V_{R/L})_2, \dots (V_{R/L})_j \dots (V_{R/L})_{nf}]$ is the discretized Rayleigh (or Love) dispersion function. Finally, **G** is a nonlinear vector-valued function of V_S. Section 6.4.3 will show an explicit representation of this vector function. The symbols *nl* and *nf* denote (respectively) the number of layers (including the half-space) of the stratified medium and the set of experimental frequencies at which the Rayleigh (or Love) phase (or group) velocity (modal or apparent) has been measured. Equation 6.2 represents the formal statement of the Rayleigh (or Love) *forward* problem in linear elastic media. As was demonstrated in Chapter 2, this equation is clearly nonlinear. If the medium is viscoelastic, Equation 6.2 still holds; however, the vectors V_S and $V_{R/L}$ become complex valued (see Section 6.4.3).

Under the assumption of weak dissipation, a procedure can be established to determine the transversal (or shear) material damping ratio profile $D_S(x_2)$ from the inversion of an experimental attenuation curve $\alpha_R(\omega)$. Details of the procedure are reported in Chapter 2, specifically Equation 2.133. As noted in Section 2.5.3, an important feature of this procedure is that it is *linear*. This is in contrast with the problem of obtaining the $V_S(x_2)$ profile from the inversion of $V_{R/L}(\omega)$, which is highly nonlinear. The statement of the corresponding Rayleigh (or Love) *forward* problem for determining material damping ratio can therefore be written as follows

$$G \cdot D_S = \alpha_{R/L} \qquad (6.3)$$

where $D_S = [(D_S)_1, (D_S)_2, \dots (D_S)_i \dots (D_S)_{nl}]$ is the vector comprising the shear damping ratio of the individual strata and $\alpha_{R/L} = [(\alpha_{R/L})_1, (\alpha_{R/L})_2, \dots (\alpha_{R/L})_j \dots (\alpha_{R/L})_{nf}]$ is the discretized Rayleigh (or Love) attenuation function. Finally, **G** is an *nf* by *nl* matrix composed by terms that include the partial derivatives of the modal (or apparent) phase velocity of surface waves with

---

[14] In principle, additional or alternative sets of model parameters may also be the *velocity of propagation* of *P-waves*, the *mass density*, and most importantly, *the thickness* of the layers. If the medium is assumed to be viscoelastic, further parameters may include the longitudinal and transversal *material* damping ratio of the strata.

respect to the model parameters $(V_P)_i$ and $(V_s)_i$ of the various layers of the soil deposit. The precise definition is given by Equation 2.133. (The "dot" operator at the left-hand side of Equation 6.3 denotes matrix multiplication.)

Summing up, the *forward problems* that can be conceived in surface wave testing may be of *three* types:

I. The problem of predicting the Rayleigh (or Love) dispersion vector function $\mathbf{V}_{R/L} = [(V_{R/L})_1, (V_{R/L})_2, ... (V_{R/L})_j .... (V_{R/L})_{nf}]$ by knowing the profile of shear wave velocity $\mathbf{V}_S = [(V_S)_1, (V_S)_2, ... (V_S)_i .... (V_S)_{nl}]$ of the multilayered medium.

II. The problem of predicting the Rayleigh (or Love) attenuation vector function $\boldsymbol{\alpha}_{R/L} = [(\alpha_{R/L})_1, (\alpha_{R/L})_2, ... (\alpha_{R/L})_j .... (\alpha_{R/L})_{nf}]$ by knowing the profile of shear damping ratio $\mathbf{D}_S = [(D_S)_1, (D_S)_2, ... (D_S)_i .... (D_S)_{nl}]$ of the multilayered medium.

III. The problem of predicting the complex-valued Rayleigh (or Love) dispersion vector function $\mathbf{V}_{R/L}^* = \left[\left(V_{R/L}^*\right)_1, \left(V_{R/L}^*\right)_2, ... \left(V_{R/L}^*\right)_j .... \left(V_{R/L}^*\right)_{nf} \right]$ by knowing the profile of complex-valued shear wave velocity $\mathbf{V}_S^* = \left[\left(V_S^*\right)_1, \left(V_S^*\right)_2, ... \left(V_S^*\right)_i .... \left(V_S^*\right)_{nl} \right]$ of the multilayered medium. The complex-valued shear wave velocity can be constructed from the profile of shear wave velocity and shear damping ratio using the formalism of complex variable theory (Equation 2.125). In a similar fashion, the Rayleigh (or Love) dispersion and the attenuation functions can be recovered from the corresponding complex-valued dispersion vector functions (Equation 2.124).

Problems I and III are nonlinear forward problems,[15] whereas problem II is linear. Consequently, in investigating the characteristics of the corresponding inverse problems and of the algorithms for their solution, it is necessary to investigate the features of both the *linear* and the *nonlinear* inverse problem.

6.3 SURFACE WAVE INVERSION BY EMPIRICAL METHODS

A simple, empirical procedure for estimating the shear wave velocity profile directly from the experimental dispersion curve was introduced in the early applications of surface wave testing for near-surface site characterization

---

[15] Problem III is further complicated by the complex nature of the mapping (Equation 6.2). It turns out, however, that owing to the elegance and efficiency of complex variable theory, the complications that arise are not severe. This result carries over even in the solution of the corresponding inverse problem as will be shown in Section 6.4.3.

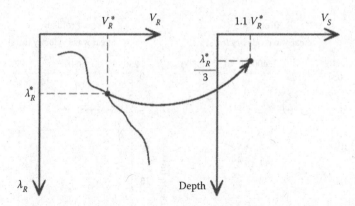

Figure 6.5 Empirical inversion procedure introduced in the early applications of surface wave testing in a test known as the Steady-State Rayleigh Method. (From Foti, S., Multi-station methods for geotechnical characterisation using surface waves, PhD Dissertation, Politecnico di Torino, 2000.)

(Jones 1958). At that time, the test was called the SSRM. Despite the inversion procedure adopted in the SSRM was very rough, it allowed a rapid, preliminary estimation of the shear wave velocity profile at a site. The essence of the empirical interpretation of surface wave data used in the SSRM will be briefly reviewed.

As discussed in Chapter 2, the ground motion induced by a traveling surface wave is confined in the uppermost part of the soil deposit (i.e., the *skin depth*). It can be assumed that most of the strain energy associated with surface wave motion is confined within a depth of about a wavelength from the ground surface (Achenbach 1984). At the same time, in a homogeneous half-space, the Rayleigh phase velocity assumes values that are close to the shear wave velocity of the medium (see Figure 2.15). A rough estimate is $V_S \approx 1.1 \cdot V_R$. To a first approximation, this estimate of shear wave velocity may be considered representative of the value of V_S at a depth equal to one-half/one-third of the wavelength. This interpretation may be viewed as a mapping from the $\{V_R, \lambda_R\}$ domain to the $\{V_S, \text{depth}\}$ range (Figure 6.5).

A rough estimate of the shear wave velocity profile at a site may be determined by repeating the procedure for the whole set of experimental data.

6.3.1 Numerical example

Figure 6.6 shows the results obtained with the empirical inversion of the SSRM for three different synthetic profiles of soil deposits. The Figure clearly shows that the method gives a reasonable estimate of the shear wave velocity profile only for case A, which represents a *normally* dispersive medium with a gradual increase of V_S with depth. This estimate can be considered

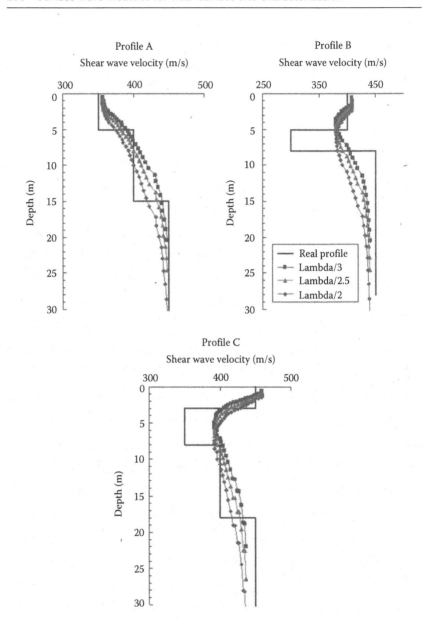

Figure 6.6 Application of the empirical inversion procedure of the SSRM to estimate the shear wave velocity at a site. (From Foti, S., Multi-station methods for geotechnical characterisation using surface waves, PhD Dissertation, Politecnico di Torino, 2000.)

acceptable in some applications without the need for implementing a rigorous inversion procedure. Moreover, it can be used as a good first guess in an iterative inversion algorithm. However, Figure 6.6 shows that the procedure is not acceptable in *inversely* dispersive soil profiles (cases B and C). Yet, it is remarked that the change of curvature exhibited by the V_S profile caused by the presence of a soft, intermediate layer (case B) or alternatively by a stiff top layer (case C) is still captured by the method.

6.3.2 Manual inversion

Empirical methods for surface wave inversion also include *trial-and-error procedures* that are in essence manual approaches to the inversion of the experimental dispersion curve. In these methods, the model parameters are successively adjusted in order to obtain a theoretical dispersion curve that matches the experimental data as best as possible. Their implementation requires the availability of an algorithm for the solution of the forward problem. In practice, a limited number of forward simulations are performed with the model parameters adjusted each time in an attempt to visually minimize the error misfit between the numerical and the experimental dispersion curves. The procedure is strongly subjective (i.e., operator dependent) and requires a certain degree of experience in order to achieve acceptable results in a reasonable amount of time.

Sometimes trial-and-error procedures represent the only viable approach in the inversion of "pathological" dispersion curves. In such situations, automatic algorithms may, in fact, get stuck and fail to converge due to instabilities in the calculation of the Jacobian needed to carry out the inversion, particularly if the latter is not performed analytically but using a discrete, finite difference scheme (Section 6.4.3).

6.4 SURFACE WAVE INVERSION BY ANALYTICAL METHODS

6.4.1 Measures of fitting goodness

A linear regression is the simplest parameter identification problem that can be conceived. This consists of finding a slope and an intercept of a straight line passing through a set of experimental data. Because there are only two model parameters (i.e., the slope and the intercept), a minimum of two measurements are needed to solve the problem. However, in practice, there is a redundancy of experimental data and the linear regression problem is *overdetermined,* which means that no exact solution can be found and the interpolation problem is converted into a *fitting* problem. The number of data is, in fact, greater than the number of model parameters, and the solution can be found only in an *approximate* sense.

The standard approach for solving this problem is that of the *least-squares method*, which consists of finding that particular value of the slope and intercept of the straight line that minimizes the sum of the squares of the individual errors. The latter are defined by the difference between the measured data and the data predicted by the linear model. The linear regression problem is then converted into a *parameter optimization problem* of finding a set of model parameters that minimize the overall misfit between the measured data and the predictions of the model represented by a straight line. If Equation 6.1 is specialized for a linear model, it becomes

$$G \cdot m = d \tag{6.4}$$

In this case, the vector-valued function operator G degenerates into a matrix, and the forward problem is represented by a linear system of algebraic equations. The error misfit (or prediction error) may then be written as follows

$$Er = \sum_{i=1}^{N} \left[d_i - \sum_{j=1}^{M} (G_{ij} m_j) \right]^2 \tag{6.5}$$

where N is the number of experimental data and G_{ij} are the elements of an N by M matrix (in this particular case, $M = 2$) where the first column is constituted of ones and the second column of the N values of the independent variable x_i at which the experimental data are measured. Finally, in this case, m_j is the component of an M-dimensional vector containing the intercept and slope of the linear model. Equation 6.5 may be interpreted as the definition of the Euclidean *norm* of a vector with components $\left[d_i - \sum_{j=1}^{M} (G_{ij} m_j) \right]$ and $i = 1, N$. So the error misfit Er may be viewed as the square of a Euclidean *distance*, and the method of least squares may be viewed as the procedure of finding the slope and the intercept of the straight line that minimizes a specific measure of distance between the measured data and the model predictions. The solution to this problem is provided by the following equation (Menke 1989)

$$m = (G^T G)^{-1} G^T d \tag{6.6}$$

Fitting a straight line is a particular type of linear regression problem characterized by only two model parameters. By allowing vector m to have an arbitrary finite size M, Equation 6.6 still represents the solution of an *over determined*, linear, discrete, inverse problem by means of the least-squares method. The issue of *stability* of the solution provided by Equation 6.6 will be investigated in Section 6.4.2.

The least-squares solution of the linear regression problem is by no means unique. Other solutions are possible introducing alternative definitions of error misfit and thus of Equation 6.5. Recalling that the L_p norm of a vector with components $\left[d_i - \sum_{j=1}^{M} (G_{ij} m_j) \right]$ is given by the relation

$$\left\| d_i - \sum_{j=1}^{M} (G_{ij} m_j) \right\|_p = \left\{ \sum_{i=1}^{N} \left[d_i - \sum_{j=1}^{M} (G_{ij} m_j) \right]^p \right\}^{\frac{1}{p}} \tag{6.7}$$

it is possible to introduce other linear models fitting a dataset in the sense of being defined by an error misfit different from the L_2 norm. Figure 6.7 shows an example of a set of measurements fitted alternatively by using the L_2 and L_1 norms. The line denoted by L_p norm corresponds to the expected trend with higher values of p. As can be seen from the figure, the dataset includes the presence of an *outlier* that is a datum that is severely discordant with the rest of the measurements; thus, it is likely to be affected

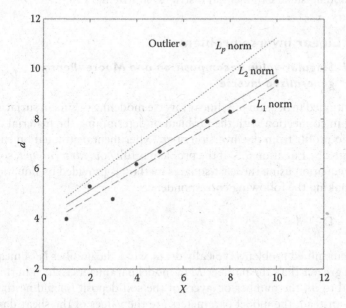

Figure 6.7 Linear regression using different measures of error misfit.

by a gross error. According to Equation 6.7, the higher the order of the norm, the greater is the weight attributed to large errors. Therefore, larger-order norms are suitable for adoption in experimental data characterized by great precision because the fact that an individual datum falls distant from a general trend is important. On the contrary, lower-order norms should be used if the data are characterized by a large uncertainty about the model, and the presence of few outliers should not affect the goodness of the fitting that much. Lower-order norms tend to give similar weight to errors of different sizes, and in this sense, they are less sensitive to the presence of outliers (Menke 1989). Figure 6.7 clearly shows this. Parameter estimation algorithms of this type are said to be *robust* (Claerbout and Muir 1973). Solution of the linear regression problem using the L_1 norm is far more involved than the standard least-squares problem. For instance, it can be carried out using the iteratively reweighted least-squares (IRLS) algorithm (Barrowdale and Roberts 1974). In general, this complexity also holds when using norms of an order higher than $p = 2$.

According to the *maximum likelihood principle*, it can be shown that the solution to the linear regression problem under the L_2 norm corresponds to the most likely solution if the data measurements are normally distributed, that is if they follow Gaussian statistics.

Likewise, by the same principle, it is possible to demonstrate that the L_1 norm solution of the linear regression problem represents the maximum likelihood estimator for data characterized by errors that are distributed like a double-sided exponential distribution (Menke 1989).

6.4.2 Linear inverse problem

6.4.2.1 Singular-value decomposition and Moore–Penrose generalized inverse

As mentioned in Section 6.2, linear inverse modeling occurs in surface wave testing in connection with the problem of determining the material damping ratio profile from the inversion of an experimental attenuation curve as illustrated by Equation 6.3. If the problem is fully *overdetermined,* solution of this equation using the least-squares method is provided by Equation 6.6 after making the following correspondence

$$\mathbf{m} = \mathbf{D}_S; \quad \mathbf{d} = \boldsymbol{\alpha}_{R/L} \tag{6.8}$$

Overdetermined problems typically occur when the number N of measured data is greater than the number M of model parameters, which in this case is equal to nl, the number of layers of the soil deposit (including the half-space). Instead, the model parameters are the values of the shear damping ratio of the individual layers.

In general, if Problem 6.4 is *even-determined, underdetermined,* or *mixed-determined,* the solution given by Equation 6.6 may no longer make sense. Indeed, if the problem is underdetermined and thus $N < M$, there are more unknowns than available equations and the problem will no longer have a unique solution. As a matter of fact, it will have infinite solutions. Given this state of affairs, it would be desirable to generalize Equation 6.6 so that it would be applicable to all possible circumstances and not only to overdetermined problems. An efficient and convenient way to obtain this generalization is through the concept of the *Moore–Penrose pseudo-* or *generalized inverse matrix* (Moore 1920; Penrose 1955), which formally extends the application of the rules of ordinary matrix algebra to the solution of the problem of finding **m** from the inversion of Equation 6.4.

The definition of the Moore–Penrose pseudoinverse of a matrix **G** requires a prior introduction of the *singular-value decomposition* (SVD), which is a particular type of eigenvalue factorization of a matrix (Strang 1988). Any N by M matrix **G** can be decomposed into the product of the following three matrices

$$\mathbf{G} = \mathbf{Q}_1 \Sigma \mathbf{Q}_2^T \tag{6.9}$$

where $\mathbf{Q}_1$ is an N by N square orthogonal matrix with columns that are unit basis vectors spanning the *data space* and $\mathbf{Q}_2$ is an M by M square orthogonal matrix with columns that are unit basis vectors spanning the *model parameter space.*[16] Finally, Σ is an N by M diagonal matrix whose non-negative diagonal elements, which are called *singular values.*

The SVD generalizes the standard eigenvalue–eigenvector factorization $\mathbf{Q}\Lambda\mathbf{Q}^T$ of a symmetric square matrix to rectangular matrices.[17] There are plenty of algorithms to efficiently compute the SVD of a matrix (Golub and Van Loan 1996). However, although numerically not very efficient, it is insightful to perform the SVD of matrix **G** by solving two standard eigenvalue problems for matrices $\mathbf{GG}^T$ and $\mathbf{G}^T\mathbf{G}$, respectively. It turns out that the eigenvectors of $\mathbf{GG}^T$ form the columns of matrix $\mathbf{Q}_1$, whereas matrix $\mathbf{Q}_2$ is formed by the eigenvectors of $\mathbf{G}^T\mathbf{G}$. Finally, the singular values of matrix Σ are obtained from the square roots of the nonzero eigenvalues of both $\mathbf{GG}^T$ and $\mathbf{G}^T\mathbf{G}$ (Strang 1988). The singular values of matrix Σ are usually arranged in order of decreasing magnitude. As will be shown in the next section, they play an important role in the construction of the Moore–Penrose generalized inverse matrix and also in controlling the stability of the inversion algorithm.

---

[16] To understand the rationale for these denominations of vector spaces, recall that in the discussion here M and N represent the number of model parameters and experimental data, respectively.

[17] In this case, $\mathbf{Q}$ is the matrix of eigenvectors and Λ is a diagonal matrix containing the eigenvalues.

If q is the *rank* of matrix **G**, this number also coincides with the number of the first nonzero (i.e., positive) singular values of matrix Σ, and the construction of the SVD of matrix **G** simplifies as follows (Menke 1989)

$$\mathbf{G} = \left(\mathbf{Q}_1\right)_q \Sigma_q \left(\mathbf{Q}_2^T\right)_q \tag{6.10}$$

where matrices $\left(\mathbf{Q}_1\right)_q$ and $\left(\mathbf{Q}_2\right)_q$ denote the first q columns of matrices $\mathbf{Q}_1$ and $\mathbf{Q}_2$, respectively. Once constructed, the SVD of matrix **G** can be used to compute the M by N Moore–Penrose generalized inverse matrix $\mathbf{G}^{-g}$ as follows

$$\mathbf{G}^{-g} = \left(\mathbf{Q}_2\right)_q \Sigma_q^{-1} \left(\mathbf{Q}_1^T\right)_q \tag{6.11}$$

Finally, the solution of the problem represented by Equation 6.4 is given by

$$\mathbf{m} = \mathbf{G}^{-g}\mathbf{d} = \left(\mathbf{Q}_2\right)_q \Sigma_q^{-1} \left(\mathbf{Q}_1^T\right)_q \mathbf{d} \tag{6.12}$$

where **m** and **d** may be interpreted in the sense of Equation 6.8. Equation 6.12 is often denoted as the *natural* solution of Equation 6.4. The Moore–Penrose generalized inverse $\mathbf{G}^{-g}$ formally acts on Equation 6.4 as if the latter would correspond to an even-determined problem with a square matrix **G**. However, Equation 6.12 represents the solution to Equation 6.4 for the most general mixed-determined problem; furthermore, it can be demonstrated that matrix $\mathbf{G}^{-g}$, and thus **m**, always exists (Aster et al. 2005). This is in contrast with the validity of Equation 6.6.

The following special cases are now examined:

- Problem 6.4 is *even-determined*, that is, $q = N = M$. Data space and model parameter space have the same dimension, thus $\mathbf{G}^{-g}$ is the inverse matrix of **G** in the ordinary sense of matrix algebra. The solution in this case is exact and unique.
- Problem 6.4 is *overdetermined*, that is, $q \leq N > M$. This means that the rank of matrix **G** coincides with the dimension of data space that is greater than the dimension of model parameter space. It can be shown that, in this case, Equation 6.12 coincides with Equation 6.6 (i.e., with the least-squares solution).
- Problem 6.4 is *underdetermined*, that is, $q \leq N < M$. This means that there are more unknowns than equations; thus, the solution is not unique and the problem is ill-posed. It can be shown that, in this case, Equation 6.12 provides a least-squares, minimum L_2 norm solution for **m**. The constraint of *minimum length* can be regarded as

supplementing Problem 6.4 with a priori information that is needed to single out one of the infinitely many solutions characterizing an underdetermined inverse problem. The minimum length solution does not always make sense physically, particularly in surface wave inversion. The next sections will further expand this concept of adding a priori information to find the solution of an ill-posed inverse problem.

In real practice, there are problems that are neither fully overdetermined nor fully underdetermined, and they are called *mixed-determined* parameter-estimation problems (Menke 1989). In geophysical prospecting and in surface wave testing, in particular, mixed-determined problems are quite common. They occur when certain portions of the domain to be investigated are fully illuminated by the seismic rays while others are left at "dark" (Figure 6.8).

In the attenuation measurement problem represented by Equation 6.3, this situation occurs if one attempts to determine the shear damping ratio of deep layers that have not been sampled by surface waves. The reason for this may be either because not enough energy at low frequencies has been generated by the seismic source or because at deep layers the motion was completely attenuated. A combination of both phenomena is of course another possibility. On the contrary, shallower layers are typically *overilluminated* due to the superposition of traveling surface waves having a broad frequency band.

The difficulties discussed here for attenuation measurements occur also in dispersion measurements with the additional complication that now the inverse problem is nonlinear, as is shown by Equation 6.2. To eliminate these inconveniences, in the forward models (Equations 6.2 and 6.3), including layers with unknown model parameters (i.e., shear wave velocity and material damping ratio) should be avoided at depths not reached by the passage of surface waves. This can be easily done by computing the skin

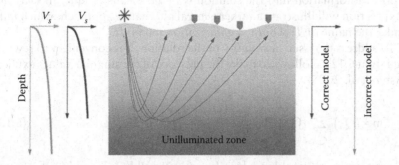

Figure 6.8 Example in geophysical prospecting of a mixed-determined problem. The shallower top layers are overdetermined because they are traversed by several seismic rays. The bottom layers are underdetermined because they are not traversed by any seismic ray. A correct model should take this into account by selecting a proper investigation depth when performing the inversion.

depth (Chapter 2) at the site under investigation from the maximum wavelength associated with the measured dispersion and attenuation curves.

The Moore–Penrose generalized inverse G^{-g} and thus Equation 6.12 make perfect sense in underdetermined as well as mixed-determined problems by providing the minimum length, least-squares solution to Problem 6.4 while properly accommodating for the various combinations of the rank of G and the size of data and model parameter spaces. However, these situations of under- and mixed-determinacy should be prevented because the solution provided by Equation 6.12, even though mathematically correct, may have undesirable features caused by the minimum length constraint appearing as a priori information to solve an otherwise indeterminate problem.

When designing a geophysical experiment and setting up the forward model for the inversion of measured data (Figure 6.8), every effort should be made to achieve redundancy and homogeneity of information throughout the spatial domain of interest so that Problem 6.4 always turns out to be overdetermined with full rank of matrix G.[18]

6.4.2.2 Instability of the solution and condition number

As stated in Section 6.1.2, according to Hadamard (1923), a mathematical problem is said to be *stable* if the solution depends continuously on the data. Existence and uniqueness are the other two conditions required by Hadamard to define a mathematical problem as well-posed. In relation to the solution of Equation 6.3, and more in general, to Equation 6.4, uniqueness is violated in underdetermined and mixed-determined problems. The strategy adopted by the Moore–Penrose generalized inverse G^{-g} to resolve this difficulty and thus enforce uniqueness in the solution represented by Equation 6.12, has been to inject a priori information into Problem 6.4 by means of the minimum length constraint.[19] This approach of introducing a priori information into the solution is by no means unique. In fact, the next section will illustrate a rather general alternative approach, which falls under the name of *Tikhonov regularization methods*.

To address the issue of *stability* of the solution, it is convenient to rewrite Equation 6.12 as follows to make the presence of the singular values explicit (Aster et al. 2005)

$$\mathbf{m} = \left(\mathbf{Q}_2\right)_q \boldsymbol{\Sigma}_q^{-1} \left(\mathbf{Q}_1^T\right)_q \mathbf{d} = \sum_{i=1}^{q} \frac{\left[\left(\mathbf{Q}_1^T\right)\right]_i \mathbf{d}}{s_i} \left[\left(\mathbf{Q}_2\right)\right]_i \tag{6.13}$$

where s_i is the ith singular value of matrix $\boldsymbol{\Sigma}$ and $[(\mathbf{Q}_1)]_i$ and $[(\mathbf{Q}_2)]_i$ denotes the ith column of matrix $\mathbf{Q}_1$ and $\mathbf{Q}_2$ respectively. The range of singular

---

[18] If N is the dimension of data space and $q = rank(G)$, then G is full rank if $q = N$.
[19] The solution obtained with this method is called the *natural* solution of Problem 6.4.

values is often called the *singular-value spectrum*, and it is important because it controls the stability of the solution of Equation 6.4 given by Equation 6.13. This equation shows that the reciprocal of the singular values s_i play the role of the coefficients of the series expansion, and the largest coefficients of the series are those associated with the smallest s_i. Thus, the solution m in Equation 6.13 is controlled de facto by the smallest singular values of matrix G. In a problem characterized by a singular-value spectrum with very small s_i, if the true data measurements $d_{true} = d$ are affected by large noise, say $d_{meas} = d_{true} + \delta$, where δ is a noise vector such that $\|\delta\|_2 \approx \|d_{true}\|_2$, then the solution m from Equation 6.13 will contain the noise terms amplified by a factor proportional to $1/s_i$.

It should be noted that the *ill-posedness*[20] and thus the instability of the solution to Problem 6.4 explicitly exhibited by Equation 6.13 is entirely independent from the accuracy of the measured data d; rather it is a property of matrix G and thus of the setting and design of the geometry of the experiment. A measure of the instability of the solution m is represented by the *condition number*, which is defined as (Aster et al. 2005)

$$cond\,(G) = \frac{s_1}{s_k} \tag{6.14}$$

where s_1 is the largest singular value of matrix G and s_k is the singular value associated with $min\,(N, M)$ that is the minimum value between the size of data and model parameter space. The larger the condition number of a G matrix, the more ill-conditioned is the problem associated with that matrix, which means the more sensitive the solution m will be to a small amount of noise in the data.

If G is not a full rank matrix,[21] then the condition number becomes infinite because s_k in Equation 6.14 would be equal to zero. However, in Equation 6.13, the solution m involves only the first nonzero singular values, which are q in number. Yet, in realistic inverse problems, there might be numerical situations where the singular values are very small but nevertheless different from zero. These singular values are the cause of severe instability in the evaluation of Equation 6.13.

A technique that can be used to mitigate this problem is to set a cutoff size by which singular values smaller than this cutoff are assumed equal to zero (Menke 1989). This operation will result in a truncation of the series represented by Equation 6.13 and as such acts like a *regularization* of the ill-posed Problem 6.4. However, dropping terms with small singular values from Equation 6.13 will determine a modification of the solution m; thus, it will worsen the fitting of m through the data d. It can be shown that

---

[20] In discrete linear problems, the *ill-posedness* is sometimes denoted as *ill-conditioning*.
[21] If G is of full rank, then $k = q = N$.

truncation of the SVD also worsens the *data and model resolution* of the problem, and this issue will be further discussed in the following section.

6.4.2.3 Tikhonov regularization methods

The SVD of matrix **G** is allowed to define a least-squares solution of Equation 6.4, which is of general validity. Simultaneously, this representation highlights the instability of this solution when the singular values s_i are very small. A "work-around" alleviation of this difficulty was the truncation of the terms in Equation 6.13 involving the smallest singular values. However, a more effective and systematic method to mitigate the ill-posedness of linear inverse problems is represented by the so-called *Tikhonov regularization* (Tikhonov and Arsenin 1977). This technique also has the advantage of very effectively addressing the nonuniqeness of the solution, which is another side of the ill-posedness of parameter identification problems.

The *zeroth-order Tikhonov regularization* consists of solving Equation 6.4 by seeking the vector of model parameter **m** and satisfying the following optimization condition

$$min \quad \left\{ \left\| \mathbf{Gm} - \mathbf{d} \right\|_2^2 + \mu^2 \left\| \mathbf{m} \right\|_2^2 \right\} \tag{6.15}$$

where μ is a *regularization parameter* also called the *Lagrange multiplier*. The solution obtained by enforcing this condition corresponds to determining the value of **m** that minimizes a combination of the *prediction error*, represented by $\left\| \mathbf{Gm} - \mathbf{d} \right\|_2^2$, and the *solution length*, represented by $\left\| \mathbf{m} \right\|_2^2$. The possible under- and/or mixed-determinacy, and therefore the lack of uniqueness in the solution of Problem 6.4, is tackled by imposing not only a reduction of the error misfit but also a property of the solution that is represented by its norm. This operation is equivalent to introducing *a priori* information to constrain the solution. In this sense, it generalizes the least-squares, minimum L_2 norm given by Equation 6.12 for underdetermined problems by allowing it to control the relative importance given to the error misfit and the simplicity of the solution represented by its L_2 norm.

As a matter of fact, the measure by which the goodness of fitting is traded in exchange for solution uniqueness is controlled by the regularization parameter μ. By setting $\mu = 0$, Equation 6.15 reduces to the same optimization condition of the standard least-squares problem whose solution is given by Equation 6.6. This solution minimizes the prediction error; however, it has all the flaws discussed earlier concerning the lack of uniqueness in under determined and mixed-determined problems. On the contrary, increasing the value of μ will significantly mitigate the ill-posedness of Problem 6.4 at the expense, however, of a simultaneous increase in the error misfit.

In practice, the choice of the optimum value to be assigned to μ is far from obvious. It is always characterized by some degree of subjectivity, although this will certainly depend on the characteristics of the specific problem at hand. An effort should be made to achieve a trade-off between two conflicting features that it is hoped the solution **m** will have: (1) the ability to fit the experimental data with the smallest possible error misfit and (2) the ability to hold a small Euclidean norm, which is a surrogate of solution simplicity.

Implementation of Equation 6.15 requires finding the minimum of an unconstrained functional; it can be achieved through the standard rules of the calculus of variations. Omitting the details, the result is (Menke 1989)

$$\mathbf{m} = (\mathbf{G}^T\mathbf{G} + \mu^2\mathbf{1})^{-1}\,\mathbf{G}^T\mathbf{d} \tag{6.16}$$

where **1** is the M by M identity matrix. Equation 6.16 is often denoted as the *damped least-squares* solution of Problem 6.4 because its potential underdeterminacy (partial or total) has been damped out. To obtain more insight into the features of the solution, it is convenient to rewrite this equation as follows using the SVD (Aster et al. 2005)

$$\mathbf{m} = \sum_{i=1}^{k} \frac{s_i^2}{s_i^2 + \mu^2} \cdot \frac{\left[\left(\mathbf{Q}_1^T\right)\right]_i \mathbf{d}}{s_i} \left[(\mathbf{Q}_2)\right]_i \tag{6.17}$$

where $k = min(N, M)$ is the minimum between the size of data space and model parameter space. It is important to note that, with this definition, all the singular values of matrix **G** are included—possibly even the very small ones.

Equations 6.17 and 6.13 differ from each other because of the presence of the *filter factors* $\frac{s_i^2}{s_i^2 + \mu^2}$ in Equation 6.17. These factors play the role of *damping* the negative influence of very small singular values that otherwise, if present, could cause a blowup of some of the terms of the expansion, as is discussed in the previous section. Clearly, from the filter factors, it is seen that the larger the regularization parameter, the greater is the damping effect and vice versa. Overall, Tikhonov regularization has restored uniqueness and stability to the solution, two properties jeopardized by the ill-posedness of Problem 6.4.

Zeroth-order Tikhonov regularization is based on minimizing a functional involving the error misfit and the square of the Euclidean norm $\mathbf{m}^T\mathbf{m}$ that was taken as an a priori measure of solution simplicity. Other measures of simplicity are possible, and in some problems they result more adequate.

Higher-order Tikhonov regularization methods are based on assuming the Euclidean norm of the first-, second-, and higher-order derivatives of **m** as a measure of solution simplicity (Aster et al. 2005). For discrete model parameters, the first and the second derivatives of **m** may be approximated by using finite difference schemes as follows

$$\begin{cases} R_1 = \|\mathbf{Lm}\|_2^2 = (\mathbf{Lm})^T \cdot (\mathbf{Lm}) \\ R_2 = \|\mathbf{L}(\mathbf{Lm})\|_2^2 = (\mathbf{L}^2\mathbf{m})^T \cdot (\mathbf{L}^2\mathbf{m}) \end{cases} \tag{6.18}$$

where **L** is an M by M real-valued matrix representing the two-point, central finite difference operator, and it is given by

$$\mathbf{L} = \begin{bmatrix} 0 & \cdots & & \\ -1 & 1 & & 0 \\ \cdots & -1 & 1 & \\ & 0 & -1 & 1 \end{bmatrix} \tag{6.19}$$

The scalars R_1 and R_2 in Equation 6.18 are two different definitions of the *roughness* of solution **m**, whereas **L** in Equation 6.19 is called the *roughening matrix*. With reference to Equation 6.3, roughness of **m** would correspond to a scalar measure of the irregularity of the shear damping ratio profile with depth. If, from geological information, the geotechnical parameters at a site are expected to vary smoothly with depth, it would make sense to solve Equation 6.3 by enforcing the minimization of roughness of the damping ratio profile.[22] In this way, solutions that satisfy the criterion of producing a small error misfit but that are too irregular will be automatically rejected by the algorithm. This is an effective way to introduce a priori information when solving a parameter identification problem. It can easily be shown that, for a continuously varying medium, the two definitions of R_1 and R_2 given by Equation 6.18 correspond to the integral over depth of the square of the first and the second derivative, respectively, of model parameter with respect to depth.

The solution to Equation 6.4 using the zeroth-order Tikhonov regularization method was found by enforcing Equation 6.15, which corresponds to determining the value of **m** that minimizes the linear functional $\|\mathbf{Gm} - \mathbf{d}\|_2^2 + \mu^2 \|\mathbf{m}\|_2^2$. In higher-order Tikhonov regularization, this condition is replaced by

$$min \quad \|\mathbf{Gm} - \mathbf{d}\|_2^2 + \mu^2 \|\mathbf{L}^n\mathbf{m}\|_2^2 \tag{6.20}$$

---

[22] Minimization of roughness of the shear damping ratio depth profile would correspond to maximization of its opposite, which is *flatness* or *smoothness*.

where $\mathbf{L}^n$ is the *nth-order roughening matrix*. For $n = 1,2$, the previous definitions of first- and second-order roughening matrices are recovered. If $n = 0$, $\mathbf{L}^0 = 1$. Then the procedure to solve Equation 6.20 becomes analogous to that used for the solution of Equation 6.15. Higher-order Tikhonov regularization methods are frequently used in geosciences because in several circumstances physical properties and geophysical and geotechnical parameters may be assumed to vary regularly with depth from the ground surface. Section 6.4.3 will describe in detail the application of first- and second-order Tikhonov regularization for the inversion of an experimental dispersion and attenuation curve to obtain the unknown shear wave velocity and shear damping ratio profiles at a site. The procedure is also known in the literature as the *Occam's algorithm*, and it will be applied to solve Equation 6.2, which is a *nonlinear* parameter identification problem.

6.4.2.4 Other regularization methods

Ill-posedness of inverse problems may be dealt with using other types of regularization methods. The objective is always the same: introduce a priori information to mitigate the instability and nonuniqeness features of the solution. Alternatives to Tikhonov regularization include *bounds constraint methods* and *total variation regularization* (Aster et al. 2005). Bounds constraint methods are based on using prior knowledge of the allowable range of variation of model parameters. For instance, with reference to Equation 6.3, damping ratio in soil layers must be nonnegative, and this poses a constraint on the lower bound that this model parameter can assume such that $\mathbf{D}_S \geq 0$. Constraint least-squares problems can effectively be solved by the *method of Lagrange* multipliers (Logan 2006). An efficient algorithm for determining the nonnegative least-squares solution is provided by Lawson and Hanson (1974).

The *total variation* regularization method is similar to first-order Tikhonov regularization in that it enforces a minimization condition similar to Equation 6.20 with $n = 1$ and the roughening matrix $\mathbf{L}$ given by Equation 6.19. However, in Equation 6.20, this method replaces the Euclidean norm with the L_1 norm. This substitution produces the effect of nonpenalizing discontinuous model parameters such as it happens with the standard Tikhonov regularization, which, in fact, favors smooth spatial variations of model parameters. Thus, the total variation method is appropriate in all those circumstances in which sharp variations and discontinuities in the model parameters are expected. One example could be represented by the presence of geological formations that have very different mechanical properties and that are separated by an abrupt interface. Like the standard Tikhonov algorithm, the total variation method still regularizes the solution of the inverse Problem 6.4; however, it performs this operation without rejecting models with possible sharp spatial variations in the model parameters. Determining $\mathbf{m}$ from Problem 6.4 by using the total variation method is not trivial because of the nondifferentiability of the L_1 norm in Equation 6.20. Yet, special algorithms have been purposely developed to overcome this difficulty (Boyd and Vandenberghe 2004).

6.4.2.5 Accuracy and resolution

Assuming the experimental data in Problem 6.4 are exempt from noise, a natural question arises about the goodness of the estimated model parameters. Let $\mathbf{G} \cdot (\mathbf{m})_{true} = (\mathbf{d})_{exp}$ denote the relation between the error-free experimental data and the *true* model parameters. Introducing this relation into the expression of the estimated model parameters $(\mathbf{m})_{est} = \mathbf{G}^{-g}(\mathbf{d})_{exp}$ yields

$$(\mathbf{m})_{est} = \mathbf{G}^{-g}(\mathbf{d})_{exp} = \mathbf{G}^{-g}\mathbf{G} \cdot (\mathbf{m})_{true} = \mathbf{R}_m(\mathbf{m})_{true} \tag{6.21}$$

where $\mathbf{R}_m = \mathbf{G}^{-g}\mathbf{G}$ identifies the M by M *model resolution matrix*. This array characterizes the bias introduced by the particular generalized inverse matrix that has been adopted to solve Problem 6.4. If $\mathbf{R}_m$ is equal to the identity matrix, then the model parameters are estimated exactly. It can be shown that in a fully overdetermined problem, $\mathbf{R}_m = 1$ (Menke 1989). In general, however, the model resolution matrix is different from the identity matrix. This is certainly true if $\mathbf{G}$ is not a full rank matrix that is $q < N$ (Aster et al. 2005). The more the diagonal elements of $\mathbf{R}_m$ are close to one, the more well resolved are the model parameters by the algorithm. Conversely, the smaller the diagonal elements, the more poorly resolved are the model parameters.

Another query that may be asked after solving Problem 6.4 is how well the estimated model parameters fit the experimental data. In other words, it is of interest to evaluate $(\mathbf{d})_{pre} = \mathbf{G} \cdot (\mathbf{m})_{est}$ using $(\mathbf{m})_{est} = \mathbf{G}^{-g}(\mathbf{d})_{exp}$. The result is

$$(\mathbf{d})_{pre} = \mathbf{G} \cdot (\mathbf{m})_{est} = \mathbf{G} \cdot \mathbf{G}^{-g}(\mathbf{d})_{exp} = \mathbf{R}_d(\mathbf{d})_{exp} \tag{6.22}$$

$\mathbf{R}_d$ is an N by N array denoted as the *data resolution matrix*. It describes the goodness of data fitting by the model. Having $\mathbf{R}_d$ equal to the identity matrix corresponds to zero prediction error, which means that the experimental data are predicted exactly. This is what happens if Problem 6.4 is fully underdetermined. In most circumstances, however, the data resolution matrix differs from the identity matrix. Thus, the prediction error is not equal to zero, and the experimental data are not perfectly resolved by the model.

Both the model and data resolution matrices describe important features of the solutions of parameter identification problems (Menke 1989). It should be noted, however, that $\mathbf{R}_m$ and $\mathbf{R}_d$ are independent from the actual data measurements and model parameters. They are only a function of the properties of $\mathbf{G}$ and $\mathbf{G}^{-g}$ and thus of the experimental setting (e.g., geometry of data acquisition system) and on the adopted inversion algorithm. Also, possible a priori information is reflected in $\mathbf{G}^{-g}$ and thus in $\mathbf{R}_m$ and $\mathbf{R}_d$. For this reason, a careful assessment of $\mathbf{R}_m$ and $\mathbf{R}_d$ may be useful in designing the experiment. With reference to surface wave testing, this would mean calculating $\mathbf{R}_m$ and $\mathbf{R}_d$ associated with attenuation measurements and thus with Equation 6.3.

The model and data resolution matrices can also be expressed in terms of the SVD of $\mathbf{G}$ and $\mathbf{G}^{-g}$. The result is (Aster et al. 2005)

$$\begin{cases} \mathbf{R}_m = (\mathbf{Q}_2)_q \cdot (\mathbf{Q}_2^T)_q \\ \mathbf{R}_d = (\mathbf{Q}_1)_q \cdot (\mathbf{Q}_1^T)_q \end{cases} \tag{6.23}$$

If $q = rank(\mathbf{G}) < M$, the model parameters are not fully resolved because $\mathbf{R}_m$ is not the identity matrix. Concerning $\mathbf{R}_d$, if $q < N$, $\mathbf{R}_d$ is not the identity matrix, and the experimental data are not perfectly resolved.

Useful scalar measures to quantify the spread of model and data resolution matrices are represented by the *Dirichlet spread functions* (Menke 1989). They are defined as the square of the Euclidean norm of the difference between $\mathbf{R}_m$ and $\mathbf{R}_d$, and the identity matrix. Formally,

$$\begin{cases} \text{spread}(\mathbf{R}_m) = \|\mathbf{R}_m - \mathbf{1}\|_2^2 \\ \text{spread}(\mathbf{R}_d) = \|\mathbf{R}_d - \mathbf{1}\|_2^2 \end{cases} \tag{6.24}$$

From this equation, the closer $\mathbf{R}_m$ and $\mathbf{R}_d$ are to the identity matrix, the smaller is the spread of these two resolution matrices.

6.4.3 Nonlinear inverse problem

6.4.3.1 Linearization by transformation of variables

Solving a nonlinear inverse problem represents a rather ordinary situation in geosciences and engineering even within the ambit of linear field and constitutive theories. The actual nonlinearity, in fact, involves the relationship between observable quantities (i.e., the measured experimental data) and a set of model parameters such as in Equation 6.1. This relationship may, of course, be nonlinear even in the context of otherwise fully linear theories. In surface wave testing, this situation occurs with the problem of determining the shear wave velocity profile from the inversion of an experimental dispersion curve in linear elastic half-spaces, as illustrated by Equation 6.2.

There are certain categories of nonlinear inverse problems that can be easily converted into corresponding linear problems through a *transformation of variables*. A typical example occurs in surface wave testing when performing attenuation measurements of particle displacement spectra (see Section 5.2).

In the uncoupled approach where dispersion and attenuation curves are measured independently, the problem of determining the experimental attenuation curve is based on the solution of the following nonlinear regression (Rix et al. 2001)

$$\left|T\left(\mathbf{r},0,\omega\right)\right|=\left|T\left(\mathbf{r},\omega\right)\right|\frac{C}{\sqrt{\mathbf{r}}}e^{-\alpha_R(\omega)\cdot\mathbf{r}} \tag{6.25}$$

where $\left|T\left(\mathbf{r},\omega\right)\right|$ is the amplitude of vertical displacement transfer function between the source and the receiver that is determined experimentally, C is a constant, $\mathbf{r}$ is the vector of receiver offsets, and $\alpha_R(\omega)$ is the unknown Rayleigh attenuation function. In Equation 6.25, higher modes of propagation are neglected.

This equation is nonlinear in the model parameter α_R and for every discrete value of the angular frequency ω_j ($j = 1$, nf), $\alpha_R(\omega_j)$ can be determined through a nonlinear regression of the measured transfer function amplitudes $\left|T\left(\mathbf{r},\omega_j\right)\right|$. Figure 6.9a shows an example of such calculation at a real testing site. The goodness of the fitting is excellent, which shows that at the frequency of 69.5 Hz the attenuation response at this site is dominated by the fundamental mode of propagation.

A quick inspection of Equation 6.25 suggests that by applying a logarithm transformation to both sides of this equation, the problem of determining the attenuation curve can be reduced to that of performing a linear regression

$$\log\left(\sqrt{\mathbf{r}}\cdot\left|T\left(\mathbf{r},\omega\right)\right|\right)=\log C-\alpha_R\left(\omega\right)\cdot\mathbf{r} \tag{6.26}$$

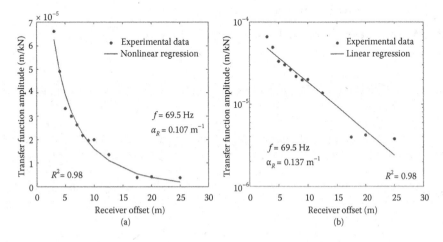

Figure 6.9 (a) Nonlinear and (b) linearly transformed regression analyses in surface wave attenuation measurements.

If r_k denotes the discrete receiver offset (with $k = 1$, np and np the number of receiver positions), Equation 6.26 can be easily recasted like Equation 6.4, which corresponds to the standard linear problem. The equivalence is shown here:

$$
G \cdot m = d = \begin{bmatrix} 1 & -r_1 \\ 1 & -r_2 \\ \cdots & \cdots \\ \cdots & -r_i \\ 1 & \cdots \\ 1 & -r_{np} \end{bmatrix} \cdot \begin{bmatrix} \log C_j \\ \alpha_R(\omega_j) \end{bmatrix} = d \qquad (6.27)
$$

where d is the data measurement vector and is given by

$$
d = \left[\sqrt{r_1} \cdot |T(r_1, \omega_j)| \ \ \sqrt{r_2} \cdot |T(r_2, \omega_j)| \ \cdots \ \sqrt{r_i} \cdot |T(r_i, \omega_j)| \ \cdots \ \sqrt{r_{np}} \cdot |T(r_{np}, \omega_j)| \right]^T
$$
$$(6.28)$$

Figure 6.9b shows the linear regression of the same data illustrated in Figure 6.9a after performing the linearization described by Equations 6.27 and 6.28. The attenuation coefficient obtained with the linear regression is about 30% greater than the one calculated using Equation 6.25. Obviously, the two procedures are not equivalent because they lead to different results. Also the goodness of the fitting is slightly dissimilar.

This example shows that although it sometimes may be convenient to apply the method of transformation of variables to linearize a nonlinear inverse problem, this procedure should be used with caution because it always introduces a bias in the solution (Menke 1989). This is particularly relevant if the uncertainty of the data has to be taken into account (Section 6.5). In linear regressions, the rigorous use of the least-squares method in the presence of uncertainty requires that data measurements are assumed to be uncorrelated, normally distributed (or Gaussian) random variables with uniform variance (Section 6.5.1). If the statistics of the data, represented, for instance, by Equation 6.28, is back-transformed into the original variables due to the nonlinearity of Equation 6.25, the original measurements will no longer be normally distributed. In addition, the fact that the data in Equation 6.28 were assumed characterized by uniform variance implies that the transfer function amplitudes were measured with an accuracy that increased with the distance from the source. This is exactly the opposite of what actually happens in reality. Because the signal-to-noise ratio tends to deteriorate with the increase of r due to the spatial, exponential decay of the displacement amplitude spectra, the accuracy of the measurements progressively decreases with the increase in the distance from the source.

6.4.3.2 LS iterative methods and GS techniques

A natural procedure for solving Equation 6.1 is to expand this equation in a Taylor's series about an initial guess of model parameters $\mathbf{m}_0$

$$\mathbf{G(m)} = \mathbf{G(m_0)} + \mathbf{J(m)}_{\mathbf{m}_0} \cdot (\mathbf{m} - \mathbf{m}_0) + o \|(\mathbf{m} - \mathbf{m}_0)\|_2^2 \qquad (6.29)$$

where $\mathbf{G(m_0)}$ denotes the vector-valued function $\mathbf{G}$ of Equation 6.1 calculated for $\mathbf{m} = \mathbf{m}_0$ and $\mathbf{J(m)}_{\mathbf{m}_0}$ is the N by M *Jacobian matrix* calculated for $\mathbf{m} = \mathbf{m}_0$, and it is defined by

$$\mathbf{J(m)}_{\mathbf{m}_0} = \left[grad\, \mathbf{G(m)} \right]_{\mathbf{m}_0} \qquad (6.30)$$

where $i = 1, N$ and $j = 1, M$. By neglecting terms higher than the first order, Equation 6.29 simplifies as follows

$$\mathbf{J(m)}_{\mathbf{m}_0} \cdot \mathbf{m} = \mathbf{J(m)}_{\mathbf{m}_0} \cdot \mathbf{m}_0 + \left[\mathbf{d} - \mathbf{G(m_0)} \right] \qquad (6.31)$$

Equation 6.31 represents a linear problem analogous to Equation 6.4 because all terms but $\mathbf{m}$ are known. Thus, it could be used as a basis for the application of any of the methods discussed in Section 6.4.2 for the solution of the linear inverse problem, including the least-squares algorithm.

The result will be a new estimate $\mathbf{m}_1$ of the unknown vector of model parameters. The procedure can then be repeated, and Equation 6.31 will provide a sequence $\{\mathbf{m}_0, \mathbf{m}_1, \mathbf{m}_2, \dots \mathbf{m}_k, \dots \mathbf{m}_n\}$ of successive approximations of the model parameters that under appropriate circumstances and conditions will converge toward the desired vector $\mathbf{m}_{true}$ of *true* model parameters. For a fully nonlinear, overdetermined problem, the recursive equation for the least-squares method can be written as follows

$$\mathbf{m}_{k+1} = \left(\mathbf{J}_k^T \mathbf{J}_k \right)^{-1} \mathbf{J}_k^T \cdot \left\{ \mathbf{J}_k \cdot \mathbf{m}_k + \left[\mathbf{d} - \mathbf{G(m_k)} \right] \right\} \qquad (6.32)$$

More refined iterative methods exist to solve the least-squares nonlinear inverse problem including the downhill simplex method and various types of gradient techniques such as the method of steepest descent, conjugate gradient, the Gauss–Newton method, the Levenberg–Marquardt algorithm. The latter is the method of choice for small- to medium-size nonlinear least-squares problems (Aster et al. 2005).

Although a nontrivial issue related to iterative methods has to do with the definition of a criterion for deciding when to terminate the iterations, the two major difficulties associated with the iterative solution of inverse problems are convergence and uniqueness of the solution. For instance, there is no guarantee that the iterative scheme specified by Equation 6.32 converges, for a sufficiently large k, to the "true" solution. Even if it does, the solution may not be unique even in overdetermined problems.

In linear inverse problems (Section 6.4.3), the situation was far more favorable. First, there was no need to introduce iterative strategies to determine the solution. Second, depending on the degree of indeterminacy of the problem at hand, a unique solution could always have been identified, if needed, by supplementing the problem with a priori information on the solution's simplicity. All these approaches are not guaranteed to work in nonlinear inverse problems. The ill-posedness that characterizes almost any inverse problem becomes severe and sometimes pathological if the problem is nonlinear.

Figure 6.10 shows a conceptual plot of the prediction error, defined according to Equation 6.5, as a function of model parameter for the case of a nonlinear (panel a) and linear (panel b) inverse problem. Solution uniqueness for the linear problem is guaranteed by the existence of a single minimum of the prediction error function the shape of which, in the most general case, is a multidimensional paraboloid.[23] If the problem is nonlinear, the error hypersurface may be nonconvex; thus, it may have multiple minima, and there is a need to distinguish between a *local* minimum and a *global* minimum (Figure 6.10a).

The "true" solution would correspond to the global minimum of the error hypersurface; however, depending on the initial guess introduced in the iterative scheme, the algorithm may actually converge and provide a solution corresponding to a local minimum. Thus, the main problem

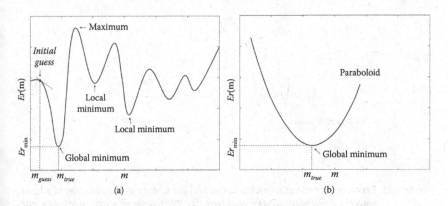

Figure 6.10 Prediction error as a function of model parameter in the solution of a least-squares (a) nonlinear and (b) linear inverse problem.

---

[23] For ease of graphical representation, in Figure 6.10, the prediction error function has been assumed to be a function of only a single model parameter.

associated with *LS iterative methods* (Figure 6.3) is the choice of an initial guess that is sufficiently close to the global minimum. This practically may be very difficult due to the extension of the solution space coupled with the objective difficulty of anticipating the position of a neighborhood of a global minimum. A wrong initial guess may lead to a failure of the iterative process represented, for instance, by Equation 6.32, and even a local minimum may not be found. As a consequence, the convergence of an LS iterative method will strongly depend on the geometry of the prediction error hypersurface (Menke 1989). Figure 6.11 shows four examples of increasing degrees of complexity that may be encountered in solving nonlinear inverse problems, from a single, well-defined

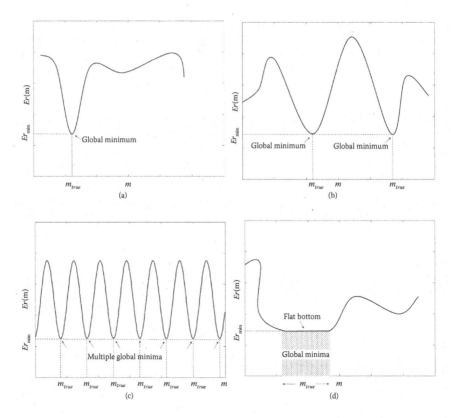

Figure 6.11 Prediction error as a function of model parameter in the solution of a least-squares nonlinear inverse problem. (a) Well-defined single minimum corresponding to a unique solution. (b) Two well-separated minima with lack of uniqueness in the solution. (c) Infinite well-separated countable minima with severe lack of uniqueness in the solution due to periodicity of the prediction error function. (d) "Flat bottom" having an uncountable finite range of solutions with severe lack of uniqueness due to ill-conditioning of the prediction error function.

global minimum (Figure 6.11a) to a "flat bottom" with an uncountable, finite range of solutions (Figure 6.11d).

GS methods (Figure 6.3) have been purposely developed to overcome these situations; in recent years, a great deal of research that is still very active has been carried out in this field. The overall subject falls under the name of global optimization (Sen and Stoffa 1995; Horst et al. 2000). Robust deterministic—and for large-size problems, stochastic—algorithms have been created to find the "true" solution of nonlinear inverse problems after scanning the whole error hypersurface in a search for the global minimum.

A conceptually simple GS method is represented by the so-called multistart strategy (Aster et al. 2005), which consists of randomly generating a large number of initial guess solutions to any one, of which a LS iterative algorithm such as the Gauss–Newton or the Levenberg–Marquardt is applied. The set of the local minima found by the method is then compared to find the solution corresponding to the global minimum. For the surface wave inverse problem represented by Equation 6.2, a set of initial guess solutions would be given by a series of initial profiles of shear wave velocity.

Multistart methods are efficient GS procedures because they can take advantage of the typical fast converging properties of LS iterative algorithms. Other global optimization techniques include simulated annealing, evolutionary algorithms, fractal inversion, enumerative methods, and, of course, Monte Carlo simulation. Although the "philosophy" adopted in each of these techniques is peculiar and different from each of the others, all aim at finding the global minimum of the prediction error function without being trapped in local minima.

GS methods have already been successfully applied in solving the inverse problem associated with surface waves. Yamanaka and Ishida (1996) have developed a genetic algorithm for the inversion of surface wave dispersion data. Although the application is concerned with the definition of the Earth crust model from the inversion of dispersion curves obtained from observed earthquake data, the algorithm could also be used for near-surface site characterization. By mimicking population genetics and processes such as selection, crossover, and mutation, genetic algorithms can simultaneously search globally and locally for the minimum of the prediction error function in the solution space by using several forward models. The authors of this paper developed a genetic algorithm to invert short- and intermediate-period surface wave dispersion data to obtain a shear wave velocity profile for a deep sedimentary basin.

Direct applications of genetic algorithms (Pezeshk and Zarrabi 2005; Dal Moro et al. 2007), simulated annealing (Beaty et al. 2002), and Monte Carlo simulations (Socco and Boiero 2008; Maraschini and Foti 2010; Bergamo et al. 2011) for near-surface site characterization using surface waves have also been done. It is an active field of research; the contributions are countless and continue to be produced. Only a few of them have been cited here.

Although artificial neural network inversion cannot be classified as a GS method, it also represents an alternative to LS techniques. Meier and Rix (1993) have proposed the use of an artificial neural network as an expeditious alternative to the trial-and-error and least-squares surface wave inversion techniques. About 99,000 synthetic dispersion curves were calculated for randomly generated two-layer shear wave velocity profiles using a theoretical wave propagation algorithm. An artificial neural network was then "taught" to map these dispersion curves back into their respective shear wave velocity profiles. After the network was successfully trained on these synthetic dispersion curves, simulated experimental dispersion curves were inverted by passing them through the neural network. Because the neural network required only a single forward pass of the data, the algorithm performed the inversion much more quickly than the standard iterative procedures. Other applications of the artificial neural network algorithm to solve the surface wave inverse problem include the works by Kim and Xu (2000) and Shirazi et al. (2009).

Similar to linear problems, the nonuniqeness of nonlinear inverse problems may also be due to the underdeterminacy or mixed-determinacy of Equation 6.1. This happens when the number of model parameters increases considerably with respect to the number of the experimental data (Aster et al. 2005). These situations may be resolved by applying regularization methods, specifically the zeroth and higher-order Tikhonov regularization that were introduced for the linear case. Formally, Equations 6.15 and 6.20 remain valid as long as the matrix product $\mathbf{Gm}$ is replaced by the nonlinear vector-valued function operator $\mathbf{G(m)}$. The same occurs for the solutions represented by Equations 6.16 and 6.17 on the condition that they are intended in a recursive sense such as it was in Equation 6.32 for the least-squares method. Once the regularization has been implemented using, for instance, the variational formulation, the actual solution can then be obtained using standard least-squares, LS iterative methods (e.g., the Levenberg–Marquardt algorithm), or GS techniques.

6.4.3.3 Analytical versus numerical Jacobian

Most least-squares, LS iterative methods require the computation of the *Jacobian* of $\mathbf{G(m)}$ with respect to model parameters $\mathbf{m}$ as shown by Equation 6.30. Depending on the problem, the partial derivatives involved in the calculation of the Jacobian may be available in closed-form, that is, analytically, or they must be computed numerically using a finite difference scheme. Numerical differentiation itself is known to be an ill-conditioned problem that introduces inaccuracies and instabilities in the implementation of the iterative scheme represented by equations such as Equation 6.32.

A critical issue in the calculation of finite difference partial derivatives is the correct choice for the step of the perturbed model parameters δm. It must be not too small to avoid round-off errors in the term at the numerator $G(m + \delta m) - G(m)$ of the finite difference, and simultaneously, it must not be too large or the calculation may become too inaccurate. A guiding principle is to set $\delta m_j = \sqrt{\varepsilon}$ ($j = 1, M$), where ε is the accuracy in the evaluations of $G(m)$ (Aster et al. 2005). Instability and lack of convergence of an algorithm implementing a nonlinear inversion may be caused by imprecise calculation of the Jacobian matrix of $G(m)$. The other problem related to computing partial derivatives numerically is that it is computationally very inefficient and time-consuming if compared with using the corresponding explicit formulas.

In light of this, it is obviously preferable when solving nonlinear inverse problems with LS iterative methods to use closed-form, analytical expressions of the partial derivatives of $G(m)$ with respect to model parameters m entering the Jacobian matrix. However, they are not always available, and sometimes the recourse to numerical differentiation is compulsory. With regard to the solution of the nonlinear surface wave problem represented by Equation 6.2, the situation is favorable because exact formulas for computing the partial derivatives of $G(V_S)$ with respect to V_S do exist. They can be derived from the variational principles of Love and Rayleigh waves (Aki and Richards 2002), and results have been obtained for computing the Jacobian of modal and apparent surface wave phase velocity with respect to medium parameters V_S and V_P. For a stratified medium composed of a finite number of homogeneous layers overlaying a homogeneous half-space, the partial derivatives of the Rayleigh phase velocity V_R with respect to changes of $(V_S)_i$ and $(V_P)_i$ of the ith layer ($i = 1, nl$) are given by the following relations

$$
\begin{cases}
\left(\dfrac{\partial V_R}{\partial V_S} \right)_i = \dfrac{(\rho V_S)_i}{2k^2 U_R I_R} \cdot \int_{x_2^i}^{x_2^{i+1}} \left[\left(k r_2 - \dfrac{dr_1}{dx_2} \right)^2 - 4k r_1 \dfrac{dr_2}{dx_2} \right] dx_2 \\[6mm]
\left(\dfrac{\partial V_R}{\partial V_P} \right)_i = \dfrac{(\rho V_P)_i}{2k^2 U_R I_R} \cdot \int_{x_2^i}^{x_2^{i+1}} \left(k r_1 + \dfrac{dr_2}{dx_2} \right)^2 dx_2
\end{cases}
\tag{6.33}
$$

where the brackets with a subscript i denote quantities evaluated in the ith layer. The limits of integration indicate the lower and upper bounds of the depth variable x_2 for the ith layer. See Chapter 2 for the definitions of the other terms appearing in Equation 6.33. The derivatives of the eigenfunctions r_1 and r_2 with respect to depth, appearing inside the integral sign, can be computed directly from the knowledge of the eigenfunctions (see Equation 2.66). Details for the derivation of Equation 6.33 are reported in Aki and Richards (2002) and Lai (2005).

The remarkable feature of these equations, which makes them so important in the solution of the Rayleigh inverse problem, is that the partial derivatives of V_R with respect to medium parameters can be computed using Rayleigh wave parameters referred to the original and not the perturbed V_S and V_P profiles. Conversely, it would be very expensive to compute these partial derivatives numerically with, say, a four-point central finite difference scheme; a single computation of a derivative would require the solution of four Rayleigh eigenproblems instead of just one eigenproblem using Equation 6.33.

It can be shown (Lee and Solomon 1979; Ben-Menahem and Singh 2000) from a detailed assessment of Equation 6.33 that the phase velocity of Rayleigh waves is relatively insensitive to changes in the parameter V_P; thus, the corresponding partial derivative is small if compared to the derivative of V_R with respect to V_S. This result is important for the solution of the Rayleigh inverse problem because it implies that inverting an equation analogous to Equation 6.2 for V_P would be a severe ill-posed problem. In fact, relatively large changes in V_P would correspond to small changes in V_R and, thus, in the prediction error function of Equation 6.5, a situation graphically described by the "flat bottom" shown in Figure 6.11d. For this reason, either the V_P profile or the Poisson ratio profile at a site is typically assumed as known in the inversion of an experimental dispersion curve.

Equation 6.33 has been derived and, therefore, is valid for a stratified medium, that is, for a discrete model of the subsoil. Similar relations have been derived by Rix and Lai (2014) for a subsurface model in which the material properties vary continuously with depth. Once the explicit formulas for the partial derivatives of V_R with respect to the medium parameters have been calculated, they can be introduced in Equation 6.30 to form the Jacobian matrix required to solve the inverse Rayleigh problem[24] with any of the LS iterative methods discussed in the previous section.

6.4.3.4 An example of a LS iterative method: Occam's algorithm

This section will describe in detail the application of Tikhonov regularization for the inversion of an experimental dispersion and attenuation curve to obtain the unknown shear wave velocity and shear damping profiles at a site. The procedure is known in the literature as the *Occam's algorithm*, and it has been introduced in geophysics by Constable et al. (1987) for the inversion of electromagnetic sounding data.

The name of the algorithm refers to William of Ockham, a fourteenth-century English philosopher considered the father of a philosophical

---

[24] Explicit formulas analogous to Equation 6.33 also hold for the Love waves. For details, see Aki and Richards (2002).

principle called the *Occam's razor* according to which parsimony and conciseness should be the guiding principles to be adopted in the interpretation of natural phenomena and in the formulation of new theories. He wrote that "it is vain to do with more what can be done with fewer" (Russell 1946). Occam's razor also has become an inspiring principle in today's science because the assumptions of a theory should not be unnecessarily numerous and complicated.

Constable et al. (1987) have adopted this principle to formulate what in essence is a constrained, damped least-squares, LS iterative method for the solution of a nonlinear inverse problem. This algorithm will now be applied to the inversion of Equation 6.2 written in this form

$$G^* \left(V_S^* \right) = V_R^* \tag{6.34}$$

where the "asterisk" notation refers to the complex-valued nature of the variables involved. It was shown in Section 2.5.3 that in linear dissipative media, phase velocity and attenuation of viscoelastic waves are not independent; furthermore, they are frequency dependent as a result of material dispersion. Thus, a correct inversion procedure of surface wave data should determine these two parameters jointly, at the same frequency of excitation and for arbitrary values of material damping ratio (Lai et al. 2002).

A natural way to construct an inversion algorithm satisfying these requirements is to start from the solution of the forward problem of surface waves in strongly dissipative, layered media illustrated in Section 2.5.3. In the frequency domain, a linear viscoelastic, arbitrary dissipative, isotropic material has its constitutive parameters completely defined by the complex-valued phase velocity of bulk P- and S-waves given by Equation 2.125 and rewritten here for convenience

$$V_\chi^*(\omega) = \frac{V_\chi(\omega)}{\sqrt{[1+4D_\chi^2(\omega)]}} \cdot \left[\frac{1 + \sqrt{[1+4D_\chi^2(\omega)]}}{2} + i \cdot D_\chi \right] \tag{6.35}$$

with $\chi = P, S$ and by the Kramers–Krönig relations (Equations 2.130, 2.131, or 2.132) specifying the functional dependence between material damping ratio $D_\chi(\omega)$ and phase velocity $V_\chi(\omega)$. In Equation 6.35, $i = \sqrt{-1}$ is the imaginary unit. The formalism of complex-valued bulk wave velocities allows two fundamental model parameters of the medium (the shear wave velocity and the shear damping ratio) to group together in a single parameter.

An algorithm for the strongly coupled, joint inversion of Rayleigh dispersion and attenuation data may be constructed by viewing the complex-valued Rayleigh phase velocity as a *holomorphic* (or *analytic*) *mapping* (Remmert 1997) of the complex-valued speed of propagation of shear waves of the medium. Formally, this mapping is represented by Equation 6.34, where $\mathbf{V}_S^* = \left[\left(V_S^*\right)_1, \left(V_S^*\right)_2, \ldots \left(V_S^*\right)_i \ldots \left(V_S^*\right)_{nl}\right]$ and $\mathbf{V}_R^* = \left[\left(V_R^*\right)_1, \left(V_R^*\right)_2, \ldots \left(V_R^*\right)_j \ldots \left(V_R^*\right)_{nf}\right]$. Again the symbols nl and nf denote the number of layers (including the half-space) of a stratified medium and the set of experimental frequencies at which the Rayleigh phase velocity has been measured, respectively.

Equation 6.34 represents the Rayleigh forward problem in linear visco-elastic media. The complex-valued Rayleigh phase velocity vector $\mathbf{V}_R^*$ jointly encloses the dispersion and attenuation (discrete) curves. It is computed by combining Equation 6.35 and the following relation (Lai et al. 2002)

$$D_R(\omega) = \left[\frac{\dfrac{\alpha_R \cdot V_R}{\omega}}{1 - \left(\dfrac{\alpha_R \cdot V_R}{\omega}\right)^2}\right] \tag{6.36}$$

where $\alpha_R = \alpha_R(\omega)$ is the frequency-dependent Rayleigh attenuation coefficient.

For the solution of the Rayleigh inverse problem, Equation 6.34 can be expanded in a *Taylor series* about an initial guess of the model parameter vector $\mathbf{V}_{S_0}^*$, thereby obtaining

$$\mathbf{G}^*\left(\mathbf{V}_S^*\right) = \mathbf{G}^*\left(\mathbf{V}_{S_0}^*\right) + \mathbf{J}^*\left(\mathbf{V}_S^*\right)_{\mathbf{V}_{S_0}^*} \cdot \left(\mathbf{V}_S^* - \mathbf{V}_{S_0}^*\right) + o\left\|\left(\mathbf{V}_S^* - \mathbf{V}_{S_0}^*\right)\right\|_2^2 \tag{6.37}$$

where... $\|...\|_2$ denotes the Euclidean norm of a complex-valued vector[25] and $\mathbf{G}^*\left(\mathbf{V}_{S_0}^*\right) = \mathbf{V}_{R_0}^*$ is the nf by 1 vector of Rayleigh phase velocities corresponding to the solution of Equation 6.34 with model parameter equal to $\mathbf{V}_{S_0}^*$. The term $\mathbf{J}^*\left(\mathbf{V}_S^*\right)_{\mathbf{V}_{S_0}^*}$ represents the nf by nl complex-valued Jacobian matrix calculated for $\mathbf{V}_S^* = \mathbf{V}_{S_0}^*$, and it is defined by the following expression

$$\mathbf{J}^*\left(\mathbf{V}_S^*\right)_{\mathbf{V}_{S_0}^*} = \left[grad\ \mathbf{G}^*\left(\mathbf{V}_S^*\right)\right]_{\mathbf{V}_{S_0}^*} \tag{6.38}$$

The elements of the Jacobian matrix are defined as follows

$$\left[\left(J^*\right)_{jk}\right]_{\mathbf{V}_{S_0}^*} = \left[\frac{\partial\left[\mathbf{G}^*\left(\mathbf{V}_S^*\right)\right]_j}{\partial\left(\mathbf{V}_S^*\right)_k}\right]_{\mathbf{V}_{S_0}^*} = \left[\frac{\partial\left(V_R^*\right)_j}{\partial\left(V_S^*\right)_k}\right]_{\mathbf{V}_{S_0}^*} \tag{6.39}$$

---

[25] The Euclidean norm of a complex-valued vector $\mathbf{v}^*$ is given by the square root of the product of vector $\mathbf{v}^*$ by vector $(\mathbf{v}^*)^H$, which is the transpose of the complex conjugate of $\mathbf{v}^*$. The vector $(\mathbf{v}^*)^H$ is also called the *Hermitian* transpose of *vector* $\mathbf{v}^*$.

where $j = 1, nf$ and $k = 1, nl$. The subscripts outside the brackets of Equation 6.39 indicate the value of the model parameter at which the Jacobian matrix is evaluated. The partial derivatives of the Jacobian are calculated using Equation 6.33 extended by *analytic continuation* to complex values of field variables.

Neglecting terms higher than the first order, Equation 6.37 reduces to

$$\mathbf{J}^*\left(\mathbf{V}_S^*\right)_{\mathbf{V}_{S_0}^*} \cdot \mathbf{V}_S^* = \mathbf{J}^*\left(\mathbf{V}_S^*\right)_{\mathbf{V}_{S_0}^*} \cdot \mathbf{V}_{S_0}^* + \left[\mathbf{G}^*\left(\mathbf{V}_S^*\right) - \mathbf{G}^*\left(\mathbf{V}_{S_0}^*\right)\right]$$

$$= \mathbf{J}^*\left(\mathbf{V}_S^*\right)_{\mathbf{V}_{S_0}^*} \cdot \mathbf{V}_{S_0}^* + \left(\mathbf{V}_R^* - \mathbf{V}_{R_0}^*\right)$$

(6.40)

Equation 6.40 could be used as a basis for the implementation of a standard least-squares algorithm to determine the complex-valued shear wave velocity profile of a layered medium possibly subjected, in under- and mixed-determined problems, to the minimum norm constraint. However, as discussed extensively earlier, inversions performed with algorithms where the solution is constrained to have the minimum norm are often inadequate, and they may lead to physically unreasonable profiles of model parameters (Constable et al. 1987). The inadequacy of this class of algorithms often can be attributed to a lack of physical justification for assuming the minimum norm constraint.

A more reasonable approach for constraining the solution is represented by the *Occam's inversion* (Parker 1994), the strategy of which is to find the *smoothest* profile of model parameters subjected to the constraint that the error misfit between observed and predicted data cannot exceed a prescribed value. In the current problem, the profile of model parameters is represented by the vector $\mathbf{V}_S^*$ containing the complex-valued shear wave velocities of the individual layers. Because the number of layers nl is generally assumed, the inverted profile of model parameters will depend on the a priori assumption about nl, and it may contain large discontinuities or other features that are not essential for matching the experimental dispersion and the attenuation curves. By enforcing maximum smoothness and regularity in the solution, it is possible to mitigate its dependence upon the assumed number of layers and at the same time reject solutions that are unnecessarily complicated (Constable et al. 1987).

Implementation of the algorithm requires a quantitative definition of *smoothness* of a profile of model parameters. This definition, actually its opposite *roughness*, has already been introduced in Section 6.4.2 when discussing higher-order Tikhonov regularization methods. Equations 6.18 and 6.19 provide two scalar definitions, R_1 and R_2, of the *roughness* of a discrete profile of model parameters. However, they are applicable only for real-valued model parameters. For complex-valued model parameters such as $\mathbf{V}_S^*$, roughness may be defined by either one of the two following expressions (Menke 1989; Lai 2005)

$$\begin{cases} R_1 = \left\| \mathbf{L}\mathbf{V}_S^* \right\|_2^2 = \left(\mathbf{L}\mathbf{V}_S^*\right)^H \cdot \left(\mathbf{L}\mathbf{V}_S^*\right) \\[2mm] R_2 = \left\| \mathbf{L}\left(\mathbf{L}\mathbf{V}_S^*\right) \right\|_2^2 = \left\| \mathbf{L}^2\mathbf{V}_S^* \right\|_2^2 = \left(\mathbf{L}^2\mathbf{V}_S^*\right)^H \cdot \left(\mathbf{L}^2\mathbf{V}_S^*\right) \end{cases} \tag{6.41}$$

where $\mathbf{L}$ is the nl by nl real-valued matrix representing the two-point central finite difference operator defined in Equation 6.19, and the symbol $(...)^H$ denotes the Hermitian transpose of a complex-valued matrix.

The prediction error misfit Er between the measured and the predicted complex-valued Rayleigh phase velocities may be written as follows

$$Er = \left[\mathbf{W}^*\overline{\mathbf{V}}_R^* - \mathbf{W}^*\mathbf{G}^*\left(\mathbf{V}_S^*\right) \right]^H \cdot \left[\mathbf{W}^*\overline{\mathbf{V}}_R^* - \mathbf{W}^*\mathbf{G}^*\left(\mathbf{V}_S^*\right) \right] \tag{6.42}$$

where $\overline{\mathbf{V}}_R^*$ is a complex-valued vector of experimental Rayleigh phase velocities, and $\mathbf{W}^*$ is a complex-valued diagonal nf by nf *weight matrix* defined by

$$\mathbf{W}^* = \begin{bmatrix} 1/\sigma_1^* & ... & 0 & ... & ... & 0 \\ 0 & 1/\sigma_2^* & ... & 0 & ... & 0 \\ 0 & ... & ... & ... & 0 & 0 \\ 0 & 0 & ... & 1/\sigma_j^* & ... & 0 \\ 0 & ... & 0 & ...^\bullet & ... & 0 \\ 0 & ... & ... & 0 & ... & 1/\sigma_{nf}^* \end{bmatrix} \tag{6.43}$$

with σ_j^*, $j = 1, nf$ being the *uncertainties* associated with the experimental data $\overline{\mathbf{V}}_R^*$ (see Section 6.5.2 to learn how they can be estimated). Equation 6.42 defines a *weighted* measure of the error prediction misfit. This definition is able to accommodate the fact that, in real practice, some experimental measurements may be more accurate than others; thus, it would make more sense to assign to these measurements a greater weight in the overall estimate of the error misfit (Menke 1989). If $\mathbf{W}^* = 1$, that is the weight matrix is equal to the identity matrix, then Equation 6.42 reduces to the standard definition of error misfit given by Equation 6.5.

Solution of the *linearized* Rayleigh inverse problem represented by Equation 6.40 with the Occam's algorithm consists of finding a vector $\mathbf{V}_S^*$ that minimizes R_1 and R_2 defined by Equation 6.41 while constraining the residual error to be equal to $\hat{E}r$, an acceptable value in light of the uncertainties. This setting of the problem corresponds to the iterative application of the first- and second-order Tikhonov regularization (Section 6.4.2) to a locally linearized nonlinear inverse problem. The method of Lagrange multipliers is employed to solve this complex-valued, constrained optimization problem resulting in

$$
\mathbf{V}_S^* = \left\{ \mu\left(\mathbf{L}^T\mathbf{L}\right) + \left[\mathbf{W}^* \cdot \left(\mathbf{J}^*\right)_{\mathbf{V}_{S0}^*}\right]^H \cdot \left[\mathbf{W}^* \cdot \left(\mathbf{J}^*\right)_{\mathbf{V}_{S0}^*}\right] \right\}^{-1}
$$
$$
\cdot \left[\mathbf{W}^* \cdot \left(\mathbf{J}^*\right)_{\mathbf{V}_{S0}^*}\right]^H \cdot \mathbf{W}^* \cdot \left[\left(\mathbf{J}^*\right)_{\mathbf{V}_{S0}^*} \cdot \mathbf{V}_{S0}^* + \left(\bar{\mathbf{V}}_R^* - \mathbf{V}_{R0}^*\right)\right]
$$

(6.44)

In Equation 6.44, μ is the Lagrange multiplier, which may be viewed as a *smoothing parameter* and must be determined with the additional constraint that the specified residual error $\hat{E}r$ is matched with a vector $\mathbf{V}_S^*$ composed only of negative imaginary parts. This condition will insure that the shear damping ratio $\mathbf{D}_S$ obtained from the inversion algorithm always turns out to be a positive quantity.

Equation 6.44 is used iteratively to refine the estimated complex-valued shear wave velocity profile $\mathbf{V}_S^*$ until convergence. Once the unknown complex-valued vector $\mathbf{V}_S^*$ is estimated, the real-valued shear wave velocities $(V_S)_k$ and shear damping ratios $(D_S)_k$, $(k = 1, nl)$ of the nl-layer medium can be recovered using the following relations, obtained from Equation 6.35

$$
\begin{cases}
(V_S)_k = \left[V_S(x_S, y_S)\right]_k = \left[\dfrac{(x_S^2 + y_S^2)}{x_S}\right]_k \\[4mm]
(D_S)_k = \left[D_S(x_S, y_S)\right]_k = \left[\dfrac{x_S \cdot y_S}{(x_S^2 - y_S^2)}\right]_k
\end{cases}
$$

(6.45)

where $(x_S)_k = \left[\Re\left(V_S^*\right)\right]_k$ and $(y_S)_k = \left[\Im\left(V_S^*\right)\right]_k$ are the real and the imaginary parts of $\left(V_S^*\right)_k$, respectively, which is the complex-valued shear wave velocity of the kth layer.

In this derivation, no specific assumption was made about the frequency-dependence law of the complex-valued vector $\mathbf{V}_S^*$. A standard technique used to account for material dispersion is to carry out the inversion procedure at a prescribed *reference frequency* often assumed equal to $\omega_{ref} = 2\pi$ (Lee and Solomon 1979; Herrmann 2007). The procedure can then be used to perform a *causal inversion* of the Rayleigh dispersion and attenuation curves provided that the partial derivatives appearing in Equation 6.39 are computed with respect to $\left(V_{S(ref)}^*\right)_k$ and the vector $\mathbf{V}_S^*$ in Equation 6.44 is replaced by $\mathbf{V}_{S(ref)}^* = \mathbf{V}_S^*\left(\omega_{ref}\right)$.

Figures 6.12 and 6.13 show the results obtained from the application of the Occam's algorithm to a pair of synthetic dispersion and attenuation curves generated using the medium parameters of Table 6.1 (Lai 2005).

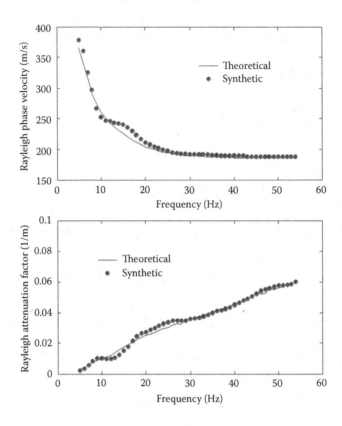

Figure 6.12 Comparison between synthetic Rayleigh dispersion and attenuation curves generated using the medium parameters of Table 6.1 and the corresponding theoretical curves (fundamental mode of propagation) obtained by employing the constrained, damped least-squares Occam's algorithm. (From Lai, C. G., Surface waves in dissipative media: Forward and inverse modelling, *Surface Waves in Geomechanics—Direct and Inverse Modelling for Soil and Rocks*, Springer-Verlag, New York, 2005.).

The synthetic response functions were computed by solving the Rayleigh inhomogeneous forward problem for a vertical, time-harmonic point source applied at the free boundary of the layered medium of Table 6.1. The displacement field was calculated at 24 receiver locations, with the closest receiver located at a distance equal to one wavelength from the source. The frequency range considered in the analysis varied from 5 to 54 Hz. The techniques used to determine dispersion and attenuation curves from the synthetic displacement field are not discussed here because they have been thoroughly explained in Chapters 4 and 5.

The algorithm converged after only three iterations. Figure 6.12 shows the agreement between *synthetic* and *theoretical* Rayleigh dispersion and

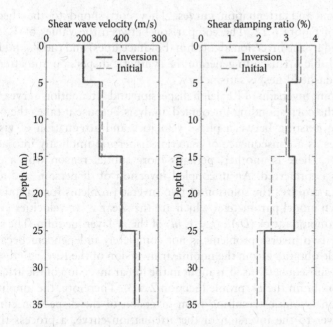

Figure 6.13 Shear wave velocity and shear damping ratio profiles obtained from the joint inversion of Rayleigh synthetic dispersion and the attenuation curves of Figure 6.12 using the constrained, damped least-squares Occam's algorithm (boldface line). The dashed lines denote the V_S and D_S profiles of Table 6.1. (From Lai, C. G., Surface waves in dissipative media: Forward and inverse modelling, *Surface Waves in Geomechanics—Direct and Inverse Modelling for Soil and Rocks*, Springer-Verlag, New York, 2005.)

Table 6.1 Medium parameters used to generate Rayleigh synthetic dispersion and attenuation curves*

Layer	Thickness (m)	V_P (m/s)	V_S (m/s)	D_P (–)	D_S (–)	ρ (Mg/m³)
1	5.0	400	200	0.020	0.035	1.7
2	10.0	600	300	0.015	0.030	1.8
3	10.0	800	400	0.010	0.025	1.8
Half-space	∞	1000	500	0.010	0.020	1.8

* Jointly inverted by employing the constrained, damped least-squares Occam's algorithm (Lai 2005).

attenuation curves at the last iteration. The theoretical curves were computed with reference to the *fundamental* mode of propagation, and the good agreement with the synthetic curves can be attributed to the fact that the layered medium of Table 6.1 is *normally dispersive*.

Figure 6.13 illustrates the shear wave velocity V_S and shear damping ratio D_S profiles resulting from the *joint* inversion of Rayleigh synthetic dispersion and attenuation curves. They correspond to the theoretical curves of Figure 6.12. The comparison between the values of V_S and D_S predicted by the inversion algorithm (boldface lines) and the V_S and D_S profiles of Table 6.1 used to generate the synthetic dispersion and attenuation curves (dashed lines) is satisfactory.

The joint inversion of Rayleigh dispersion and attenuation curves is superior to the corresponding uncoupled analysis because it takes the inherent coupling existing between phase velocities and attenuation of viscoelastic waves as a consequence of material dispersion implicitly into account. However, there is another, perhaps more subtle, reason why a coupled analysis is preferred. An uncoupled inversion of dispersion and attenuation data requires the solution of two inverse problems for a total of $2 \cdot nl$ unknown model parameters, which are the shear wave velocities $(V_S)_k$ and shear damping ratios $(D_S)_k$, $(k = 1, nl)$ of the nl-layer medium. The solution of these two inverse problems is not completely independent because the V_S profile obtained from the nonlinear inversion of the Rayleigh dispersion curve is subsequently used as input in the linear inversion of the attenuation curve to obtain the D_S profile (Section 2.5.3). Therefore, the amplification of the uncertainties resulting from inversion of the dispersion curve will carry over to the inversion of the attenuation curve, a process that will eventually add its own uncertainty. In other words, the uncoupled inversion suffers from a negative cross-coupling effect resulting from the solution of two inverse problems where the input data of one problem comes from the solution of the other.

Conversely, the joint inversion of both dispersion and attenuation curves is freed from this negative cross-coupling because both sets of experimental data are inverted simultaneously using the elegant formalism of complex variable theory. Furthermore, the coupled analysis of surface wave data takes advantage of a built-in *internal constraint* embedded in the algorithm that makes the inversion a *better-posed* mathematical problem. This internal constraint is represented by the *Cauchy–Riemann equations* that are satisfied by the Rayleigh phase velocity V_R^* if viewed as an *holomorphic* function $V_R^* = G^*\left(V_S^*\right)$ of the complex-valued shear wave velocity (Remmert 1997). In summary, application of complex variable theory for the joint inversion of surface wave data is not only an elegant procedure to account for the inherent coupling between phase velocities and dissipative properties of viscoelastic media, but it also improves the well-posedness of the associated inverse problem. An example of joint inversion of experimental data is presented in Chapter 7.

6.4.4 A priori information in surface wave inversion

The importance of a priori information as a means of mitigating the ill-posedness of linear and nonlinear inverse problems has been emphasized throughout the whole chapter. Of particular significance is the need to alleviate the problems arising from the nonuniqeness of the solution emerging from the interpretation of the experimental measurements.

In geotechnical site investigation, surface wave tests are typically conducted in association with numerous other surveys including drilling of boreholes, penetration tests, other types of geophysical prospecting, and so on. The extension of an exploration program at a construction site is usually redundant, and a considerable amount of data is available for the definition of a geological–geotechnical model of the subsurface. This situation is ideal for making good use of cross-correlations among the results of various tests and of a priori information for the inversion of geophysical measurements including surface wave data. At a larger spatial scale, detailed geological information may also be useful to constrain for instance, the position of the bedrock or to define the geometries of the contacts among different geological formations or lithostratigraphic units.

6.4.4.1 Borehole logs

If borehole data and lithostratigraphic information are available, layer thicknesses may be assumed as known a priori information to reduce the number of unknowns in setting up a 1D geotechnical model at a site. If these data are not available, a good rule of thumb is to assume layer thicknesses increasing with depth to comply with the decrease of model parameter resolution with depth, which is an intrinsic shortcoming of noninvasive geophysical methods including surface wave testing. Obviously, this modeling assumption reduces the ability of surface wave tests to locate the spatial position of layer interfaces.

6.4.4.2 P-wave refraction survey

Testing configurations for MASW testing and seismic P-wave refraction survey are basically the same, so that with a minimum effort in ensuring data acquisition compatibility between the two tests, useful information from the refraction survey may be exploited for the inversion of surface wave data (Foti et al. 2003). Two typical situations in which refraction data are particularly valuable for the inversion of surface wave data are the presence of either a shallow bedrock interface or of a shallow water table.

Furthermore, P-wave refraction can also be useful for recognizing the presence of dipping layers or other lateral variations that cannot be resolved with 1D modeling adopted in the standard interpretation of surface wave data.

In the example reported in Figure 6.14, the depth of the bedrock roof evaluated with seismic refraction has been used as a constraint in

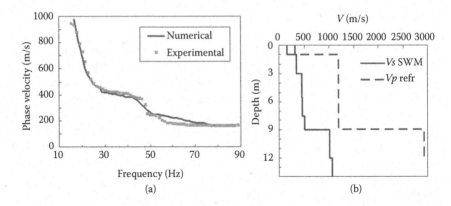

Figure 6.14 Inversion of surface wave data with constraints from P-wave refraction (a) theoretical and experimental dispersion curves. (b) Shear wave velocity profiles. (From Foti, S. et al. *Near Surface Geophys*, 1(1), 119–129, 2003.)

the inversion of surface wave data, and the final result is in good agreement with independent borehole logs at the same site.

The role that the spatial position of the water table plays in surface wave inversion is subtle, yet quite relevant. Although the presence of the water table does not have a great influence on the values of the measured shear wave velocity, which is the main goal of surface wave survey, it significantly affects the measured value of the speed of propagation of P-waves and, in turn, of the resulting Poisson ratio. Even though the dispersion curve's sensitivity to changes in the Poisson ratio is low compared to the sensitivity to changes in the shear modulus profile and layer thicknesses, the role played by the Poisson ratio during the inversion is not always negligible (Foti and Strobbia 2002). The transition from dry to wet soil, which is sharp for coarse-grained materials, is always associated with an abrupt change of Poisson ratio and mass density of the soil. If this transition is not properly recognized, it may lead to erroneous results.

The examples shown in Chapter 7 demonstrate with real data the importance of a priori information concerning the position of the water table when interpreting surface wave data. Although this information may be obtained through different types of exploration methods (e.g., electrical surveys, borehole drilling, piezometer measurements), P-wave refraction appears to be an effective and economically convenient option. When no information is available, it is necessary to take into account the influence that the uncertainty in defining the position of the water table may have on the results of surface wave measurements. A possible strategy in this regard could be to perform a few trials assuming reasonable guesses for the position of the water table and then use *engineering judgment* to appreciate the differences, if any, among the results obtained from the inversions of surface wave data.

6.4.4.3 Joint inversion of geophysical data

Synergies of surface wave testing with other geophysical exploration methods can be effective for improving the reliability of the investigation campaign and for obtaining a more consistent soil model. Hayashi et al. (2005) performed a joint analysis of surface wave data and microgravity measurements aimed at detecting the presence of underground voids. Another appealing possibility is represented by the joint inversion of surface wave measurements and vertical electrical sounding (VES) data (Hering et al. 1995; Misiek et al. 1997; Comina et al. 2002; Wisén and Christiansen 2005). Indeed, despite the marked differences between these two geophysical techniques, the interpretation of data measurements presents rather strong similarities. VES data analysis is based on the inversion of field measurements consisting of an apparent resistivity curve, which can be considered analogous of the dispersion curve for surface wave data. Moreover, the inversion of the resistivity curve is typically based on assuming a 1D layered model similar to the one adopted in surface wave testing; however, in this case, the unknown parameters are the electrical resistivities of the layers.

Assuming that layer interfaces are the same for the seismic and the electrical model, it is possible to use a joint inversion strategy, simultaneously inverting the experimental dispersion and apparent resistivity curves to obtain thickness, shear wave velocity, and resistivity of each layer. In practice, $3nl-1$ unknowns are determined from the joint inversion of the two experimental datasets, whereas each uncoupled inversion aims at determining $2nl-1$ parameters from a single dataset. From a mathematical point of view, the solution of the joint inverse problem is better-posed than the solution of two single inverse problems.

The main limitation of this approach is demonstrated by situations in which variations in electrical properties are not associated with variations in seismic velocity or vice versa. Indeed, there are some circumstances in which a discontinuity in one of the two model parameters does not correspond to a discontinuity in the other. Another potential shortcoming of these types of joint inversions is computational. In fact, not all the partial derivatives of the response functions with respect to the model parameters required to construct the Jacobian matrix can be computed analytically using explicit formulas. Thus, the recourse to inefficient and unstable finite difference calculations and numerical differentiation is unavoidable (Section 6.4.3).

The general principles and further applications of joint inversion with other geophysical data are discussed in Section 8.3.

6.5 UNCERTAINTY

6.5.1 Inverse problems and measurement errors

6.5.1.1 Linear problems with Gaussian data errors

An essential aspect of any measurement technique is the ability to (a) estimate the uncertainty of the measured data and (b) to determine how this uncertainty is projected onto the model parameters derived from them. This is particularly important in geophysical prospecting where the parameters of interest are often inferred from complex inversion procedures. In surface wave testing, the measured data are represented by the Rayleigh (or Love) dispersion and attenuation curves, whereas the derived parameters are the variation with depth of the shear modulus (or equivalently, the shear wave velocity) and shear damping ratio at small-strain levels.

The statistics of normally distributed experimental data enjoy peculiar properties such as being completely described by the expected value and variance. Furthermore, a basic theorem of probability theory states that any linear combination of normally distributed (i.e., Gaussian) random variables is normally distributed (Harr 1996). Therefore, if the relationship linking the experimental data and the model parameters is linear and, for instance, is represented by Equation 6.12 and, furthermore, if the data errors are assumed normally distributed, the model parameters obtained from the inversion are also normally distributed. Moreover, if X is a multivariate, normal, random vector with expected value $E(X)$ and covariance matrix $\mathrm{Cov}(X)$, and $Y = AX$ is a multivariate random vector linearly related through matrix A to the random variable X, Y is also multivariate normal with expected value and covariance given by the following relations (Aster et al. 2005)

$$\begin{cases} E(\mathbf{Y}) = \mathbf{A}E(\mathbf{X}) \\ \mathrm{Cov}(\mathbf{Y}) = \mathbf{A}\mathrm{Cov}(\mathbf{X})\mathbf{A}^T \end{cases} \tag{6.46}$$

Equation 6.46 is important because, in a linear inverse problem, it allows us to estimate, with a simple calculation, mean and covariance of model parameters directly from the mean and covariance of experimental measurements. This relation clearly shows that the way the uncertainty of data errors is mapped into uncertainty of model parameters is intimately connected to the algorithm adopted for the inversion represented here by matrix A. For instance, for the standard least-squares problem solved in an overdetermined situation, from Equation 6.6, matrix A would then be equal to $A = (G^T G)^{-1} G^T$.

In general, if the linear inverse problem is resolved through the formalism of the Moore–Penrose generalized inverse matrix G^{-g} (Section 6.4.2) as shown by Equation 6.12. Equation 6.46 can be specialized as follows

$$\begin{cases} E(\mathbf{m}) = \mathbf{G}^{-g}E(\mathbf{d}) \\ \mathrm{Cov}(\mathbf{m}) = \mathbf{G}^{-g}\mathrm{Cov}(\mathbf{d})\left(\mathbf{G}^{-g}\right)^T \end{cases} \qquad (6.47)$$

where $E(\mathbf{d})$ and $\mathrm{Cov}(\mathbf{d})$ are, respectively, the expected value and covariance matrix of the experimental measurements viewed as a multivariate, normal, random vector. In the very special, yet relevant, case in which $\mathrm{Cov}(\mathbf{d}) = \sigma^2 \cdot \mathbf{1}$, where σ^2 is the variance of data vector and $\mathbf{1}$ is the identity matrix, Equation 6.47 simplifies to

$$\begin{cases} E(\mathbf{m}) = \mathbf{G}^{-g}E(\mathbf{d}) \\ \mathrm{Cov}(\mathbf{m}) = \sigma^2 \cdot \mathbf{G}^{-g} \cdot \left(\mathbf{G}^{-g}\right)^T \end{cases} \qquad (6.48)$$

6.5.1.2 Normality assessment

The validity of Equation 6.47 rests on the assumption that the data errors are normally distributed. Prior to its adoption, it is therefore of primary importance to assess the validity of this hypothesis with respect to multivariate vector $\mathbf{d}$ of experimental measurements. A standard test that can be performed to validate the assumption that a set of experimental data follows a Gaussian distribution (also called a *normality test*) is the χ^2-test ("*chi-squared test*"). Given a random variable of interest, say x, the test is based on grouping a set of discrete, observed realizations x_j ($j = 1, Nc$) of the variable into Nc classes followed by the calculation of the absolute observed f_j^O and expected frequency f_j^E for each class. The latter is computed as if the sample came from a Gaussian distribution with mean and variance determined from the observed data. The following quantity is finally calculated

$$X^2 = \sum_{j=1}^{Nc} \frac{\left(f_j^O - f_j^E\right)^2}{f_j^E} \qquad (6.49)$$

It can be shown that the random variable X^2 defined in Equation 6.49 approximately approaches the χ^2 distribution, the probability density function of which is given by the following relation (Bain and Engelhardt 1992):

$$f_{\chi^2}(x|\nu) = \frac{x^{\left(\frac{\nu-2}{2}\right)} \cdot e^{-x/2}}{2^{\nu/2}\Gamma(\nu/2)} \qquad (6.50)$$

where ν denotes the degrees of freedom of the χ^2 distribution and for the Gaussian distribution is equal to ν = Nc–3; Γ(...) defines the gamma function. An assessment can be made about the goodness of the assumption of normal distribution for x by comparing the value of X^2 in Equation 6.49 and the value of χ^2 given by Equation 6.50; χ^2 is small if the observed and the expected frequencies are nearly the same. Thus, a small value of χ^2 is indicative of a good correspondence between measured and normal probability distributions. It is standard practice to assume that the data do not follow a normal distribution if values greater or equal than X^2 occur with a frequency of less than 5% (Menke 1989). The threshold upper bound of X^2 is established from the χ^2 distribution.

Figure 6.15 shows an example of the relative frequency distribution of Rayleigh phase velocities computed using the results of an actual surface wave test performed at a site in Italy (Lai et al. 2005). The relative and cumulative frequency distributions of the experimental phase velocities were computed for each frequency and were arranged in five homogeneous classes. The elements of each class were then compared with those obtained for a Gaussian distribution with mean and standard deviation

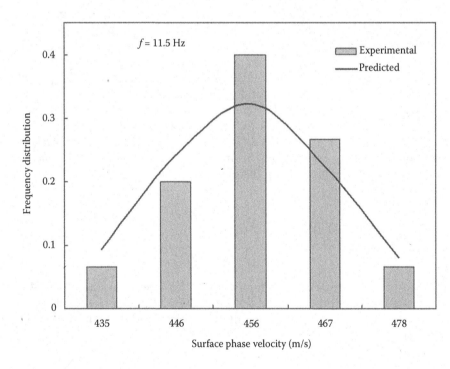

Figure 6.15 Relative frequency distribution of experimental phase velocities measured in a real surface wave test. Comparison with the values predicted by a Gaussian distribution. (From Lai, C. G. et al., *J Environ Eng Geophys*, 10, 219–228, 2005.)

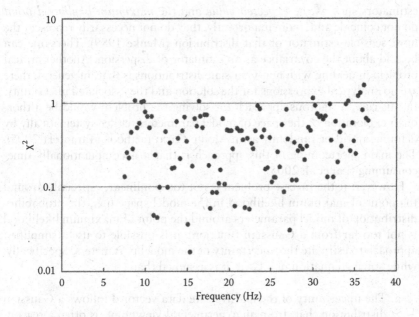

Figure 6.16 Results of χ^2-test using frequency distribution of surface phase velocities measured in a real surface wave test over the whole frequency range. (From Lai, C. G. et al., *J Environ Eng Geophys*, 10, 219–228, 2005.)

derived from the experimental data. The comparison between the experimental and the theoretical distributions appears reasonable. For the same example, Figure 6.16 shows the plot of the quantity X^2 computed with the experimental dataset as a function of frequency. From the plot, it can be noted that all the points fall below $\chi^2_{0.05}$ confirming that Gaussian distribution is a reasonable assumption for this dataset over the whole frequency range of interest.

In addition to the χ^2-test, there are a number of other statistical trials that can be used for normality assessment including the Kolmogorov–Smirnov test, the Anderson–Darling test, the Lilliefors test, and the Q–Q plot (Bain and Engelhardt 1992; Aster et al. 2005).

6.5.1.3 Nonlinear problems with Gaussian data errors

It can be shown that if an inverse problem is nonlinear, a Gaussian distribution of the errors in the data generally will be mapped into a non-Gaussian distribution of the errors in the model parameters (Tarantola 2005). In principle, this fact poses serious difficulties in the estimation of the uncertainty of model parameters derived from nonlinear inversion of data affected by measurement errors. First, in non-Gaussian probability distributions, central estimators such as the *expected value* and the *maximum likelihood point* do not coincide and, more importantly, they do not necessarily represent the most sensible estimator of that distribution (Menke 1989). The same can be said about the covariance as an estimator of dispersion. Another critical problem in dealing with non-Gaussian distributions is that, in general, there are no analytical expressions for the solution and the associated uncertainty. In this case, a rigorous approach for solving the problem would be a thorough exploration of the space of model parameters, either systematically by defining a grid or randomly as in Monte Carlo methods (Tarantola 2005). The main shortcoming of this approach is that it is computationally time-consuming (Lai et al. 2005).

However, if the inverse problem is not too nonlinear, especially around the point of maximum likelihood in the model space (i.e., the probability distribution of model parameters around the point of maximum likelihood is not too far from a Gaussian function), it is possible to use a simplified approach to estimate the uncertainty of the model parameters. Specifically, while solving equation $\mathbf{m} = \mathbf{G}^{-g}\mathbf{d}$, it is assumed that

a. The uncertainty of the multivariate data vector $\mathbf{d}$ follows a Gaussian distribution that, from an experimental viewpoint, is often a reasonable assumption.
b. The inverse problem $\mathbf{m} = \mathbf{G}^{-g}\mathbf{d}$ is only moderately nonlinear around its solution.
c. Relation $\mathbf{G}(\mathbf{m}) = \mathbf{d}$ is inverted using the *method of maximum likelihood*.

then the uncertainty associated with the expected value of model parameter vector $\mathbf{m}$ can be approximately calculated using the following formula (Menke 1989)

$$
\begin{aligned}
\mathrm{Cov}[\mathbf{m}] &\approx \left[\left(\mathbf{J}^{\mathrm{T}}\left(\mathrm{Cov}[\mathbf{d}]\right)^{-1}\mathbf{J}\right)^{-1}\mathbf{J}^{\mathrm{T}}\left(\mathrm{Cov}[\mathbf{d}]\right)^{-1}\right]_{\mathrm{last\,\#}} \\
&\cdot\mathrm{Cov}[\mathbf{d}]\cdot\left[\left(\mathbf{J}^{\mathrm{T}}\left(\mathrm{Cov}[\mathbf{d}]\right)^{-1}\mathbf{J}\right)^{-1}\mathbf{J}^{\mathrm{T}}\left(\mathrm{Cov}[\mathbf{d}]\right)^{-1}\right]_{\mathrm{last\,\#}}^{\mathrm{T}}
\end{aligned}
\tag{6.51}
$$

where the covariance of data vector Cov(d) has been assumed to be a diagonal matrix composed of *statistically uncorrelated* measurements and $\mathbf{J}$ is the Jacobian matrix calculated from *grad* $\mathbf{G}(\mathbf{m})$. The subscript last # outside the brackets of Equation 6.51 denotes that the terms inside the parentheses (essentially the Jacobian matrix) have to be computed with respect to the *last iteration* in the solution of the nonlinear problem $\mathbf{G}(\mathbf{m}) = \mathbf{d}$ using a LS recursive technique (for instance, the standard least-squares method).

Because Equation 6.51 is not exact, the " = " sign has been replaced by "≈" to denote the approximated nature of this relationship. For most problems, if the nonlinearity of $G(m) = d$ is not too severe at the point near the optimal values of the model parameter where the function $G(m)$ has been replaced by the Jacobian linearization, this approximation turns out to be adequate unless data measurements are exceptionally noisy (Aster et al. 2005). Finally, from Equation 6.51, it is noted that the availability of the analytical Jacobian would be beneficial not only for the solution of the nonlinear inverse problem itself (Section 6.4.3) but also for the estimate of the uncertainty of the calculated model parameters.

6.5.2 Uncertainty in surface wave measurements

Uncertainties in any experimental test are related to a number of different sources. In geophysical measurements such as surface wave tests, it is important to distinguish between uncertainties in the measurements (also called *aleatory*) and uncertainties related to the model used for the interpretation of the experimental results (also called *epistemic*). The latter are caused, for instance, by near-field effects or by the inadequacy of the horizontally layered model to capture lateral heterogeneities of the soil deposit. Strobbia and Foti (2006) proposed a statistical procedure to detect the presence of epistemic uncertainty based on the regression of phase versus offset. This section quantifies aleatory or data uncertainties in surface wave measurements.

Aleatory uncertainties are mainly, although not exclusively, due to noise in the recorded signals and to geometrical uncertainties related to the configuration and possible tilting of the receivers. The influence of several sources of uncertainty occurring in surface wave testing has been studied by O'Neill (2003) using Monte Carlo numerical simulations and repeated field tests. The author reports minimal influence from geophone tilting and ground coupling, although positional errors, static shift, and additive Gaussian noise introduce larger uncertainties in the measured experimental dispersion curve. Overall, he found *coefficients of variation*[26] for Rayleigh wave phase velocities increasing nonlinearly with the decrease of frequency and proposed that the errors be described by a *Cauchy–Lorentz* distribution for the low-frequency range. Uncertainties in recorded signals are associated with coherent and uncorrelated noise. The latter is externally generated noise (i.e., environmental noise) and can be studied via the statistical distribution of the recorded signals if many repetitions of the test in a given configuration are available (Lai et al. 2005). Coherent noise may be due, for instance, to events generated by the seismic source (e.g., near-field effects, lateral variations, higher modes; see Chapter 3).

---

[26] The coefficient of variation of a random variable is the ratio between the square root of the variance (i.e., the standard deviation) and the expected value (i.e., the mean).

In general, the estimation of the uncertainty associated with surface wave measurements is not a trivial exercise also because it is strongly dependent upon the particular technique that is adopted to calculate the experimental dispersion and attenuation curves (Chapters 4 and 5). Technical difficulties arise in properly quantifying the influence that inevitable errors affecting the measured quantities propagate through a series of complicated steps of data-processing and curve-fitting procedures, from the acquisition of the signals in the time-offset domain, to the calculation of dispersion and attenuation curves. (These errors may be due to factors such as the inaccuracies of the instrumentation, testing configuration, modality of data acquisition, and the even subjective behavior of the operator running the test.) The following sections will describe simplified procedures that can be used to quantify the uncertainty in surface wave measurements.

6.5.2.1 Experimental dispersion curve

6.5.2.1.1 Spectral analysis of surface waves

In conventional SASW measurements, the experimental dispersion curve is determined using the *two-station method* (see Sections 3.4.3.2 and 4.3). The data are collected with two receivers located at distances x_1 and x_2 from the source. Typically, data are recorded in the frequency domain, and the Rayleigh phase velocity is computed from Equation 4.7

$$V_R(\omega) = \frac{\omega(x_2 - x_1)}{\arg\left[S_{12}(\omega)\right]} \tag{6.52}$$

where arg $[S_{12}]$ is the phase of the particle velocity *cross-power spectrum* of the signals detected by the two receivers (see Section 4.3). This is the quantity measured experimentally. By making the simplifying, though reasonable, assumption that the experimental data are normally distributed, the measurements represent an estimate of the expected value of S_{12}, hereinafter denoted by $E[S_{12}(\omega)]$. The *variance* of S_{12} can be computed using the following relations (Bendat and Piersol 2010)

$$\begin{cases} \mathrm{Var}\left[\,\left|S_{12}(\omega)\right|\,\right] = \dfrac{\left|S_{12}(\omega)\right|^2}{n_d \cdot \gamma_{12}^2} \\[4mm] \mathrm{Var}\left[\arg\left\{S_{12}(\omega)\right\}\right] \approx \dfrac{\left(1 - \gamma_{12}^2\right)}{2\gamma_{12}^2} \end{cases} \tag{6.53}$$

where n_d is the number of independent averages used to estimate $S_{12}(\omega)$ and $\gamma_{12}^2(\omega)$ is the *ordinary coherence function* of the signals (see Section 4.3).

Equation 6.53 allows us to compute the uncertainty of the measured cross-power spectrum in a *single* realization of a test. It represents the uncertainty associated with the spectral estimate of arg $[S_{12}(\omega)]$ inferred from the signals detected by the two receivers.

A rapid inspection of Equation 6.52 also shows that $V_R(\omega)$ is a nonlinear function of $\arg\left[S_{12}(\omega)\right]$; thus, $V_R(\omega)$ is generally non-Gaussian, even with Gaussian data. However, applying the FOSM method (Harr 1996; Baecher and Christian 2003) to Equation 6.52 yields the following relations for the expected value and variance of $V_R(\omega)$

$$\begin{cases} E\left[V_R(\omega)\right] \approx \dfrac{\omega(x_2 - x_1)}{E\left[\arg\left(S_{12}(\omega)\right)\right]} \\[4mm] Var\left[V_R(\omega)\right] \approx \dfrac{Var\left[\arg\left(S_{12}(\omega)\right)\right] \cdot \left[\omega(x_2 - x_1)\right]^2}{\left\{E\left[\arg\left(S_{12}(\omega)\right)\right]\right\}^4} \end{cases} \tag{6.54}$$

Because the experimental dispersion curve is constructed by averaging Rayleigh phase velocities $[V_R(\omega)]_j$ ($j = 1, n_{SP}$) obtained over n_{SP} different receiver spacings (see Section 4.3), and because every linear combination of normal distributions is itself a normal distribution, the expected value and variance of the *composite* experimental dispersion curve $V_R(\omega)$ are given by

$$\begin{cases} E\left[V_R(\omega)\right] = \dfrac{1}{n_{SP}} \displaystyle\sum_{j=1}^{n_{SP}} E\left[V_R(\omega)\right]_j \\[4mm] Var\left[V_R(\omega)\right] = \dfrac{1}{n_{SP}^2} \displaystyle\sum_{j=1}^{n_{SP}} Var\left[V_R(\omega)\right]_j \end{cases} \tag{6.55}$$

In Equation 6.55, the variance of $V_R(\omega)$ has been computed under the assumption that the individual dispersion curves $[V_R(\omega)]_j$ are statistically uncorrelated.

Equations 6.54 and 6.55 can equivalently be used in either a *single* or a *multiple* realization of a SASW test. It all depends on how the quantity $Var[\arg\{S_{12}(\omega)\}]$ is defined. If it is calculated by means of Equation 6.53, then Equations 6.54 and 6.55 allow us to determine the effects yielded onto the experimental dispersion curve by the uncertainty of $\arg[S_{12}(\omega)]$ in a single realization of a test. Alternatively, $Var\{\arg[S_{12}(\omega)]\}$ can be determined from the inherent variability of results in a multiple realization of a test; in that case, Equations 6.54 and 6.55 allow us to estimate how this uncertainty is projected onto uncertainty of the experimental dispersion curve (Tuomi and Hiltunen 1996).

6.5.2.1.2 Multichannel analysis of surface waves

In recent years, the number of investigators that have adopted multichannel techniques to determine the experimental dispersion curve has grown considerably (Gabriels et al. 1987; Tokimatsu 1995; Tselentis and Delis 1998; Park et al. 1999; Foti 2000). These techniques use linear arrays of multiple receivers for active measurements and a 2D array for passive measurements. Multichannel methods comprise a variety of techniques depending on how the signals detected at the receivers are processed to obtain the dispersion curve (Chapter 4). However, they share the common feature that the signals detected at the receiver array are processed jointly as an ensemble. As a result, the surface wave dispersion curve determined with multichannel methods is generally smoother and more regular than that obtained with the conventional two-station method.

The simplest procedure for determining the apparent dispersion curve from the signals recorded by an array of receivers is to perform a linear regression of the displacement (or particle velocity, acceleration, or transfer function) phases measured at each receiver location (Lai 1998; Strobbia and Foti 2006). The method (see also Section 4.4) consists of determining the Rayleigh wavenumber $k_R(\omega)$ from a linear regression involving the experimental displacement phases for each of the experimental frequencies

$$k_R(\omega) \cdot \mathbf{r} = -\arg[\mathbf{u}_2(\mathbf{r},0,\omega)] \qquad (6.56)$$

where $\arg[\mathbf{u}_2(\mathbf{r},0,\omega)] = \arg[\mathbf{u}_2(\mathbf{r},\omega)]$ is the phase of the vertical component of particle displacement and $\mathbf{r}$ is the vector of receiver offsets. The Rayleigh phase velocity is then obtained from $V_R(\omega) = \omega / k_R(\omega)$. By estimating $E[k_R(\omega)]$ using a standard least-squares algorithm, it is a straightforward matter to calculate the uncertainty associated with $k_R(\omega)$ using Equation 6.47, with $\mathbf{G}^{-g} = (\mathbf{G}^T\mathbf{G})^{-1}\mathbf{G}^T$ and $\mathrm{Cov}(\mathbf{d}) = -\arg[\mathbf{u}_2(\mathbf{r},\omega)]$ assuming the data $\arg[\mathbf{u}_2(\mathbf{r},\omega)]$ to be uncorrelated. The term $\mathrm{Cov}(\mathbf{m})$ in Equation 6.47 denotes a 2 by 2 diagonal matrix whose two nonzero elements are the variances of the intercept and the slope (i.e., $Var[k_R(\omega)]$) of the linear regression. Finally, matrix $\mathbf{G}$ has the following expression

$$\mathbf{G} = \begin{bmatrix} r_1 & r_2 & \cdots & r_{np} \\ 1 & 1 & \cdots & 1 \end{bmatrix}^T \qquad (6.57)$$

where np defines the number of receiver positions. Having defined $E[k_R(\omega)]$ and $Var[k_R(\omega)]$, the expected value and variance of $V_R(\omega) = \omega/k_R(\omega)$

can be determined using the FOSM method (Harr 1996; Baecher and Christian 2003), which consists of expanding $V_R(\omega) = \omega/k_R(\omega)$ in a Taylor's series[27] about $E[k_R(\omega)]$ and truncating it to first-order terms only. If it is further assumed that $Var[\omega] = Var[r] = 0$ (i.e., frequency of excitation and receiver offsets are considered deterministic rather than random variables), using Equation 6.47 it is possible to obtain the following result

$$\begin{cases} E\big[V_R(\omega)\big] \approx \dfrac{\omega}{E\big[k_R(\omega)\big]} \\[2em] Var\big[V_R(\omega)\big] \approx \dfrac{\omega^2 Var\big[k_R(\omega)\big]}{\big\{E\big[k_R(\omega)\big]\big\}^4} \end{cases} \tag{6.58}$$

Other techniques for determining the apparent dispersion curve include the so-called *transform-based methods* that are based on transforming the data measured in the time-offset domain into a different domain where the identification of the loci of points defining the dispersion curves is simpler (Section 4.6).

The estimation of the uncertainty becomes more involved when the experimental dispersion curve is determined using transform-based procedures such as the f–k method. The problem arises from the difficulties of properly quantifying the propagation of the uncertainty in each step of the rather complicated sequence of data processing operations leading to the calculation of the dispersion curve. It is for this reason that the uncertainty associated with the experimental dispersion curve determined in MASW testing with transform-based methods is more easily calculated by a direct measurement of the statistical distribution of primitive and derived surface wave data. Even though this approach requires considerably more time and effort to be implemented, it does not involve simplifying assumptions (e.g., small variance of the raw data); furthermore, it is exempt from major technical difficulties (Lai et al. 2005).

6.5.2.2 Experimental attenuation curve

The experimental setting adopted in attenuation measurements of surface waves is that associated with the multichannel technique. Both controlled harmonic sources (e.g., hydraulic vibrators and electromechanical shakers)

---

[27] Because $V_R(\omega)$ is a nonlinear function of $k_R(\omega)$, $V_R(\omega)$ generally is *non-Gaussian* distributed. However, assuming that the experimental data, that is, $\arg[u_2(r,\omega)]$ are characterized by small variances, it is possible to linearize $V_R(\omega) = \omega/k_R(\omega)$ about the expected value of $k_R(\omega)$ and to use Equation 6.47 without committing large errors.

and transient impulsive sources (e.g., sledgehammers and dropped weights) are used. The source term can be measured using controlled sources. The Rayleigh attenuation coefficients $\alpha_R(\omega)$ are obtained from measurements of the vertical displacement amplitudes $|u_2(r,\omega)|$ at several receiver offsets over a prescribed frequency range (see Section 5.2).

The relevant spectral quantity is now the particle velocity autopower spectrum $S_{rr}(\omega)$ calculated at each receiver spacing. Because ambient noise may be important, particularly at large receiver offsets, the experimental particle velocity spectra are corrected to account for the noise effects (Rix et al. 2000)

$$S_{rr}(\omega) = \tilde{\gamma}_{sr}^2(\omega) \cdot \tilde{S}_{rr}(\omega) \tag{6.59}$$

where $\tilde{S}_{rr}(\omega)$ is the measured autopower spectrum that is presumed to contain noncoherent noise. The quantity $\tilde{\gamma}_{sr}^2(\omega)$ is the ordinary coherence function between a harmonic source and the measured vertical particle velocity at the receiver. Computation of $\tilde{\gamma}_{sr}^2(\omega)$ requires the use of an accelerometer at the source to monitor the motion of the harmonic oscillator. From Equation 6.59, the experimental vertical particle displacement spectrum is computed from the following relation

$$|u_2(r,0,\omega)| = |u_2(r,\omega)| = \frac{|v_2(r,\omega)|}{\omega \cdot C(\omega)} = \frac{\sqrt{S_{rr}(\omega)}}{\omega \cdot C(\omega)} \tag{6.60}$$

where $C(\omega)$ is a frequency-dependent *calibration factor* that converts the output of the velocity transducer (volts) into engineering units (e.g., cm/s); $v_2(r, \omega)$ is the Fourier transform of the vertical particle velocity recorded at the receivers located at distance r from the source. Using the FOSM method illustrated here for $V_R(\omega)$, namely expanding Equation 6.60 in a truncated Taylor's series about $E[S_{rr}]$ and assuming $Var[C(\omega)] = 0$, it is possible to obtain

$$\begin{cases} E\Big[|u_2(r,\omega)|\Big] \approx \dfrac{\sqrt{E\big[S_{rr}(\omega)\big]}}{\omega \cdot C(\omega)} \\[4mm] Var\Big[|u_2(r,\omega)|\Big] \approx \dfrac{Var\big[S_{rr}(\omega)\big]}{4E\big[S_{rr}(\omega)\big] \cdot \big[\omega \cdot C(\omega)\big]^2} \end{cases} \tag{6.61}$$

Once the vertical displacement amplitudes $|u_2(r,\omega)|$ have been calculated, the Rayleigh attenuation coefficients $\alpha_R(\omega)$ can be determined from

$$|u_2(r,0,\omega)| = |u_2(r,\omega)| = F \cdot \Upsilon_2(r,0,\omega) \cdot e^{-\alpha_R(\omega) \cdot r} \tag{6.62}$$

where $\Upsilon_2(r,0,\omega)$ is the vertical component of the Rayleigh geometric attenuation function and F is the amplitude of the harmonic vertical point load (Chapter 2). In homogenous half-spaces, $\Upsilon_2(r,0,\omega) \propto 1/\sqrt{r}$. Sometimes this may be a reasonable approximation even in heterogeneous geomaterials.[28]

Equation 6.62 forms the basis of a nonlinear regression analysis to determine the frequency-dependent attenuation coefficients $\alpha_R(\omega)$ from the experimental displacement amplitudes $|u_2(r,\omega)|$. The geometric spreading function $\Upsilon_2(r,0,\omega)$ is known because it is obtained from the solution of the elastic Rayleigh forward problem (Section 2.5.3). Alternatively, Equation 6.62 may be substituted with Equation 6.25 where $|u_2(r,\omega)|$ has been replaced by the amplitude of vertical displacement transfer function $|T(r,\omega)|$ between the source and the receiver.

The expected value and variance of the Rayleigh attenuation coefficient $\alpha_R(\omega)$ can be obtained from a regression of either Equation 6.62 or Equation 6.25. Due to the nonlinearity of the relationship between either $|u_2(r,\omega)|$ or $|T(r,\omega)|$ and $\alpha_R(\omega)$, only estimates of $E[\alpha_R(\omega)]$ and $Var[\alpha_R(\omega)]$ can be obtained. By solving Equation 6.62 using a standard nonlinear least-squares algorithm, the uncertainty associated with the attenuation coefficient can be approximately calculated with the following relation

$$\mathrm{Cov}\left[\alpha_R(\omega)\right] \approx \left[\left(\mathbf{J}_{\alpha_R}^T \mathbf{J}_{\alpha_R}\right)^{-1} \mathbf{J}_{\alpha_R}^T\right]_{\mathrm{last\#}} \mathrm{Cov}\left[|u_2(r,\omega)|\right] \left[\left(\mathbf{J}_{\alpha_R}^T \mathbf{J}_{\alpha_R}\right)^{-1} \mathbf{J}_{\alpha_R}^T\right]_{\mathrm{last\#}}^T \quad (6.63)$$

where $\mathrm{Cov}\left[|u_2(r,\omega)|\right]$ is an np by np matrix containing the covariances of the experimental displacement amplitudes at the np different receiver positions at a given frequency ω. Because it is assumed that the data $|u_2(r,\omega)|$ are uncorrelated, the data covariance matrix is diagonal with the nonzero elements equal to the variances of $|u_2(r,\omega)|$ given by Equation 6.61. The Jacobian $\mathbf{J}_{\alpha_R}$ is an np by 1 vector whose components $(\mathbf{J}_{\alpha_R})_k$ $(k = 1, np)$ are obtained from Equation 6.62

$$(\mathbf{J}_{\alpha_R})_k = \left[\mathbf{J}_{\alpha_R}(\omega)\right]_k = \frac{\partial |u_2(r_k,\omega)|}{\partial \alpha_R} = -r_k \cdot |u_2(r_k,\omega)| \quad (6.64)$$

The subscript last # outside the brackets of Equation 6.63 indicates that the terms inside the parentheses, essentially the Jacobian $\mathbf{J}_{\alpha_R}$, refer to the *last iteration* in the solution of the nonlinear regression (Equation 6.62).

---

[28] This occurs, for instance, in *normally dispersive* soil deposits namely shallow geological formations where the mechanical impedance increases gradually with depth. In this case, the wave field is dominated by the fundamental mode of propagation. As a result, the geometric spreading function $\Upsilon_2(r,0,\omega)$ is well approximated by a term proportional to $1/\sqrt{r}$ (Chapter 2).

6.5.2.3 Joint experimental dispersion and attenuation curves

In the standard approach of surface wave measurements, experimental dispersion and attenuation curves are determined separately using different source-to-receiver configurations and different interpretation methods. However, Rix et al. (2001) proposed a methodology based on measuring the displacement transfer functions in which dispersion and attenuation curves are determined simultaneously from a single set of measurements using the same source and linear array of receivers. The multichannel simultaneous estimation of velocity and attenuation is also possible (Chapter 5). This approach has been introduced after recognizing that, in dissipative media, Rayleigh phase velocity and attenuation are not independent due to material dispersion. Therefore, a coupled analysis of dispersion and attenuation data is a more consistent and fundamentally correct approach. Also, the coupled measurement is compatible with the joint inversion algorithm that can be used to obtain the shear wave velocity and shear damping ratio profiles of a soil deposit (Section 6.4.3).

In coupled measurements of dispersion and attenuation data, the quantity measured experimentally is the particle velocity transfer function $H(r,\omega)$ between source and receiver and is defined by $H(r,\omega) = v_2(r,\omega)/F \cdot e^{i\omega t} = i\omega \cdot u_2(r,\omega)/F \cdot e^{i\omega t} = i\omega T(r,\omega)$. The expected value and variance of the modulus and phase of $T(r,\omega)$ can be computed from the following relations (Bendat and Piersol 2010)

$$E\big[\big|T(r,\omega)\big|\big] = \omega \cdot E\big[\big|H(r,\omega)\big|\big]; \quad E\big[\arg(T(r,\omega))\big] = E\big[\arg(H(r,\omega))\big] + \frac{\pi}{2}$$

$$Var\big[\big|T(r,\omega)\big|\big] \approx \omega^2 \cdot \frac{(1-\gamma_{sr}^2) \cdot \big|H(r,\omega)\big|^2}{2n_d \cdot \gamma_{sr}^2}; \quad Var\big[\arg(T(r,\omega))\big] \approx \frac{(1-\gamma_{sr}^2)}{2\gamma_{sr}^2}$$

$$(6.65)$$

where $\gamma_{sr}^2(\omega)$ represents the ordinary coherence function between the harmonic source and the receiver output signal. Because the complex Rayleigh wavenumber

$$k_R^*(\omega) = \left[\frac{\omega}{V_R(\omega)} + i \cdot \alpha_R(\omega)\right] \tag{6.66}$$

is determined from the nonlinear regression

$$T(r,\omega) = \Upsilon_2(r,0,\omega) \cdot e^{-i \cdot k_R^*(\omega) \cdot r} \tag{6.67}$$

it is first necessary to calculate $E[T(r,\omega)]$ and $Var[T(r,\omega)]$ in order to compute the expected value and variance of $k_R^*(\omega)$. For this purpose, it is convenient to write $T(r,\omega) = [T_1(r,\omega) + i \cdot T_2(r,\omega)]$ from which it is simple to obtain (Lai 1998)

$$\begin{cases} E\left[\left|T_1\left(r,\omega\right)\right|\right] \approx E\left[\left|T\left(r,\omega\right)\right|\right] \cdot \cos\left\{E\left[\arg\left(T\left(r,\omega\right)\right)\right]\right\} \\ E\left[\left|T_2\left(r,\omega\right)\right|\right] \approx E\left[\left|T\left(r,\omega\right)\right|\right] \cdot \sin\left\{E\left[\arg\left(T\left(r,\omega\right)\right)\right]\right\} \\ Var\left[T\left(r,\omega\right)\right] = Var\left[T_1\left(r,\omega\right)\right] - Var\left[T_2\left(r,\omega\right)\right] \end{cases} \quad (6.68)$$

where

$$\begin{cases} Var\left[T_1\left(r,\omega\right)\right] \approx \cos^2\left[\arg\left(T\left(r,\omega\right)\right)\right]Var\left[\left|T\left(r,\omega\right)\right|\right] \\ \qquad + \left|T\left(r,\omega\right)\right|^2 \sin^2\left[\arg\left(T\left(r,\omega\right)\right)\right]Var\left[\arg\left(T\left(r,\omega\right)\right)\right] \\ Var\left[T_2\left(r,\omega\right)\right] \approx \sin^2\left[\arg\left(T\left(r,\omega\right)\right)\right]Var\left[\left|T\left(r,\omega\right)\right|\right] \\ \qquad + \left|T\left(r,\omega\right)\right|^2 \cos^2\left[\arg\left(T\left(r,\omega\right)\right)\right]Var\left[\arg\left(T\left(r,\omega\right)\right)\right] \end{cases} \quad (6.69)$$

The uncertainty of the complex wavenumber $k_R^*(\omega)$ is finally computed from the nonlinear regression represented by Equation 6.67 following a procedure that is formally identical to that used for determining the uncertainty of the Rayleigh attenuation coefficient $\alpha_R(\omega)$. The result is

$$\mathrm{Cov}\left[k_R^*\left(\omega\right)\right] \approx \left[\left(\mathbf{J}_{k_R^*}^H \mathbf{J}_{k_R^*}\right)^{-1}\mathbf{J}_{k_R^*}^H\right]_{\text{last \#}} \cdot \mathrm{Cov}\left[\mathbf{T}(\mathbf{r},\omega)\right]\left[\left(\mathbf{J}_{k_R^*}^H \mathbf{J}_{k_R^*}\right)^{-1}\mathbf{J}_{k_R^*}^H\right]_{\text{last \#}}^H \quad (6.70)$$

where $\mathrm{Cov}[\mathbf{T}(\mathbf{r},\omega)]$ is an np by np matrix representing the covariances, at a given frequency ω, of the experimental displacement transfer functions at the np receiver positions. Because it is assumed that the data $\mathbf{T}(\mathbf{r},\omega)$ are uncorrelated, matrix $\mathrm{Cov}[\mathbf{T}(\mathbf{r},\omega)]$ is diagonal with the nonzero elements equal to the variances of $\mathbf{T}(\mathbf{r},\omega)$ and given by Equations 6.68 and 6.69. The term $\mathbf{J}_{k_R^*}$ is an np by 1 complex-valued vector the components $\left(\mathbf{J}_{k_R^*}\right)_j$, $j = 1, np$ of which are obtained by differentiating Equation 6.67, and they are equal to

$$\left(\mathbf{J}_{k_R^*}\right)_j = \left[\mathbf{J}_{k_R^*}\left(\omega\right)\right]_j = \frac{\partial T\left(r_j,\omega\right)}{\partial k_R^*} = -i \cdot \mathbf{r}_j \cdot T\left(r_j,\omega\right) \quad (6.71)$$

Again the subscript last # outside the brackets of Equation 6.71 indicates that the terms inside the parenthesis refer to the *last iteration* in solving Equation 6.67. Once $E\left[k_R^*(\omega)\right]$ and $Var\left[k_R^*(\omega)\right]$ are computed, the expected value and variance of the complex-valued Rayleigh phase velocity $V_R^*(\omega)$ are the complex-valued counterparts of Equation 6.58, namely

$$
\begin{cases}
E\left[V_R^*(\omega)\right] \approx \dfrac{\omega}{E\left[k_R^*(\omega)\right]} \\[3mm]
Var\left[V_R^*(\omega)\right] \approx \dfrac{\omega^2 \cdot Var\left[k_R^*(\omega)\right]}{\left\{E\left[k_R^*(\omega)\right]\right\}^4}
\end{cases}
\tag{6.72}
$$

Equation 6.72 completes the statistical analysis of surface wave measurements when dispersion and attenuation data are determined simultaneously from the particle velocity transfer functions $H(r,\omega)$. However, it may also be important to compute $E\left[V_R^*(\omega)\right]$ and $Var\left[V_R^*(\omega)\right]$ when surface wave data, namely $V_R(\omega)$ and $\alpha_R'(\omega)$, are obtained independently. This happens, for instance, in carrying out a joint inversion of dispersion and attenuation curves to determine the shear wave velocity and shear damping ratio profiles at a site where $V_R(\omega)$ and $\alpha_R(\omega)$ have been measured following an uncoupled procedure.

The solution to the problem is obtained by applying the FOSM method to Equation 2.127. In particular, this equation can be viewed as a mapping[29] that assigns a complex random variable V_R^* to a pair of independent random variables V_R and D_R. If the variables V_R and D_R are normally distributed and the mapping is sufficiently smooth, it may be expanded in a Taylor series about the point $\{E(V_R), E(D_R)\}$. Again, assuming small variances for variables V_R and D_R, this series is truncated to first-order terms only, yielding

$$
\begin{cases}
E\left[V_R^*(\omega)\right] \approx \dfrac{E\left[V_R(\omega)\right]}{1+E^2\left[D_R(\omega)\right]} \cdot \left(1+i\cdot E\left[D_R(\omega)\right]\right) \\[4mm]
Var\left[V_R^*(\omega)\right] \approx \left[\dfrac{1+i\cdot E\left[D_R(\omega)\right]}{1+E^2\left[D_R(\omega)\right]}\right]^2 \cdot Var\left[V_R(\omega)\right] \\[4mm]
\qquad + \left[\dfrac{E\left[V_R(\omega)\right]\cdot\left[2E\left[D_R(\omega)\right]+i\left(1+E^2\left[D_R(\omega)\right]\right)\right]}{\left(1+E^2\left[D_R(\omega)\right]\right)^2}\right]^2 \\[4mm]
\qquad \cdot Var\left[D_R(\omega)\right]
\end{cases}
\tag{6.73}
$$

---

[29] The frequency-dependence of $V_R^*(\omega)$ in this mapping does not constitute a major problem in this discussion because it can be considered parametric.

where

$$
\begin{cases}
E\big[D_R(\omega)\big] \approx \dfrac{E\big[\alpha_R(\omega)\big]E\big[V_R(\omega)\big]}{\omega} \\[4mm]
Var\big[D_R(\omega)\big] \approx \left[\dfrac{E\big[V_R(\omega)\big]}{\omega}\right]^2 \cdot Var\big[\alpha_R(\omega)\big] \\[4mm]
\qquad\quad + \left[\dfrac{E\big[\alpha_R(\omega)\big]}{\omega}\right]^2 \cdot Var\big[V_R(\omega)\big]
\end{cases}
\tag{6.74}
$$

In deriving Equations 6.73 and 6.74, it was assumed that the random variables V_R and D_R are uncorrelated.

In conclusion, despite the fact that the relationship among *directly* measured quantities $S_{rr}(\omega)$, $S_{12}(\omega)$, and $H(r,\omega)$ and *derived* surface wave data, namely $V_R(\omega)$, $\alpha_R(\omega)$, and $V_R^*(\omega)$, is nonlinear in most circumstances, the assumption of small variances allowed us to obtain explicit results for the expected values and variances of $V_R(\omega)$, $\alpha_R(\omega)$, and $V_R^*(\omega)$.

6.5.3 Estimate of variance of model parameters

Once the uncertainty of the experimental dispersion and attenuation curves has been estimated, the remaining goal is to determine the projection of this uncertainty into the expected values of the model parameters, which are the shear wave velocity and shear damping ratio profiles at a site. The solution of this problem is not unequivocal because it depends on the specific algorithm adopted for the inversion. Furthermore, whereas determining the shear wave velocity is a nonlinear inverse problem, the calculation of the shear damping ratio is a linear problem, as is shown by Equations 6.2 and 6.3, respectively. Of course, in the case of a joint inversion, the problem is nonlinear and also complex valued (Section 6.4.3).

If the dispersion curve is represented by an *nf* by 1 array $\mathbf{V}_R$ of Rayleigh phase velocities associated with *nf* discrete frequencies and it is viewed as a multivariate, normal, random vector[30] with expected value $E(\mathbf{V}_R)$ and covariance *nf* by *nf* matrix $Cov(\mathbf{V}_R)$, then Equation 6.51 may be used to estimate $Cov(\mathbf{V}_S)$, where $\mathbf{V}_S$ is an *nl* by 1 vector, the components of which

---

[30] It is further assumed that the components of data vector $\mathbf{V}_R$ are statistically uncorrelated; thus, the matrix $Cov(\mathbf{V}_R)$ is diagonal with the nonzero elements equal to the variances of $\mathbf{V}_R$.

are the unknown shear wave velocities of a stratified soil deposit having nl layers (including the half-space). The result is

$$\mathrm{Cov}\left[V_S\right] \approx \left[\left(J_{V_S}^T\left(\mathrm{Cov}\left[V_R\right]\right)^{-1}J_{V_S}\right)^{-1}J_{V_S}^T\left(\mathrm{Cov}\left[V_R\right]\right)^{-1}\right]_{\text{last }\#}$$

$$\cdot\mathrm{Cov}\left[V_R\right]\cdot\left[\left(J_{V_S}^T\left(\mathrm{Cov}\left[V_R\right]\right)^{-1}J_{V_S}\right)^{-1}J_{V_S}^T\left(\mathrm{Cov}\left[V_R\right]\right)^{-1}\right]^{T}_{\text{last }\#}$$

(6.75)

where J_{V_S} is the nf by nl Jacobian matrix calculated from $grad$ $G(V_S)$ in Equation 6.2. It can be computed analytically using Equation 6.33. The subscript last # outside the brackets of Equation 6.75 denotes again that the terms inside the parentheses are computed with respect to the final iteration when solving Equation 6.2 by using, for instance, the standard least-squares method. Finally, $\mathrm{Cov}(V_S)$ is an nl by nl diagonal matrix, the elements of which are the variances of the estimated shear wave velocities of the nl layers.

If Equation 6.2 is inverted using the Occam's algorithm (Section 6.4.3), it is possible to show that the uncertainty associated with the estimated profile of shear wave velocities can be computed with the following relationship (Lai et al. 2005)

$$\mathrm{Cov}\left[V_S\right] \approx \left[\left(\mu L^T L + \left(W J_{V_S}\right)^T W J_{V_S}\right)^{-1}\left(W J_{V_S}\right)^T W\right]_{\text{last }\#}$$

$$\cdot\mathrm{Cov}\left[V_R\right]\cdot\left[\left(\mu L^T L + \left(W J_{V_S}\right)^T W J_{V_S}\right)^{-1}\left(W J_{V_S}\right)^T W\right]^{T}_{\text{last }\#}$$

(6.76)

where μ is the Lagrange multiplier, W is a diagonal nf by nf *weight matrix* defined by Equation 6.43 where the uncertainties are now real valued, and L is the nl by nl real-valued matrix representing the two-point central finite difference operator defined in Equation 6.19.

Figure 6.17 shows the shear wave velocity profile and associated uncertainty obtained from the inversion of an experimental dispersion curve measured at a real site in Italy.

Figure 6.18 illustrates the comparison between the experimental dispersion curve and the corresponding theoretical curve (fundamental mode) obtained using the Occam's inversion algorithm after the ninth and final iteration (Lai et al. 2005). The error bars in Figure 6.17 attached to the V_S profile represent the standard deviation of the expected shear wave velocity profile with a coefficient of variation ranging from 0.2% to about 4%. The experimental dispersion curve of Figure 6.18 appears to be split into two regions separated by a threshold frequency of about 11 Hz: a low-frequency region

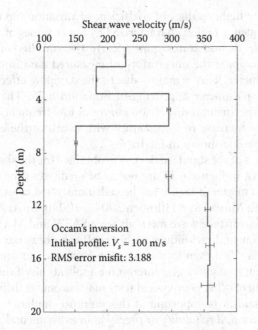

Figure 6.17 Expected value and standard deviation of the shear wave velocity profile from the inversion of the experimental dispersion curve shown in Figure 6.18, obtained at a real testing site in Italy. (From Lai, C. G. et al., *J Environ Eng Geophys*, 10, 219–228, 2005.)

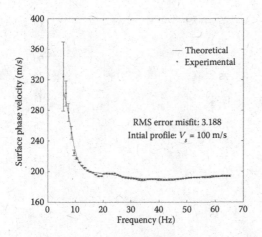

Figure 6.18 Experimental dispersion curve measured at a real testing site in Italy. Expected value and standard deviation of surface phase velocity versus frequency. The continuous line represents the theoretical dispersion curve (fundamental mode) at the final iteration of the inversion algorithm. (From Lai, C. G. et al., *J Environ Eng Geophys*, 10, 219–228, 2005.)

characterized by higher values of coefficient of variation (up to about 14%) and a high-frequency region characterized by lower values of uncertainty.

This example demonstrates quite clearly the stability of the Occam's algorithm in mapping the uncertainty of measured data into a covariance of model parameter. This is mainly due to the damping effect produced by the smoothing parameter appearing in Equation 6.76. The larger uncertainty of the experimental dispersion curves at low frequencies contributes to the observed increase of uncertainty with depth of the expected shear wave velocity profile shown in Figure 6.17.

The low values of the standard deviation obtained for the shear wave velocity profile shown in Figure 6.17 are not to be credited to the Occam's algorithm only. As a matter of fact, it has been demonstrated by other researchers (Xia et al. 2002; Marosi and Hiltunen 2004a, 2004b; Moss 2008; Cox and Wood 2011) that surface wave methods (e.g., SASW and MASW) are rather accurate experimental techniques with a coefficient of variation ranging on the order of 5%–10% even in the presence of strong ambient uncorrelated noise. This result is of particular interest for applications of these methods in urban areas where other geophysical tests may encounter difficulties.

That being said, it is important at this point to emphasize the difference between *accuracy* and *reliability* or *precision* in experimental measurements. Accuracy may be defined as the probability that the expected value of a measurement, in this case $E(V_S)$, determined with a certain technique is equal to the *true* value. Accuracy of a measurement may be affected by bias and systematic errors (Tuomi and Hiltunen 1996). Figure 6.19 shows a classical

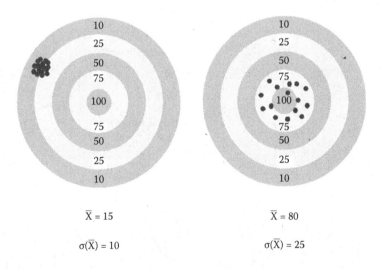

$$\bar{X} = 15 \qquad\qquad \bar{X} = 80$$

$$\sigma(\bar{X}) = 10 \qquad\qquad \sigma(\bar{X}) = 25$$

Figure 6.19 Difference between *accuracy* and *reliability* of two sets of experimental measurements. The symbol $\bar{X}$ denotes the expected value of the measurement, whereas $\sigma(\bar{X})$ represents its standard deviation.

example that is often used to explain the difference between reliability and accuracy of the outcome of an experiment.

The set of measurements on the left is characterized by a small spread around the mean (i.e., low standard deviation). However, if 100 represents the *true* value of the parameter (target), the mean is off by a factor almost equal to 7. The set of measurements on the right has a mean that is equal to 80, so relatively close to the true value, yet the measurements are grossly scattered around the mean and, as a result, the standard deviation is large. Figure 6.19 can be qualified by saying that the set of measurements on the left is *reliable* but not *accurate*, whereas the set on the right is *accurate* but not *reliable*.

Sometimes, it may be difficult to assess accuracy because the true value of a property or a physical parameter may not be known with certainty. The present discussion concerns with the reliability of surface wave measurements that represent an estimate of the scatter of V_S around its expected value and can be calculated by using Equation 6.76. The accuracy of surface wave methods in predicting the "true" V_S profile at a site may be addressed by comparing the results of the inversion with measurements of V_S profile obtained with other geophysical techniques (see Chapter 7).

The determination of the uncertainty associated with the estimated *damping ratio profile* is simpler than determining $\mathrm{Cov}(V_S)$ because the inversion of the experimental attenuation curve $\alpha_R(\omega)$ to obtain the damping ratio profile is *linear* as shown by Equation 6.3. If the attenuation curve is represented by an *nf* by *1* array $\boldsymbol{\alpha}_R$ of Rayleigh attenuation coefficients associated with *nf* discrete frequencies and it is viewed as a multivariate, normal, random vector with expected value $E(\boldsymbol{\alpha}_R)$ and covariance *nf* by *nf* matrix $\mathrm{Cov}(\boldsymbol{\alpha}_R)$,[31] then Equation 6.47 may be used to estimate $\mathrm{Cov}(\mathbf{D}_S)$, where $\mathbf{D}_S$ is an *nl* by *1* vector, the components of which are the unknown shear damping ratios of a stratified soil deposit having *nl* layers (including the half-space). If a standard least-squares algorithm is used, the result is

$$\mathrm{Cov}(\mathbf{D}_S) = \left(\mathbf{G}^T\mathbf{G}\right)^{-1}\mathbf{G}^T\mathrm{Cov}(\boldsymbol{\alpha}_R)\left[\left(\mathbf{G}^T\mathbf{G}\right)^{-1}\mathbf{G}^T\right]^T \tag{6.77}$$

where $\mathbf{G}$ is a matrix composed by terms that include the partial derivatives of the modal phase velocity of Rayleigh waves with respect to the model parameters $\mathbf{V}_P$ and $\mathbf{V}_S$ of the various layers of the soil deposit. The precise definition of this matrix can be obtained from Equation 2.133. Finally, in Equation 6.77, $\mathrm{Cov}(\mathbf{D}_S)$ is an *nl* by *nl* diagonal matrix, the elements of which are the variances of the estimated shear damping ratios of the

---

[31] It is further assumed that the components of data vector $\boldsymbol{\alpha}_R$ are statistically uncorrelated; thus, the matrix $\mathrm{Cov}(\boldsymbol{\alpha}_R)$ is diagonal with the nonzero elements equal to the variances of $\boldsymbol{\alpha}_R$.

nl layers. Of course, other algorithms besides the standard least-squares method can be used to invert the Rayleigh attenuation curve $\alpha_R(\omega)$, including the Occam's algorithm (Rix et al. 2000).

The shear damping ratio profile $\mathbf{D}_S$ can also be obtained through a joint, nonlinear inversion of dispersion and attenuation curves (Section 6.4.3). The procedure for estimating the uncertainty associated with the inverted complex-valued shear wave velocity profile $\mathbf{V}_S^*$ is identical to that described for the uncoupled inversion to obtain $\mathrm{Cov}(\mathbf{V}_S)$, and thus Equations 6.75 and 6.76 are still valid. The only difference is that the computation of $\mathrm{Cov}\left(\mathbf{V}_S^*\right)$ requires a systematic use of the formalism of complex variables.

6.5.4 Trade-off between model resolution and uncertainty

As shown in the previous section, the mode in which the uncertainty associated with an experimental dispersion or attenuation curve is projected into uncertainty of model parameters that are the shear wave velocity, and shear damping ratio profiles at a site generally will depend upon the specific algorithm adopted for the inversion. Inspection of Equations 6.75 and 6.76 makes this statement obvious. However, in addition to the algorithm, there is another factor affecting the amplification/deamplification of data errors. This factor is the spatial resolution of the geometric model assumed for the soil deposit. Suppose that at a construction site the sequence of layers and corresponding thicknesses are well-defined from independent borehole logs until a depth of 20 m. Suppose also that a surface wave test has been executed at the site and the low-frequency range of the measured dispersion curve covers depths up to 20 m. A variety of different geometrical models could be adopted for the inversion of the experimental dispersion curve not only because sharp interfaces among the layers may not exist at all but also because the low-strain shear modulus typically increases with depth due to the increase of overburden stress, even in a homogeneous soil deposit (Santamarina et al. 2001). Thus, a natural question arises regarding what would be the ideal geometrical model to be used for the inversion of the dispersion curve. Would a five-layer model be adequate? Or would a 10-layer model would be preferable? What should be the guiding principle or the rationale for making a correct choice with reference to the number of layers? The question is well-posed and relevant in virtually any field of exploration geophysics and seismology. In seismic tomography, what would be the ideal coarseness of the grid? Of course, the finer the grid, the more details may be resolved. However, a small cell (i,j) is likely to be illuminated by few seismic rays; thus, the corresponding model parameter [e.g., $(V_S)_{ij}$ and/or $(V_P)_{ij}$] determined by inverting the travel times will be characterized by large uncertainty.

In very fine grids, there could be cells that are completely at "dark" (that is, they are not sampled by any ray). In such situations, no algorithm can resolve the model parameters of these cells no matter how sophisticated it is. Again, quoting Lanczos (1961) "A lack of information in the solution of a problem cannot be remedied by any mathematical trickery." Of course, this does not mean that very fine grids should always be avoided. It all depends on the objectives of the geophysical investigation. If a high-resolution tomogram is desired, then the tomography should be carried out using large numbers of shots, and the seismograms should be detected by a large number of closely spaced receivers. The wavelength of the generated waves should be short enough to guarantee a sufficient physical resolution. In short, every cell of the spatial grid should be illuminated by a large number of rays, making sure that there is a redundancy of data measurements so that the problem of inverting the travel times becomes *overdetermined,* Of course, this is costly and not always justifiable. With large cells, the spatial resolution decreases; however, the chances of having a sufficient number of rays illuminating the cell increase even with a limited number of shots and receivers. In these cells, the uncertainty of the model parameters is small (Menke 1989). In summary, the mapping of data errors into uncertainty of model parameters is generally dependent on

- Magnitude of experimental errors
- Algorithm used to invert the experimental data
- Spatial resolution of the model

The issue of model resolution has already been discussed in Section 6.4.2 in conjunction with accuracy and data resolution. Model resolution matrix $\mathbf{R}_m$ and Dirichlet spread function $\|\mathbf{R}_m - \mathbf{1}\|_2^2$ have been introduced and specified by Equations 6.21 and 6.24, respectively. In fully overdetermined problems, $\mathbf{R}_m = \mathbf{1}$ and the model parameters are perfectly resolved by the algorithm. However, for an under- or mixed-determined problem, in general, the model resolution matrix is different from the identity matrix. Because $\mathbf{R}_m = \mathbf{G}^{-g}\mathbf{G}$, the structure of $\mathbf{R}_m$ will depend on the structure of the particular generalized inverse matrix adopted to solve the inverse problem.

The discussion in Section 6.4.2 was conducted under the assumption that the experimental data were exempt from errors. If noise is introduced in the measurements, the treatment of the problem becomes more complicated. A milestone contribution in this regard is presented by the Backus and Gilbert (1970) paper. After introducing a new definition of *resolution spread function* that better captures the spread of model resolution matrix if compared with Equation 6.24, these researchers calculated a generalized inverse matrix by minimizing the spread of the model resolution function in the presence of uncorrelated, Gaussian data errors. They found out that a unique solution to the inverse problem does not exist because resolution spread and variance of model parameters are not two independent variables. For instance, model resolution can only be improved at the expense of increasing the uncertainty of model parameters (Parker 1977). The notion of the *trade-off curve* helps to explain the nature of the problem, as shown in Figure 6.20. The above qualitative discussion has been substantiated quantitatively by Backus and Gilbert (1970). The ideal solution does not exist, and the acceptable values of model resolution and precision should be set case by case depending on the characteristics of the specific problem at hand. This also answers the initial question posed at the beginning of this section about what is the ideal geometrical model that should be used in surface wave testing for the inversion of an experimental dispersion curve.

Although the Backus and Gilbert method is rigorous only for linear inverse problems, it can be extended to a certain degree to nonlinear problems as long as the nonlinearity is not too severe (Parker 1977).

Figure 6.21 shows the results of an inversion of a dispersion curve conducted after assuming two different geometrical models. Figure 6.21a has been obtained adopting a five-layer model, whereas a ten-layer model was used in producing Figure 6.21b. The overall thickness of the two models is 20 m. Apart from the different number of layers, all the other parameters of the two models are the same. The experimental dispersion

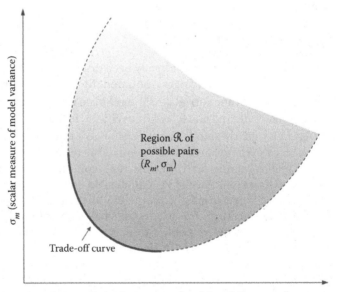

Figure 6.20 Conceptual diagram showing the trade-off curve of model resolution spread and variance in the solution of an inverse problem. (Modified from Parker, R. L., *Ann Rev Earth Planet Sci* 5, 35–64, 1977.)

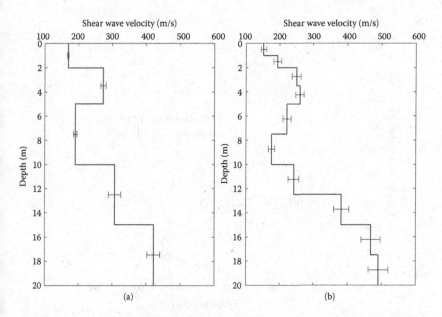

Figure 6.21 Expected value and standard deviation of the shear wave velocity profile obtained using two geometric models with different number of layers with all other parameters being the same: (a) five-layer model; (b) ten-layer model.

curve was assumed to be contaminated with uncorrelated, Gaussian noise characterized by a coefficient of variation equal to 3%, independent from frequency. Figure 6.22 shows the experimental and theoretical dispersion curves (fundamental mode) at the final iteration as obtained from the inversion algorithm. They correspond to the shear wave velocity profiles illustrated in Figure 6.21. The theoretical dispersion curves of the five- and ten-layer model are almost identical as also the error misfit. The general trend of the corresponding shear wave velocity profiles of Figure 6.21 is similar, although there are some differences.

However, Figure 6.21 clearly shows that the uncertainty associated with the values of shear wave velocity calculated for the ten-layer model is greater than the corresponding uncertainty of the five-layer model. This is consistent with the expectations of Figure 6.20 that the greater the desired spatial resolution (i.e., the greater the number of layers), the greater the variance will be. As expected, although the model parameters of the shallowest layers are estimated with greater accuracy with respect to deeper layers in both models, in the shallower layers, the uncertainty of V_S in the ten-layer model is larger than the corresponding uncertainty of the five-layer model. Recently, Xia et al. (2010) conducted an interesting study concerned with the trade-off between model resolution and covariance, specifically devoted to surface wave inversion.

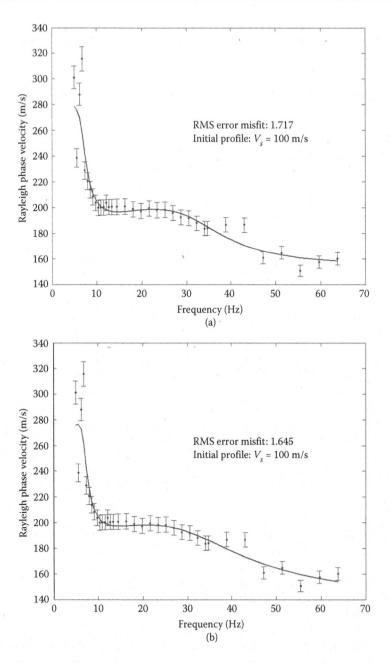

Figure 6.22 Experimental (with uncorrelated Gaussian noise) and theoretical dispersion curves (fundamental mode) at the final iteration of the inversion algorithm corresponding to the shear wave velocity profiles of Figure 6.21: (a) five-layer model; (b) ten-layer model.

6.5.5 Bayesian approach

This section briefly describes an alternative approach to the solution of inverse problems that is known in the literature as the Bayesian probability approach. The name comes from Thomas Bayes, an eighteenth-century British mathematician, who proposed a different interpretation of probability theory based on a theorem of conditional probability that he proposed. Although the theorem has a general validity, two different schools of statistics have originated from it. They are the *frequentist* (classical) and the *Bayesian* schools, and they differ in the interpretation given to the Bayes' theorem. The use of Bayesian statistics in various fields of science and engineering has been growing, particularly in recent years. This is not exempt from controversies raised by the followers of the frequentist school who claim that Bayesian probability introduces a degree of subjectivity and arbitrariness in the interpretation of the postulates of probability theory. However, it has been shown that when Bayesian probability theory is applied to inverse problems, it often leads to similar solutions (Aster et al. 2005). The monograph by Tarantola (2005) is perhaps the most comprehensive and rigorous work currently available on the application of Bayesian probability theory to the solution of parameter estimation and inverse problems.

In using the Bayesian approach to solve an inverse problem, the model parameters are assumed as random variables (more precisely, multivariate random vectors). On the contrary, in the classical approach, model parameters are viewed as coefficients to be determined from the inversion algorithm. Possible errors in the data are accounted for by projecting them into uncertainty of the model parameters. In the Bayesian approach, the solution to a parameter estimation problem consists of determining the probability distribution of the model parameters. This is a radically different method, and it constitutes an essential feature of the Bayesian approach. Its practical implementation requires an *a priori* definition of the probabilistic distribution of model parameters; this is controversial because this choice is subjective, and selecting different distributions will in general result in obtaining different outcomes.

In the Bayesian approach, the result of an inversion is the *a posteriori* probabilistic distribution of model parameters. If no information is available, then the "white" or uniform probability distribution is adopted where any value of a model parameter is equally likely, perhaps within a certain prescribed range (if known a priori). In the Bayesian vocabulary, this distribution is also known as *uninformative*, although some researchers find this terminology controversial (Tarantola 2005). Alternative probability distributions can be adopted if this is justifiable on the basis of availability of a priori information. In this sense, the Bayesian approach allows us to naturally incorporate a priori information about the model parameters (Aster et al. 2005). The maximum entropy method (Rietch 1977) can be used to properly define the a priori distribution of model parameters. This method is based on the assumption that the "best" distribution is the one with the largest information (or entropy) content defined according to the so-called *Shannon's measure* (Tarantola 2005).

Bayesian theory allows computing the a posteriori probabilistic distribution of model parameters from knowing the a priori distribution and the experimental data. From the a posteriori distribution, one can then compute standard moments such as expected value and variance. In the most general situation, the implementation of the Bayesian approach in solving an inverse problem is computationally demanding because it requires the numerical evaluation of integrals with large dimensions. It can be shown that if the Bayesian approach is used to solve a nonlinear inverse problem with Gaussian, uncorrelated data errors associated with an a priori uninformative distribution of model parameters **m**, the a posteriori largest value of **m** corresponds to the nonlinear classical least-squares solution of the problem (Aster et al. 2005).

Applications of Bayesian statistics to the inversion of surface wave data have been recently proposed by Bodin et al. (2012) and Shen et al. (2012).

Chapter 7

Case histories

This chapter presents some applications of surface wave tests for site characterization. The aim is to give an overview of the processes described in previous chapters with some examples, which will be presented step by step. First, simple datasets will be presented, for which it is possible to compare different processing and inversion techniques. This is not always the case; in some circumstances, it is not easy to process the experimental data because of several reasons:

- Inadequate signal-to-noise ratio
- Influence of higher modes, which are difficult to identify
- Influence of body waves
- Discontinuous retrieval of the dispersion curve over the frequency range of interest
- Ambiguities due to lack of resolution and spatial aliasing

These aspects necessitate some experience and engineering judgment and can cause the results to be less reliable (increasing relevance of the nonuniqueness of the solution). It is almost impossible to provide a comprehensive picture of complex datasets, and there is no substitute for experience and engineering judgment in the analysis of experimental data. Accurate design of the experimental setup and the use of adequate equipment and field procedures are prerequisites for obtaining reliable results.

The first section of the chapter provides an example of surface wave analysis of active-source data for the evaluation of the shear wave velocity profile, comparing different processing approaches. The inverse problem is subsequently solved, and there is a discussion of the relevance of a *priori* assumptions on model parameters, with specific references to the existence and position of a water table.

Section 7.2 reports an example of a combined dataset for which active-source data and passive-source data are available. The combination allows an extension of the frequency range from which the experimental dispersion curve can be retrieved and hence the evaluation of the shear wave velocity profile to larger depths, with no need for heavy and expensive active sources. Several approaches for the solution of the inverse problem are compared for this dataset.

In Section 7.3, two examples of surface wave analysis for the evaluation of the shear wave velocity profile and the shear damping ratio profile are reported.

In Section 7.4, a case in which higher modes of Rayleigh wave propagation play a relevant role is presented to discuss the interpretation of more complex datasets.

Surface wave data also can be extracted from datasets that have been collected for other purposes. In Section 7.5, the exploitation of the so-called *ground-roll* in seismic reflection records for the analysis of surface is presented using two case histories.

7.1 COMPARISON AMONG PROCESSING TECHNIQUES WITH ACTIVE-SOURCE METHODS

The test site is located in Saluggia in the northern part of Italy, close to the Dora Baltea River; it is part of a large flat area of fluvial sediments. The subsoil is basically composed of gravels and gravelly sands, with the presence of fine sand and clayey silt in the form of lenses. The water table is very shallow, between 2 and 3 m below the ground surface. The results of a cross hole (CH) test at the site are available from a previous geotechnical survey. Details on the site are reported by Foti (2000).

Data have been collected using 24 vertical geophones (4.5 Hz natural frequency) with two different setups having receiver spacing equal to 1 and 3 m, respectively, and with the source-first geophone offset equal to the inter-receiver spacing. Impact sources have been used for both testing configurations: a 6 kg sledgehammer and a 130 kg weight-drop system height 3 m above the ground level respectively. The two configurations have been chosen to investigate the high- and the low-frequency ranges, respectively.

Examples of the experimental data are reported in Figure 7.1, which shows the vertical stacking of the 15 shot gathers recorded for each of the test setups. The difference in frequency content is evident in the seismograms: the central frequency of the trace at 3 m from the shot is about 25 Hz with the weight drop and 50 Hz with the sledgehammer. Also, a higher mode appears with the larger relative amplitude in the short offset gather with the lighter source.

7.1.1 Two-station (spectral analysis of surface waves)

For the two-station procedure, the signal pairs at the following inter-receiver distance have been used: 3, 6, 12, 18, and 30 m. Figure 7.2 shows an example of the spectral quantities needed for the phase velocity evaluation (see Section 4.3). A set of traces from seven repetitions of the impact has been

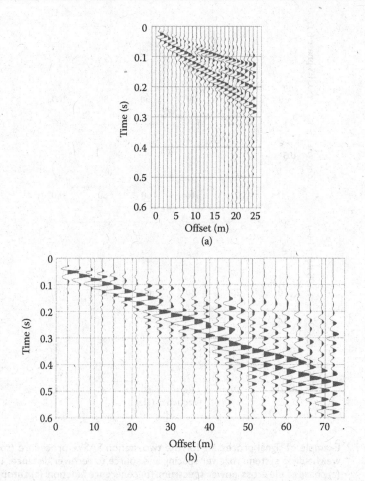

Figure 7.1 Experimental data: (a) 1 m spacing and 6 kg sledgehammer; (b) 3 m spacing and 130 kg weight-drop system (from 3 m height).

used to calculate all the average quantities. The cross-power spectrum phase is used to estimate the frequency-dependent time delay among the receivers and, because the inter-receiver distance is known, the phase velocity. The other quantities are used to locate the distribution of energy (through the autopower spectrum at the two receivers) and the frequency ranges with a high signal-to-noise ratio (SNR) (corresponding to a coherence function close to 1). This information is used to recognize the frequency range over which the cross-power spectrum is reliable. The useful frequency range selected for this pair of receivers goes from about 10 to 35 Hz.

A crucial aspect is related to phase unwrapping; indeed, the low SNR makes it difficult to clearly identify possible phase jumps in the frequency range below 10 Hz. An error in unwrapping the phase would badly affect

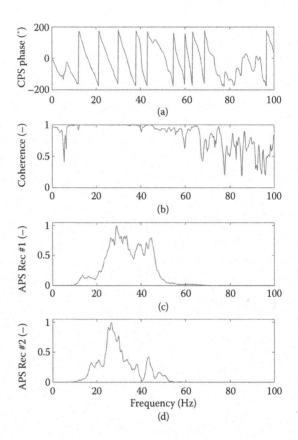

Figure 7.2 Example of signal processing for the two-station SASW procedure (source: weight-drop system; receiver spacing and source to receiver distance: 18 m): (a) phase of the cross-power spectrum; (b) coherence function; (c) autopower spectrum for receiver #1; (d) autopower spectrum for receiver #2.

the evaluation of the dispersion curve. A careful inspection of the comparison of the branch of the experimental dispersion curve obtained with different spacing is of paramount importance. Figure 7.3 shows the different branches of the experimental dispersion curve as retrieved from the different receiver couples. A possible strategy for data processing starts from closely spaced receiver pairs, which typically provide reliable information in the high-frequency range with high SNR. Then, when analyzing data from widely spaced receiver couples, it is possible to check the consistency among the different branches of the experimental dispersion curve to identify possible errors due to phase unwrapping. With good quality data, a significant overlap of information from different receiver pairs is typically obtained. In any case, the evaluation of the experimental dispersion curve requires experience and engineering judgment.

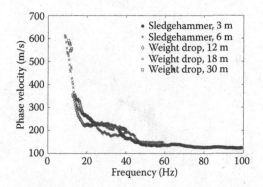

Figure 7.3 Two-station SASW experimental dispersion curve obtained from the analysis of several receiver pairs extracted from the dataset of Figure 7.1.

7.1.2 Frequency–wavenumber analysis

The analysis of multistation data is typically performed using transform-based methods in which the experimental data are transformed from the time–space domain in which they have been collected into a domain in which phase shifts associated with the propagation can be identified. In Figure 7.4a, the amplitude of the frequency–wavenumber spectrum of the experimental data of Figure 7.1b is reported. As detailed in Section 4.6.1, several options are available for the evaluation of the transform. In this case, the spectrum has been obtained with the application of a double Fourier transform. Although this is likely the simplest option, its application is limited to datasets collected with equally spaced receivers. The typical acquisition parameters allow for a sufficient resolution in the frequency domain because this is associated with the length of the acquisition window, which is constrained by the necessity of also recording the whole wave train at the farthest receiver. However, the resolution in the wavenumber domain is limited because the number of available receivers and logistic space constraints typically limit the length of the array. For this reason, a zeropadding procedure is necessary in space domain to obtain an adequate resolution in the wavenumber domain. For example, the spectrum in Figure 7.4a has been evaluated with a zero padding to 1024 sample points in space (i.e., 1000 signals with zero amplitude have been appended to the experimental dataset). The experimental dispersion curve is evaluated from the location of spectral density maxima in Figure 7.4a. From this dataset, the dispersion curve is retrieved consistently in the frequency range between 8 and 43 Hz. The limitation in the high-frequency range is imposed by the length of the array and by receiver spacing. Indeed, on the one hand, the attenuation of high-frequency components prevents the possibility of getting high SNR for the farthest

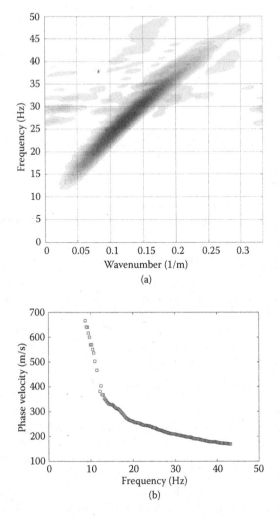

Figure 7.4 Analysis in the *f–k* domain of experimental data from Figure 7.1b: (a) amplitude of the frequency–wavenumber spectrum; (b) experimental dispersion curve evaluated from the location of the maxima of the spectrum.

receivers; on the other hand, receiver spacing limits the maximum available wavenumber.

On a complementary basis, the experimental data from the 1 m spacing array (Figure 7.1a) allow the experimental dispersion curve to be retrieved in the high-frequency range (Figure 7.5). The agreement between the two experimental datasets in the frequency range in which the relative

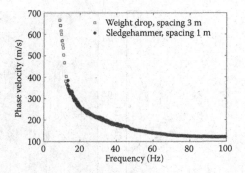

Figure 7.5 Experimental dispersion curve obtained from the combination of the two datasets of Figure 7.1 with *f–k* analysis.

dispersion curve overlaps is very good, showing that the same experimental mode is retrieved.

A key advantage of multistation processing is that, in principle, the dispersion curve is obtained from the location of the maxima without a need for engineering judgment on the part of the operator. For more complex datasets, the influence of higher modes and presence of frequency ranges with unclear maxima due to low SNR lead to the necessity of some engineering judgment for the evaluation of reliable branches of the dispersion curve.

As seen in Section 4.6, other transform-based methods are available. A tool widely used is the slant-stack or τ-p transform, which allows for the set of signals to be represented as the superposition of straight-line events. The two steps of the *f–p* transform of the experimental data in Figure 7.1b are shown in Figure 7.6. On the basis of the subsequent application of the τ-p transform (Figure 7.6b) and a 1D Fourier transform (Figure 7.6c), it is possible to transform the data in the frequency–slowness domain where the experimental dispersion curve is retrieved as the location of the maxima.

A similar procedure is implemented in the multichannel analysis of surface waves (MASW) method proposed by Park et al. (1999). For the same dataset of Figure 7.1b, the frequency–wavenumber, frequency–slowness, and frequency–velocity spectra are plotted in Figure 7.7. The frequency–velocity domain is obtained straightforwardly from the frequency–slowness domain because the velocity is the inverse of the slowness. Its use has the advantage that the experimental dispersion curve is directly visualized as the spectral maxima.

A comparison of the average experimental dispersion curve obtained with the two-station spectral analysis of surface waves (SASW) procedure

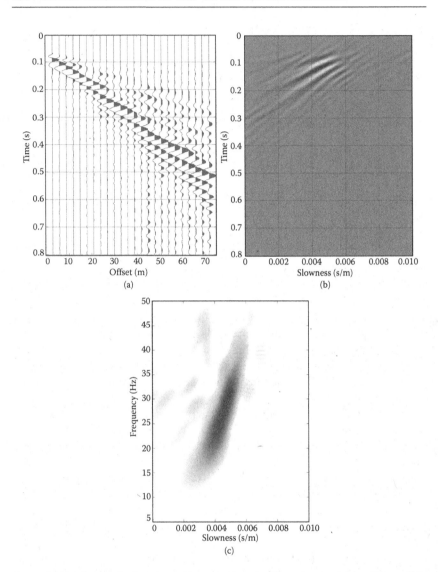

Figure 7.6 Analysis in the frequency–slowness domain of experimental data of Figure 7.1b. (a) experimental data, (b) τ-p spectrum, and (c) f–p spectrum.

with the experimental dispersion curves obtained with multistation procedures is reported in Figure 7.8, showing that the different procedures lead to comparable results in the same dataset. In particular, the dispersion curves obtained with frequency–wavenumber analysis and frequency-slowness analysis are indistinguishable.

The inversion of the experimental dispersion curve obtained from the analysis in the frequency–wavenumber domain is reported below.

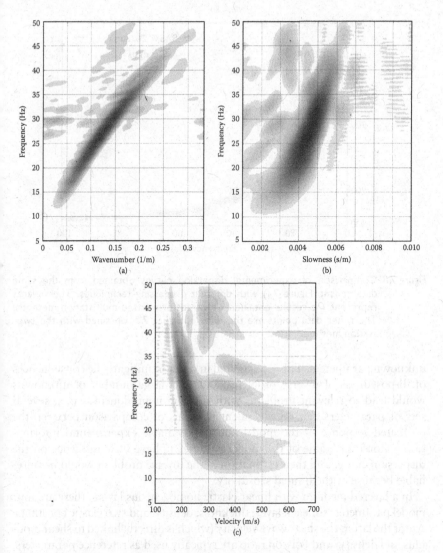

Figure 7.7 Amplitude spectra in different domains of the experimental data of Figure 7.1b: (a) frequency–wavenumber domain; (b) frequency–slowness domain; (c) frequency–velocity domain.

The inversion has been performed using a local search approach based on a weighted and damped least squares algorithm (Herrmann 1994). A comparison among different inversion schemes will be presented in the next section on a different experimental dataset.

Inversion of surface wave data is typically performed with a priori assumptions on some model parameters in order to reduce the number of

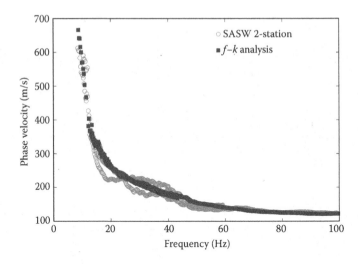

Figure 7.8 Comparison of experimental dispersion curves obtained from the same dataset (see Figure 7.1) with different processing techniques. The squares represent the results obtained with transform-based multistation methods. The other data points are the ones of Figure 7.3, obtained with the two-station method.

unknown parameters. This is suggested in order to mitigate the consequences of ill-posedness of inverse problems. Indeed a large number of unknowns would lead to relevant problems with solution nonuniqueness (i.e., several sets of parameters can be equivalent in terms of comparison between the estimated response of the model and the available experimental information). Moreover, some of the model parameters have little influence on the dispersion curve, and their estimation in the inverse problem would be unreliable because of the limited sensitivity.

In a layered medium with linear elastic homogeneous layers, there are four model parameters for each layer: thickness, density, and two elastic constants. As for the latter, the shear wave velocity (which is directly linked to shear modulus and density) and Poisson ratio are typically used as reference parameters.

Most often the inverse problem is resolved assuming a priori values for density and Poisson ratio of each layer. Material density can be estimated a priori with a sufficient level of confidence with a basic knowledge of the local geology.

The minor influence of the Poisson ratio on surface wave propagation in a homogeneous elastic half-space (Richart et al. 1970) could initially justify the choice of assuming an a priori value for this parameter when interpreting experimental data.

A parametric study was performed by Nazarian (1984) to assess the relative influence of model parameters on the dispersion curve. Considering the similarity among numerical curves obtained for different values of density and of Poisson ratio over a wide frequency range, he concluded that the influence of density and Poisson ratio was negligible compared to that of layer thickness and shear wave velocity. But in that parametric study, only simultaneous variations of the Poisson ratio for all the layers were considered. Different values for different layers can produce variations in the overall shape of the dispersion curve that may significantly influence the solution of the inverse problem.

In particular, a relevant effect may be associated with the water table position at the site. Indeed, for a dry soil, the range of variation of the Poisson ratio is about 0.1 to 0.3, whereas for a saturated porous medium, the apparent Poisson ratio is very close to 0.5 because in terms of wave propagation, an undrained behavior is expected. Substantial errors can be driven by the wrong hypothesis about the water table position, as will be shown in the following discussion.

Information on water table position at the Saluggia site can be inferred from a P-wave refraction survey that has been carried out together with surface wave tests. In fact, active-source surface wave tests and P-wave seismic refraction share a very similar testing setup. For this reason, it is interesting to evaluate the synergies between the two tests (see Section 6.4.4 for further details). P-wave refraction surveys in shallow sediments are very sensitive to the water table position because of the abrupt change in P-wave velocity associated with the passage from unsaturated to saturated conditions. At the Saluggia site, a simple interpretation of P-wave arrival times (assuming a horizontally layered system, which is reasonable for the geological context) shows an interface at a depth of 3 m, which can be assumed to be the position of the water table.

In order to show the importance of the assumption of the water table position, three different inversions of the experimental dispersion curve in Figure 7.5 are compared. The inversions have been performed with the same code, assuming the same starting V_S profile (constant V_S with depth). Number and thicknesses of the layers have been assumed a priori on the basis of a previous interpretation of the same surface wave dataset (Foti 2000).

Inversion #1 has been conducted assuming the correct position of the water table (3 m below the ground surface), with values of density and Poisson ratio of a saturated medium (2100 kg/m³ and 0.49, respectively) below this level. Inversion #2 (no water table) has been conducted assuming a constant value of density and Poisson ratio with depth (1800 kg/m³ and 0.2, respectively). Inversion #3 assumes an incorrect position for the water table (7 m below the ground surface).

In terms of shear wave velocity profiles, the final results are reported in Figure 7.9 and are compared to the independent results of a cross-hole test at the same location. The dispersion curves (Figure 7.10) show the perfect

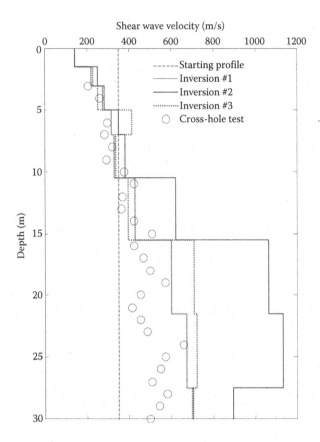

Figure 7.9 Relevance of the hypothesis on water table position on the inversion of surface wave data. Results of the inversion are compared to the shear wave velocity profile obtained with a cross-hole test. (From Foti, S., and C. Strobbia, Some notes on model parameters for surface wave data inversion. In: Proceedings of SAGEEP 2002, Las Vegas, USA, February 10–14, 2002, CD-Rom.)

equivalence among the results of the different inversions and a good match with the experimental dispersion curve. The shear wave velocity profiles show great discrepancies. Inversion #1, with a priori assumptions on the density and the Poisson ratio consistent with available information on the water table position, provides a shear wave velocity profile with a trend of gradual increase with depth that is in agreement with the stratigraphy of the site and with the results of the cross-hole test. The hypothesis of the absence of the water table (Inversion #2) is strongly misleading and results in severe inaccuracies in the final results. The wrong hypothesis on the position of the water table (Inversion #3) leads to an overestimation of the shear wave velocity profile and hence of the soil stiffness at depth. These errors

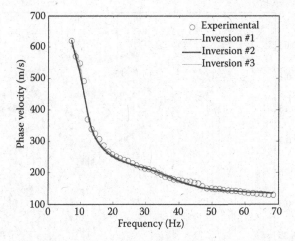

Figure 7.10 Experimental dispersion curve compared to the theoretical dispersion curves for the three profiles in Figure 7.9. (From Foti, S., and C. Strobbia, Some notes on model parameters for surface wave data inversion. In: Proceedings of SAGEEP 2002, Las Vegas, USA, February 10–14, 2002, CD-Rom.)

are driven by the necessity of following the trend of increasing Rayleigh wave velocity for decreasing frequency. This trend is partially generated by the higher Poisson ratio, but it is interpreted only as a difference in shear wave velocity with depth for Inversion #2 and partially for Inversion #3.

This example shows the importance of a proper selection of a priori values for parameters that are not considered as unknown in the inversion process. Cautious and scrupulous judgment by the operator is required in this instance. All the available information on the site has to be taken into account in the selection.

7.2 COMPARISON AMONG INVERSION STRATEGIES

A comparative analysis of inversion strategies is presented in this section. The test site is part of a wide alluvial fan in the Italian Alps. In this case, experimental data for surface wave analysis have been collected using an active-source multistation setup and a circular array for microtremors. The combination of active-source and passive data is often adopted to extend the characterization to significant depths when heavy seismic sources such as truck-mounted Vibroseis are not available.

7.2.1 Experimental dataset

The active-source dataset has been collected using an 8 kg sledgehammer as the seismic source and 48 4.5 Hz vertical geophones. The spacing between

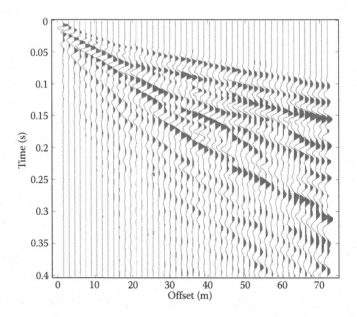

Figure 7.11 Experimental data for active-source setup (48 receivers with 1.5 m spacing and 6 kg sledgehammer).

geophones is 1.5 m for a total array length of 70.5 m. A total of 30 shot gathers have been collected for subsequent vertical stacking and analysis of experimental uncertainty. The resulting stacked shot gather is reported in Figure 7.11, and it shows surface waves preceded by large amplitude body waves.

The corresponding frequency–wavenumber spectrum is reported in Figure 7.12. From the location of spectral maxima in Figure 7.12, it is possible to obtain the experimental dispersion curve in the frequency range 10–40 Hz.

Passive data have been collected with 2 Hz vertical geophones deployed in a circular array of 75 m diameter. The spectral analysis has been performed with the frequency domain beamformer method as implemented by Zywicki (1999). Some examples of contour plots of the spectral power density are reported in Figure 7.13. Each spectrum is plotted as a function of wavenumbers in two orthogonal directions for a given frequency. The origin of the axes coincides with the center of the acquisition array. The position of the energy peak identifies the direction of the arrival of the main perturbation with respect to the center of the array and allows for the identification of the wavenumber associated with the specific frequency, which can be used to evaluate the velocity of propagation. A circle is plotted on top of each spectrum: its radius is the wavenumber corresponding to the identified phase

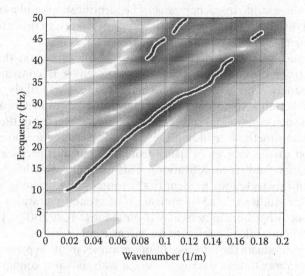

Figure 7.12 Analysis in the *f–k* domain of experimental data of Figure 7.11, amplitude of the frequency–wavenumber spectrum and its absolute maxima.

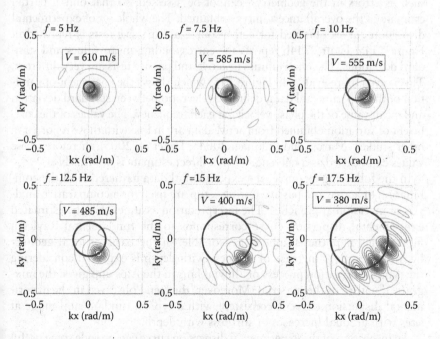

Figure 7.13 Analysis of passive data in the frequency–wavenumber domain, examples of amplitude of the spectrum for different frequencies. The circles represent the dominant wavenumber from which the phase velocity for the specific frequency is evaluated. (From Foti, S. et al., *Riv Ital Geotecnica*, 41(2), 39–47, 2007.)

velocity at the specific frequency value. The identification of phase velocities is repeated for all the frequencies in the range of interest in order to evaluate the experimental dispersion curve for Rayleigh waves.

At this site, waves are arriving from a specific direction that can be identified from the location of the maxima, which is consistent in the different panels. The main source of microtremors was likely associated with construction activities in an area a few hundreds of meters away from the site. In such conditions, special care should be exerted in the choice of the processing method, avoiding the methods that assume homogeneous distribution of the sources in space around the array (specifically spatial autocorrelation and its modification). This example also shows the great potential of this technique to identify the sources of vibrations, which can be useful for studies related to mitigation of ground vibrations caused by industrial or other human activities (Comina and Foti 2007). The source identification may also be relevant for forensic applications.

In order to evaluate the influence of uncertainties in the experimental data, independent processing of datasets collected with the same configuration has been performed for active-source and passive methods. This procedure allows for an evaluation of uncertainty related to background noise but other effects—such as errors in the geometry—cannot be assessed, so that only a partial estimate of the overall uncertainty is obtained. The whole set of experimental dispersion curves obtained for active-source and passive tests is reported in Figure 7.14a; Figure 7.14b reports the corresponding mean values and standard deviations. The uncertainties in active and passive tests are very different. This is even more evident in Figure 7.15, which shows the coefficient of variation of the experimental data, defined as the ratio between standard deviation and mean value of the phase velocity at each frequency. The values of the coefficient of variation obtained from active data are in line with those reported in other studies (Marosi and Hiltunen 2004a; Lai et al. 2005); therefore, these values can be used as a reference when a direct estimate is not feasible.

In the following example, the experimental data gathered with the combination of active and passive testing setup are used for a comparative analysis of inversion approaches. The interpretation assumes that the estimated experimental dispersion curve corresponds to the fundamental Rayleigh mode. This hypothesis appears reasonable on the basis of local geology because a gradual increase of stiffness with depth is expected considering the typical formation process of alluvial fans in the Alps and given the mix-grained nature of the material. Moreover, the trend observed in the experimental dispersion curve is consistent with a typical fundamental mode at sites with gradual increases of stiffness with depth.

An inspection of the experimental dispersion curve can provide some useful hints for the solution of the inverse problem. Taking into account maximum and minimum available wavelength, it is possible to have an indication of the investigation depth and the expected resolution close to the ground surface.

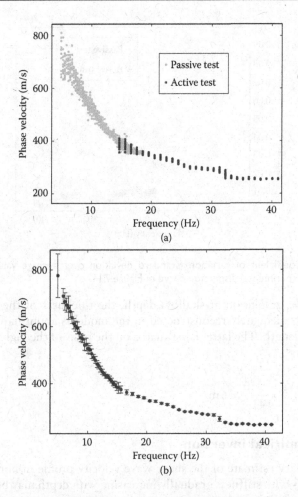

Figure 7.14 Experimental dispersion curve from the combination of active-source and passive data: (a) ensemble of data points for several repetitions of the test with the same geometrical setup; (b) average dispersion curve and its associated standard deviation evaluated with the data in (a). (From Foti, S. et al., *Riv Ital Geotecnica*, 41(2), 39–47, 2007).

The investigated depth is about half of the maximum available wavelength. Considering the data point at the lowest available frequency

$$\lambda_{max} = \frac{V_R}{f} = \frac{780}{5} = 155\,m \tag{7.1}$$

Hence, the investigated depth is about 80 m. Layers below this depth would not be sufficiently constrained by the experimental data, and it is inappropriate to include them in the inversion.

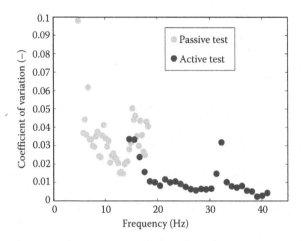

Figure 7.15 Coefficient of variation (standard deviation over average values) for the experimental dispersion curve of Figure 7.14.

As for the resolution at shallow depth, the thickness of the first layer that can be adequately reconstructed in the model is about half the minimum wavelength. The latter is estimated on the basis of the high-frequency components

$$\lambda_{\min} = \frac{V_R}{f} = \frac{250}{41.5} = 6\,\text{m} \tag{7.2}$$

7.2.2 Empirical inversion

A preliminary estimate of the shear wave velocity profile in normally dispersive sites with stiffness gradually increasing with depth may be obtained with the mapping of the points of the dispersion curve into approximate values of shear wave velocity with depth (Section 6.3).

Each point of the experimental dispersion curve is transformed into a point of the pseudo–shear wave velocity profile by associating a shear wave velocity slightly higher than the phase velocity of the Rayleigh wave at that frequency (e.g., $V_S = 1.1\ V_R$) to a depth equal to a fraction of the wavelength (e.g., 1/2.5). This procedure is derived from the observation of the wave propagation parameters in a homogeneous linear elastic half-space. Application to the experimental dispersion curve of Figure 7.14b gives the pseudo–shear wave velocity trend of Figure 7.16a.

This approach cannot be considered a substitute of a proper inversion process, but it can be useful for estimating a first tentative profile to be used as a starting profile (Figure 7.16b) in iterative procedures for automatic inversion or for trial and error procedures.

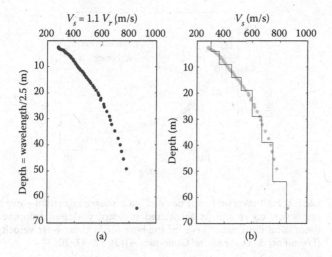

Figure 7.16 Estimate of the shear wave velocity profile with direct mapping of the point of the experimental dispersion curve: (a) pseudo–shear wave velocity profile; (b) construction of a first tentative profile for subsequent solution of the inverse problem.

7.2.3 Deterministic approach (least squares)

The deterministic inversion of the mean values of the experimental dispersion curve based on a weighted damped least square algorithm (Herrmann 1994) is reported in Figure 7.17. The profile of Figure 7.16b has been used as a starting model for the iterative procedure. The obtained shear wave velocity profile is compared to the results of down-hole tests at two sites in the same alluvial fan. The actual distance between each of the three sites is about 1 km so a direct comparison is not appropriate, but considering the nature of the sediments, the agreement appears to be meaningful anyway.

In order to obtain the standard deviation of the final shear wave velocity profile, the standard deviation of the Rayleigh wave phase velocities (Figure 7.17) can be incorporated in the inversion process using a linearization in the neighborhood of the final solution (Tarantola 2005). The corresponding error bars on the velocity profile of Figure 7.18 have been obtained using the procedure described in Section 6.5.3.

The low experimental uncertainty of active-source data in the high-frequency range (Figure 7.17) leads to very low uncertainties in the velocity estimates for shallow layers (Figure 7.18). For deeper layers, the higher uncertainty of passive data leads to higher uncertainties in the shear wave velocity. The coefficient of variation of shear wave velocity ranges from less than 1% for shallow layers to about 6% for deeper layers, showing that the inversion process is not affected by error magnification.

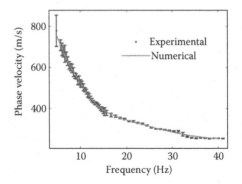

Figure 7.17 Local search inversion with damped least square algorithm—experimental dispersion curve and its associated standard deviation compared to the theoretical dispersion curve for the best-fitting shear wave velocity profile. (From Foti, S. et al., *Riv Ital Geotecnica*, 41(2), 39–47, 2007.)

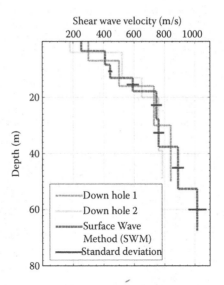

Figure 7.18 Local search inversion with damped least square algorithm—best-fitting shear wave velocity profile and its associated standard deviation compared to results of down-hole tests at locations nearby. (From Foti, S. et al., *Riv Ital Geotecnica*, 41(2), 39–47, 2007.)

7.2.4 Stochastic approach (Monte Carlo)

Deterministic inversions with local search methods provide a solution that may represent a local minimum of the misfit function (typically a norm of the difference between experimental and theoretical dispersion curves). The local minimum in which the solution is trapped depends also on the starting profile for the iterative procedure. In order to mitigate the risk of getting trapped into a local minimum, global search methods can be applied.

Moreover, a stochastic approach can provide an assessment of the solution's nonuniqueness. Indeed, for mathematically ill-posed problems, such as inverse problems, the solution is not unique, and several solutions can be considered equivalent because they provide the same or similar agreement with experimental data. In this respect, it is also necessary to account for uncertainties in experimental data when comparing them to the theoretical dispersion curve of possible solutions.

Experimental data for the La Salle site have been also interpreted with a global search method based on a Monte Carlo approach, using the code by Socco and Boiero (2008). It allows a limited population size to be used because it applies scale properties of modal curves (Socco and Strobbia 2004; Maraschini et al. 2011) to improve the efficiency in sampling the model parameter space. Moreover, the use of scale properties allows model parameters to be outside the boundaries initially selected for the generation of the population (Socco and Boiero 2008). The application of a statistical test (one-tailed Fisher test) leads to the selection of a set of acceptable shear wave velocity profiles according to data quality and model parameterization (Socco and Boiero 2008). Selected V_S velocity profiles represent the final result of the inversion for a chosen level of confidence. The results reported in Figure 7.19 have been obtained with a population of 10^5 models. Well-resolved parameters produce a small range of variations, while poorly constrained parameters can assume virtually any value within a wide range.

Figure 7.20 shows all the theoretical dispersion curves for the selected profiles of Figure 7.19 (compared to the experimental one) and also shows the equivalence of these profiles. The hierarchical relative misfit representation adopted for the equivalent profiles (Figure 7.19) is such that the darkest color corresponds to the profile having the lowest misfit. The deterministic inversion of the previous section and its associated standard deviation are also reported, for comparison, in Figure 7.19.

High resolution and low uncertainty are attained for shallow layers, although a much higher uncertainty from nonuniqueness is obtained for deeper layers. This is due to the fact that the resolution of surface waves inevitably decreases with depth and to the higher experimental uncertainty in the passive tests (Figure 7.15). The uncertainty in the first layer looks negligible, both for its velocity and its thickness, thanks to the large amount of data available to constrain these parameters in the inverse problem solution. It has to be stressed that the obtained parameters correspond to the average laterally homogeneous model. The model errors, in particular those due to the lateral variations, are not estimated.

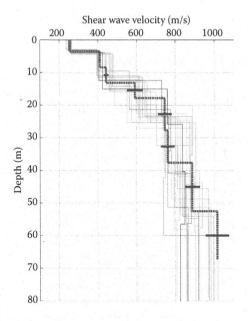

Figure 7.19 Global search inversion with Monte Carlo approach—best-fitting shear wave velocity profiles selected with a statistical test compared to the solution of Figure 7.18 obtained with the local search method (LSM). The profiles are plotted in a grayscale color range according to their associated misfit (the darker the line, the lower the misfit). (From Foti, S. et al., *Riv Ital Geotecnica*, 41(2), 39–47, 2007.)

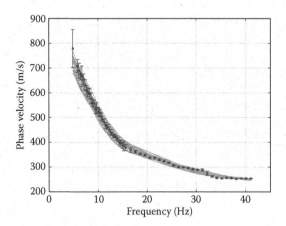

Figure 7.20 Global search inversion with Monte Carlo approach—experimental dispersion curve and its associated standard deviation compared to the theoretical dispersion curve for the best-fitting shear wave velocity profiles of Figure 7.19 selected with a statistical test. (From Foti, S. et al., *Riv Ital Geotecnica*, 41(2), 39–47, 2007.)

It can be interesting to assess the consequences of solution nonuniqueness in terms of the specific application for which site characterization is performed. In this respect, Figure 7.21 shows the response spectra obtained for the same strong motion record by assuming each one of the shear wave velocity models of Figure 7.19. Clearly, the differences are quite limited. If no a priori information is used to constrain the results, errors in the identification of a single parameter can be very high. Nevertheless, the consequences of such uncertainties are very limited for seismic site response studies, at least where a 1D approach is deemed to be reasonable. Indeed, if shear wave velocity profiles are equivalent with respect to Rayleigh wave dispersion, they are also equivalent with respect to seismic ground response. Different profiles that represent the global response of the site in terms of Rayleigh wave propagation are also representative of the seismic response of the soil deposit.

Another relevant issue is related to the estimation of $V_{S,30}$ (i.e., the average velocity of the shallowest 30 m), which is often adopted in seismic studies and in seismic codes for subsoil classification (Borcherdt 1994). The estimate of $V_{S,30}$ for each of the equivalent shear wave velocity profiles in Figure 7.19 is reported in Figure 7.22. Because $V_{S,30}$ is an average parameter, the influence of solution nonuniqueness on its estimate is limited, and the range of variation of the different estimates is very narrow.

A preliminary estimate of $V_{S,30}$ can be obtained directly from the experimental dispersion curve. The idea that the dispersion curve offers a consistent and simplified way of estimating the $V_{S,30}$ has been considered by Brown et al. (2000), who proposed the Rayleigh wave phase velocity at a

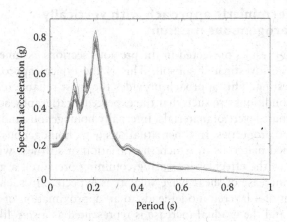

Figure 7.21 Simulation of 1D seismic ground response for the group of equivalent shear wave velocity profiles of Figure 7.19. (From Foti, S. et al., *Soil Dynam Earthquake Eng*, 29(6), 982–993, 2009.)

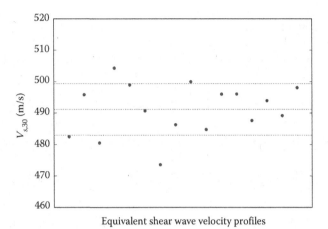

Equivalent shear wave velocity profiles

Figure 7.22 $V_{S,30}$ evaluated for each of the equivalent shear wave velocity profiles of Figure 7.20.

wavelength of 36 m as a direct estimate of $V_{S,30}$. Such an approach cannot be considered as an alternative to the solution of the inversion process because this oversimplification can lead to large errors, especially for complex stratigraphical situations. Nevertheless it can provide a useful reference for preliminary assessments. For the experimental dispersion curve of Figure 7.14, this estimate would be equal to 450 m/s, which is about 10% less than the values estimated with the shear wave velocity profiles obtained from the inversion (Figure 7.22).

7.2.5 Deterministic approach with vertically heterogeneous medium

The inversion results presented in the previous sections assumed a layered model for the investigated subsoil. This is the typical approach in surface wave testing. This approach provides the most reliable results where geological conditions are such that the expected stratigraphy actually consists of alternate layers of materials, internally homogenous and having different seismic properties. In other situations (e.g., homogenous deposits of coarse grained materials in which the variation of stiffness with depth is mainly due to the effect of increasing confining pressure), a gradual and continuous increase of shear wave velocity is expected. In such situations, the adoption of a layered model leads to an approximation of actual field conditions, and the gradual increase is represented as a step-like increase. The interpretation of surface wave data can be performed with a more appropriate model allowing for continuous variations of seismic properties (see Section 2.4).

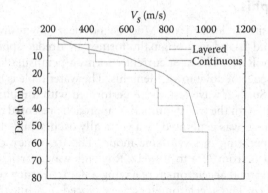

Figure 7.23 Comparison of shear wave velocity profiles obtained with local search methods using different reference models—smooth vertical heterogeneous elastic model versus stack of linear elastic layers.

As mentioned before, the La Salle site is part of a large alluvial fan, which is mainly composed of chaotic material. The inversion of the dispersion curve of Figure 7.14b, in terms of smooth variations, has been performed with the code developed by Rix and Lai (2013). The results are shown in Figure 7.23 and are compared to the results obtained with the classical layered medium approach (Figure 7.18). For the very shallow portion of the profile, which shows a marked jump of shear wave velocity likely associated with the superficial layer of residual and vegetated soil, the interpretation in terms of layered medium is likely more appropriate.

7.3 EXAMPLES FOR DETERMINING V_S AND D_S PROFILES

A crucial aspect in site characterization for seismic applications is related to the quantification of material damping. Indeed, internal energy dissipation in hysteretic cycles plays a major role in the definition of seismic ground response. Laboratory tests provide an evaluation over a wide strain range, but they are performed on a limited number of samples; hence, they represent a local estimate not representative of the behavior of the whole deposit. *In situ* seismic tests can in principle be used to characterize large volumes, but their implementation faces difficulties related to the necessity of separating geometric attenuation and material attenuation in wave propagation phenomena. Surface wave data can be interpreted to provide an estimate of material damping. The methods for the estimation of the experimental attenuation curve have been discussed in Chapter 5, whereas the inversion procedures to get the material damping profile are reported in Chapter 6. The application of the whole procedure to experimental data is reported in the following two examples.

7.3.1 Memphis

This testing site is located on Mud Island near downtown Memphis, Tennessee. Mud Island was originally formed by dredge spoil taken from the Mississippi River. The area currently forms a peninsula located on the northwestern edge of downtown Memphis. The water table is 8 m below the ground level. Surface wave tests were performed with a multistation array and interpreted with the transfer function approach presented in Section 5.3.

The wave train was generated by a vertically oscillating, electrodynamical shaker operating in swept-sine mode. The frequency range used in the field test was from 3.75 to 100 Hz. Rayleigh wave particle motion was recorded by vertical accelerometers having a flat frequency response from 0.10 to 300 Hz. The accelerometers were placed in a linear array at the following offsets from the source: 2.44, 3.05, 3.66, 4.57, 5.49, 6.71, 8.54, 10.37, 12.80, 15.24, 18.29, 21.34, 24.39, 28.96, and 33.54 m. The acceleration of the armature of the electrodynamical shaker was also measured with a piezoelectric accelerometer to allow the input force to be calculated. The experimental transfer function was obtained at each receiver location from an average in the frequency domain of 10 measurements to reduce the variance of the measured spectral quantities.

The complex-valued transfer function is used for estimating the complex wavenumber through a regression with the analytical expression of Equation 5.10. An example of the fitting of the regression process is reported in terms of phase and amplitude of the transfer function in Figure 7.24. The process is repeated over the frequency range scanned with the harmonic source in order to simultaneously obtain the experimental dispersion and attenuation curves (Figure 7.25). The inverse problem aimed at estimating

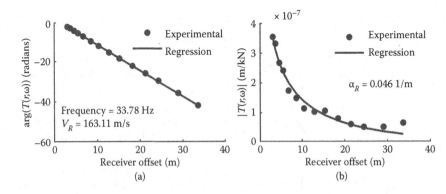

Figure 7.24 Regression of the complex-valued experimental transfer function for a sample frequency: (a) plot of the phase and (b) plot of the amplitude as a function of offset. (From Lai, C. G. et al., *Soil Dynam Earthquake Eng*, 22(9–12), 923–930, 2002.)

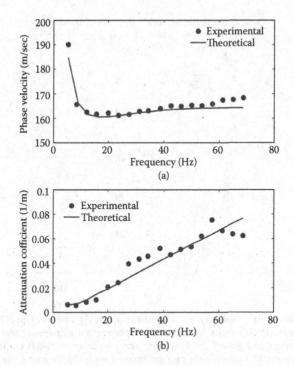

Figure 7.25 Comparison between experimental and theoretical dispersion (a) and attenuation (b) curves (fundamental mode). (From Lai, C. G. et al., *Soil Dynam
Earthquake Eng*, 22(9–12), 923–930, 2002.)

the shear wave velocity and damping ratio profiles has been solved with a
damped least square approach with coupled and simultaneous inversion
(see Section 6.4.3.4). The results are reported in Figure 7.26, where the
shear wave velocity profile is also compared to results of a down-hole survey performed with the seismic cone (Mayne 2000).

7.3.2 Pisa

The site of the Leaning Tower of Pisa has been extensively characterized over
the years for studies on the stability of the tower and for planning the remedia-
tion works for its safeguard. The local stratigraphy, from top down, is as follows. The first formation is predominantly slightly clayey and sandy silt with
lenses of sand and clay. It extends down to a depth of around 10 m. The underlying clayey formation has a thickness of approximately 30 m. In particular,
the zone between 10 and 20 m from the ground surface is a very uniform
deposit of soft clays that was the main cause of the differential settlement during the construction of the tower. The water table is shallow (2 to 3 m below
the ground surface) and is subject to marked seasonal fluctuations.

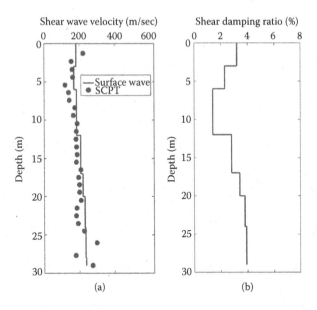

Figure 7.26 Shear wave velocity (a) and shear damping ratio (b) profiles obtained with simultaneous inversion of experimental dispersion and attenuation curves of Figure 7.25. The shear wave velocity profile is compared to the results of a down-hole test performed with the seismic cone penetration test (SCPT). (From Lai, C. G. et al., *Soil Dynam Earthquake Eng*, 22(9–12), 923–930, 2002.)

The field measurements for surface wave testing were conducted using a 24-channel seismograph and vertical geophones having a natural frequency of 4.5 Hz. Two different seismic impact sources were used to generate energy over a broad frequency range: a weight-drop system (130 kg released under self-weight from a height of 3 m) and an 8 kg sledgehammer for long and short arrays, respectively. The necessity of two different testing configurations is due to the trade-off between the length of the array and the frequency band. The test location is shown in Figure 7.27, in which it is also shown the location of a cross-hole test performed in the vicinity of the Leaning Tower. The lateral homogeneity of the soil deposit across the site allows a meaningful comparison of the results.

Because equal receivers are used at each location, their frequency response is considered equivalent, and the experimental transfer function is evaluated using the deconvolution procedure reported in Section 5.3. An example of experimental transfer function and its regression for obtaining the complex wavenumber is reported in Figure 5.6. By repeating the regression process over the usable frequency range, the experimental dispersion and attenuation curves are obtained (Figure 7.28). The inverse problem has been solved using a damped least squares approach to simultaneously obtain the shear

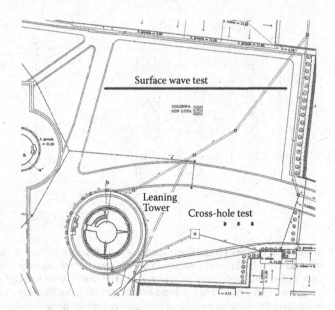

Figure 7.27 Location of surface wave test and cross-hole test at the Leaning Tower of Pisa site.

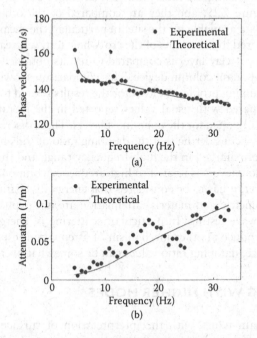

Figure 7.28 Comparison between experimental and theoretical dispersion and attenuation curves (fundamental mode). (From Foti, S., *Geotechnique*, 53, 455–461, 2003.)

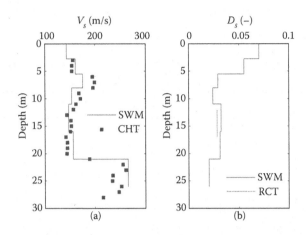

Figure 7.29 Shear wave velocity (a) and shear damping ratio (b) profiles from surface wave method (SWM) obtained with simultaneous inversion of experimental dispersion and attenuation curve of Figure 7.28. The shear wave velocity profile is compared to the results of the cross-hole test (CHT); the damping ratio profile is compared to laboratory data on high-quality samples obtained with the resonant column device (RCT). (From Foti, S. *Geotechnique*, 53, 455–461, 2003.)

wave velocity and damping ratio profiles (Herrmann 1994). The results are reported in Figure 7.29, and they are compared to the other independent experimental data available for the site. In particular, the shear wave velocity profile is compared to the results of the cross-hole test, whereas the damping ratio for the upper clay layer is compared to results obtained in the laboratory with the resonant column device. This comparison shows the reliability of the shear damping profile—at least for the results related to deeper layers. However, compared with usual values reported in the literature, relatively high values are obtained for the uppermost layers. These discrepancies can be partially justified considering that the damping ratio of soils has a frequency dependency more marked in the high-frequency range and that information related to shallow layers is related to high-frequency components of surface waves. Moreover, it has to be considered that energy dissipation phenomena other than geometric and material attenuation are not explicitly accounted for in surface wave testing. In particular, scattering of energy at heterogeneities can introduce another mechanism of attenuation, especially at high frequency; hence, damping ratio values can be somewhat overestimated.

7.4 DEALING WITH HIGHER MODES

One of the main issues with the interpretation of surface wave data is related to the effect of higher modes of propagation and with the difficulties in properly accounting for them in the inversion process. As discussed in Section 2.4, propagation of surface waves in a vertically heterogeneous medium can be interpreted as the effect of the superposition of free modes of propagation, which are characterized by different dispersion curves. The possibility of identifying individual modes of propagation in experimental data is strongly dependent on the acquisition parameters (specifically on the array length), as discussed in Chapter 3. Often the experimental dispersion curve is indeed an apparent one, arising from the combined effects of different modes of propagation. In such situations, the solution of the inverse problem is not straightforward and requires an adequate strategy to account for higher modes' contribution.

A possible approach is based on the numerical evaluation of an apparent dispersion curve, which takes into account the contribution of higher modes and the layout used in the acquisition of experimental data. The misfit function for the solution of the inverse problem is then defined as a norm of the difference between the experimental dispersion curve and the apparent dispersion curve obtained from the forward model. Two crucial issues are the time-consuming procedure required for the solution of the forward problem and the difficulties in the evaluation of partial derivatives for local search methods.

The following example is based on a different approach in which the misfit function for the solution of the inverse problem is defined in terms of the values of the determinant associated with free surface wave modes (Maraschini et al. 2010). As seen in Section 2.4, the modal curves for free surface waves in a layered (visco-)elastic medium can be obtained by solving an eigenvalue problem, which can be formulated in terms of a propagator matrix representing the dynamic interaction among layers. Examples of such procedures are the transfer matrix method (Thomson 1950; Haskell 1953), the stiffness matrix method (Kausel and Roesset 1981), and the reflection and transmission formulation (Kennet 1974). Wavenumbers that represent the modal solutions can be obtained by imposing the determinant of the matrix to be equal to zero.

If experimental wavenumbers are plugged into the determinant for a given subsoil model, the value of the determinant itself will be different from zero, unless the experimental data are an exact modal dispersion curve for that given model (Figure 7.30). In this sense, the value of the determinant can be considered a measure of the misfit between experimental data and the parameters of propagation in the specific model. With reference to Figure 7.30, the misfit is the sum of the values of the determinant for any specific data point of the experimental dispersion curve. The inverse problem can be solved using this definition of misfit rather than the usual distance between dispersion curves. The advantage of such an approach is twofold. First, it is not necessary to associate a priori the experimental data to the fundamental mode rather than to any of the higher modes. Second, the solution of the forward problem is very light from a computation point of view because time-consuming zero search procedures are avoided. For this reason, the approach is very appealing for global search

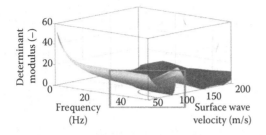
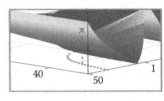

Figure 7.30 Representation of the determinant misfit function. The point indicated with X is the absolute value of the determinant of the synthetic model m* corresponding to an experimental point; this value is the misfit of m* for the single experimental point. (From Maraschini, M., and S. Foti, *Geophys J Int*, 182(3), 1557–1566, 2010.)

methods because very large populations can be used with limited computational efforts. An example of application with a multimodal Monte Carlo procedure (Maraschini and Foti 2010) is reported in the following.

The test has been performed on an unpaved track in which the top layer is an artificially compacted layer of gravels and pebbles. The stratigraphic conditions below this superficial layer report an alluvial–colluvial deposit a few meters thick that lies over weak torbiditic bedrock (interbedded silts and sands). The experimental data have been collected with an array of 24 vertical geophones with a spacing of 0.7 m because of limited area. The short array was considered sufficient as the target of the characterization was very shallow because the seismic bedrock is expected to be close to the ground surface. An 8 kg sledgehammer has been used as a source. The experimental dispersion curve has been obtained with a frequency–wavenumber analysis.

A velocity increase with frequency can be observed in the experimental dispersion curve (Figure 7.31b), suggesting the presence of a velocity inversion in the subsoil model. In this condition, the apparent dispersion curve follows the higher modes in the high-frequency band. On the contrary, the sharp stiffness jump expected at the contact between the alluvial layers and the bedrock could cause a transition to the first higher mode at low frequency as observed in other synthetic and experimental datasets (Foti 2002).

Data were inverted with a Monte Carlo approach considering a population of 10^7 random profiles. The best profiles, which have been selected with a statistical test, are reported in Figure 7.31a. The stiff top layer is coherent with available information on the site. The solutions present a velocity inversion in the shallow portion of the soil profile. The experimental dispersion curve is reported on top of the Haskell–Thomson determinant of the best-fitting profile in Figure 7.31c. Although the inversion process does not require the evaluation of modal dispersion curves, the dispersion curves of the best profiles have also been computed to show their agreement with experimental data (Figure 7.31b).

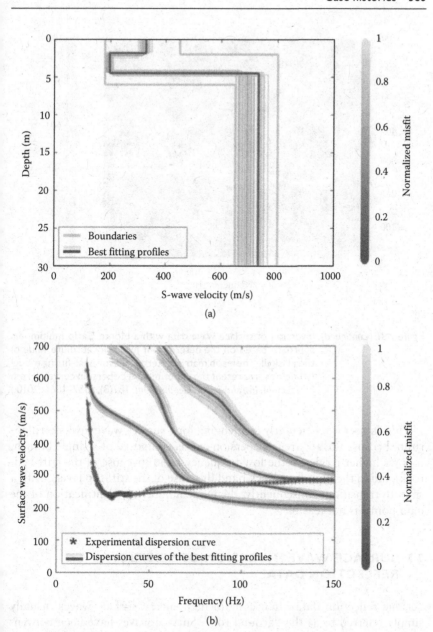

Figure 7.31 Inversion of surface wave data with a Monte Carlo multimodal approach based on the determinant misfit: (a) best-fitting profiles and boundaries for the generation of the population (decreasing misfit from clear to dark); (b) dispersion curves for best models compared with the experimental dispersion curve; same color scale as (a).

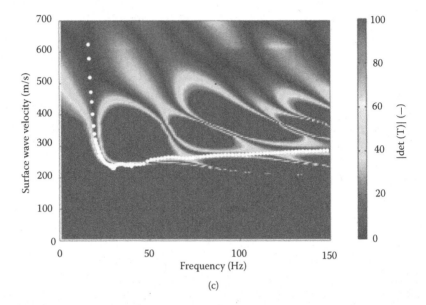

(c)

Figure 7.31 (Continued) Inversion of surface wave data with a Monte Carlo multimodal approach based on the determinant misfit: (c) absolute value of the Haskell–Thomson matrix determinant for best-fitting model (white dots represent the experimental dispersion curve). (From Maraschini, M., and S. Fot, *Geophys J Int,* 182(3), 1557–1566, 2010.)

This dataset is particularly challenging for a surface wave inversion algorithm because the apparent dispersion curve, composed of a single branch, follows higher modes in the low-frequency band (because of the stiff bedrock) and in the high-frequency band (because of the stiff top layer), with a smooth transition; consequently, the mode numbering identification of the data points is not feasible.

7.5 SURFACE WAVE INVERSION OF SEISMIC REFLECTION DATA

Seismic reflection data often contain high-energy surface waves, usually simply referred to as the "ground-roll". Surface waves have been conventionally considered as coherent noise, superimposing on the useful P-wave reflections signal and filtered out as soon as possible. The conventional reflection acquisition procedures aim at the attenuation of the surface waves in the field, using receiver arrays, sometimes high-frequency geophones, and high-frequency sources—such as vibrators with sweep signals with a rather high minimum frequency.

These acquisition practices can be quite effective in attenuating the surface waves. From the perspective of a surface wave user, these practices damage the surface wave signal, limiting the usable bandwidth and introducing phase distortions. However, if the acquisition setup is not too harsh on the surface waves, reflection data can be particularly interesting. Sources are typically very powerful and provide high signal-to-noise ratio even at far offsets; the offset range usually guarantees high wavenumber resolution and modal separation; the high-fold data can offer an incredible statistical redundancy for the surface wave dispersion extraction.

Some engineering surveys, with 2D acquisition geometry, can contain well-preserved and well-sampled surface waves. The offset ranges, the frequency of receivers, the limited size of the arrays, or the use of single-sensors (or bunched geophones) produce data that are compatible with standard surface wave analysis techniques. The data of modern, large-scale exploration three-dimensional (3D) surveys can also be used, especially with broadband point-receiver acquisition systems.

7.5.1 Application with engineering data in 2D

An example of ground-roll exploitation for the construction of a pseudo–2D shear wave velocity model is reported by Socco et al. (2008, 2009). The reflection dataset has been collected along the two survey lines shown in Figure 7.32. Down-hole tests and specific surface wave tests have also been performed at specific sites as shown in the map and will be used for a comparison. In particular, Site E in Figure 7.32 is the location of the test presented in Section 7.2.

The site is located on a wide triangular alluvial fan. The maximum thickness of the quaternary deposits is expected to be around 200 m, and the fan is mainly composed by alluvial deposits (sand, gravel, stone), polygenic slivers, stones, and blocks. The general geological evolution can be summarized as a succession of alluvial fan deposits having medium to coarse grain size on deposits of glacial environment. The deposit has been preliminary investigated by two boreholes, subsequently used for the down-hole tests. The stratigraphic log to a depth of about 50 m shows the typical chaotic sequences of gravelly soils of alpine alluvial fans, with no marked layering.

Experimental dispersion curves have been extracted from overlapping windows of the seismic reflection seismograms (Figure 7.33). The analysis has been performed with standard transforms in the frequency–wavenumber domain (see Section 4.6.1).

Redundancy of experimental data provided by the several shots of the seismic reflection survey has been used to evaluate reliable average dispersion curves and their associated uncertainty. The experimental dispersion curves for the different spatial windows along the two survey lines are reported in Figure 7.34.

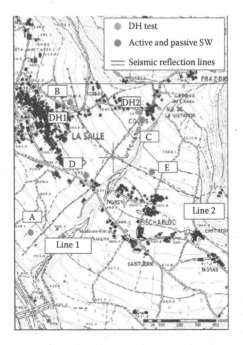

Figure 7.32 Location of seismic surveys performed for the characterization of the alluvial fan in La Salle, Italy. (From Socco, L. V. et al., *Near Surf Geophys*, 6(8), 255–267, 2008.)

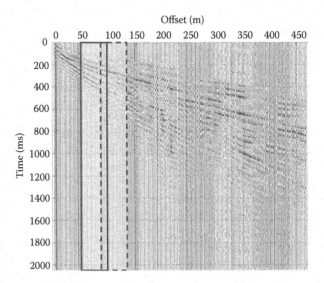

Figure 7.33 Example of experimental dataset from seismic reflection survey with the representation of the sliding window used to perform surface wave analysis.

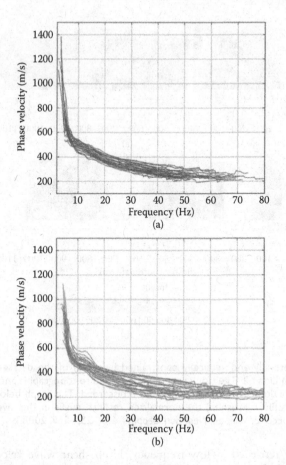

Figure 7.34 Experimental dispersion curves obtained for the different positions of the spatial window along the survey line: (a) Line L1; (b) Line L2. (From Socco, L. V. et al., *Geophysics*, 74, G35–G45, 2009.)

The inversion has been performed with a laterally constrained inversion in which the 1D shear wave velocity profiles associated with the different spatial windows are linked one to each other with lateral constrains which take also into account their distance. The results of the inversion are reported in Figure 7.35. Finally, a comparison of all the geophysical results available for the site is reported in Figure 7.36, where the V_S model obtained with laterally constrained inversion of ground-roll is plotted on top of the results of the seismic reflection survey. In the same figure, the local results obtained from down-hole tests (long arrows) and specific active–passive surface wave tests (short arrows) also are reported. The investigation depth of ground-roll inversion is limited to about 50 m because of the high-frequency geophones used for the reflection survey, which did not allow for the experimental dispersion

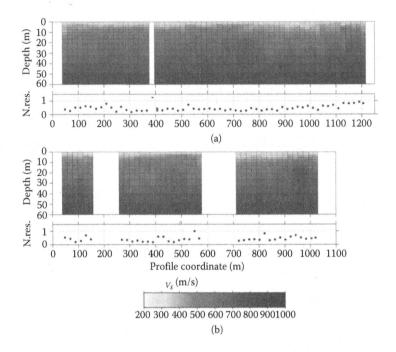

Figure 7.35 Laterally and vertically constrained inversion of the dataset: (a) Line L1; (b) Line L2. The blank sections are related to topographic anomalies where portions of the dataset had to be discarded. The graph below each single profile represents the normalized residual misfit in the inversion. (From Socco, L. V. et al., *Near Surf Geophys*, 6(8), 255–267, 2008.)

curve to be retrieved at low frequency. Deep shear wave velocity profiles obtained with the inversion of combined active and passive surface wave data confirm the deeper reflection as the position of the bedrock below the alluvial fan and allow for the extension of the model at depth (Figure 7.36).

7.5.2 Application with exploration data in 3D

In the exploration industry, acquisition practices are moving toward point-receiver acquisition, in which the wave field is better sampled in space and time. With such high-quality data, surface waves can be analyzed and inverted to characterize the near-surface.

A fine receiver spacing without geophone arrays provides data with non-aliased surface waves, with short wavelengths, and with a dense spatial sampling of the near-surface heterogeneities. Low-frequency sweeps and low-frequency receivers, besides the advantages for the reflection signals, provide the long wavelengths required to reach the large investigation depth in surface wave analysis.

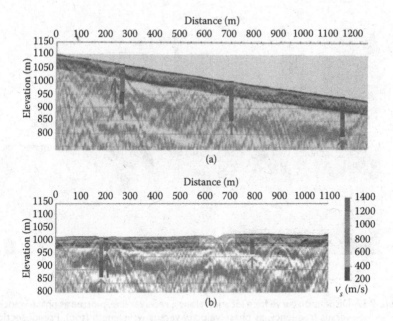

Figure 7.36 Shear wave velocity model obtained from the inversion of surface wave data in ground-roll superimposed to seismic reflection profiles: (a) Line LI and (b) Line L2. Long arrows and short arrows indicate the results of downhole tests and active plus passive surface wave tests, respectively. (From Socco, L. V. et al., *Near Surf Geophys*, 6(8), 255–267, 2008.)

For surface wave analysis and inversion, the use of 3D geometries is not a limitation but offers advantages and enables robust processing. With land 3D configurations—for example, cross-spread orthogonal shooting (Strobbia et al. 2010)—surface wave propagation is well sampled in the offset and azimuth domains. A very large statistical redundancy is available for the surface wave analysis. Local supergathers can be formed for the surface wave analysis, collecting traces corresponding to a set of receivers over an area within a defined aperture. The number of shots contributing to a single location can be very large, of the order of thousands; this extreme redundancy of high-fold land data can be exploited to enhance and extract linear events. Transform-based methods can be used to extract the dispersion. In order to process data from 3D geometries, the transforms must be able to handle unevenly spaced arrays and 2D arrays.

For source and receiver lines, the dispersion properties can be plotted as pseudosections for each individual mode. In this representation, the position along the line is the horizontal axis; the vertical axis is the wavelength (related to the investigation depth); and the color scale represents the local phase velocity. An example of continuous profiling along a 10 km receiver line is shown in Figure 7.37.

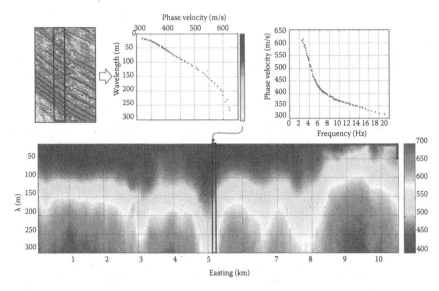

Figure 7.37 Dispersion curve for a location along a receiver line, plotted as phase velocity versus frequency, as phase velocity versus wavelength (top). Pseudosection of the phase velocity of the fundamental mode for a 10-km-long receiver line. (From Strobbia, C. et al., *First Break* 28, 85–91, 2010.)

For typical 3D geometries, the results of the analysis along source and receiver lines are merged into a volume representing the surface wave properties within a survey. The dispersion volume represents the phase velocity of a mode as a function of the geographic coordinates of the analysis point (easting and northing) and of the wavelength. The result of the first stage of analysis can be used to map the noise properties within the survey area. Due to the rapid changes from weathering, compaction, and lithology in the near-surface, large variations of the surface wave properties are often observed, even within a single receiver line. In Figure 7.38, a slice of the dispersion volume, at the wavelength 130 m, is depicted over a false-color satellite image for an area of about 300 km², from a 3D survey in Egypt (Strobbia et al. 2011).

The receiver lines are in the east–west direction and are spaced 210 m. The volume is obtained using about 30,000 dispersion curves. The fundamental mode is the highest energy event in a wide frequency range. Locally higher modes and P-guided waves are present.

The high lateral resolution of the surface waves allows mapping lithological boundaries and geological structures printing through the near-surface. For example, the high-velocity zones in the south correspond to wadi beds. The low-velocity zones in the southeast to northwest direction are fault zones, parallel to the main fault system (the Gulf of Suez, visible on the right). More details of the geological interpretation are provided by

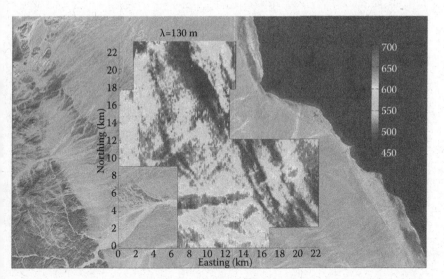

Figure 7.38 Wavelength slice of a dispersion volume for an area of about 300 km² showing the phase velocity of the Rayleigh waves for a wavelength of 130 m.
(From Strobbia, C. et al., *Near Surf Geophys* 9, 503–514, 2011.)

(Laake et al. 2010). Mapping the surface wave properties of large areas
allows the identification of large-scale geological features of the near-
surface. The surface wave propagation properties can be merged with other
data to build an integrated near-surface geological model. The identification of structural elements can be precious geological information, not only
in frontier exploration projects.

The surface wave velocity has to be inverted to infer a near-surface model
for data-processing applications. For each location, the adaptive surface
wave analysis provides a unique set of local surface wave properties, which
correspond to the local average properties. To maximize the lateral resolution in a case where the data quality is high, the inversion can be run with a
local 1D approach. In some cases, 2D or 3D schemes are used. Essentially,
they add a lateral regularization term because most 2D and 3D inversion
approaches are still based on 1D forward modeling.

The quality and the lateral smoothness of dispersion volumes do not generally require further lateral smoothing or constraints. A laterally smooth
inversion could be obtained directly from the dispersion volume. When
boundaries and sharp lateral variations are present, a high-resolution
image of the surface wave properties must be used, and sharpening can
be suitable.

When a grid-based inversion is run, the vertical dimension (depth) is
discretized with variable thickness layers. The maximum depth and the

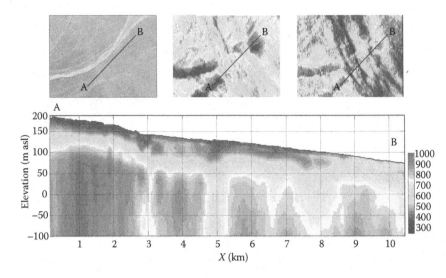

Figure 7.39 Shear wave velocity model obtained from the inversion of the pseudosection of Figure 7.37. The position of the section is shown on a satellite image and on two wavelength slices in the top part of the figure. (From Strobbia, C. et al., *Near Surf Geophys* 9, 503–514, 2011.)

cell size in the shallow section are the main model parameters. The near-surface model obtained via surface wave inversion provides geometric information about the near-surface layers, the geology, the velocity, and the attenuation distribution down to the investigation depth. With low-frequency sources and receivers, the investigated depth can reach hundreds of meters. In Figure 7.39, the results of a modal inversion are plotted for a receiver line, the pseudosection of which has been plotted in Figure 7.37. The model is a 2D velocity grid, with space variant cell size. The height of the cells increases with depth.

In geotechnical applications, the shear wave velocity is often the parameter of interest. On the contrary, in seismic reflection processing for velocity modeling and statics computation, a P-wave velocity model is required. The conversion from V_S to V_P can be done with lithological and hydrogeophysical information extracted from the near-surface integrated model. This can be relatively straightforward in arid areas with sediments where the Poisson ratio is known to have a very narrow range of values, and it becomes more critical in a context where the hydrogeophysical setting has important variations. Calibration with refraction data and up-hole or down-hole data allows the validation of the V_S to V_P conversion. Alternative approaches include the joint inversion of refraction travel times (Section 8.3.2.3).

Chapter 8

Advanced surface wave methods

The most common surface wave testing on land consists of the acquisition of vertical component Rayleigh wave data and the inversion of the dispersion to obtain a shear wave velocity model. The extension to the estimation of dissipative properties has been discussed in previous chapters.

In this chapter, we present some alternative approaches and applications. The objective is not to provide an exhaustive overview of the recent surface wave research but rather to provide an insight into advanced topics and further developments. Analogies and differences with respect to the standard surface wave testing are also discussed.

The following topics will be presented in particular: the use of Love waves on land and Scholte waves in marine applications; joint inversion of surface wave data with other geophysical measurements, with examples on the joint inversion of critically refracted body waves and surface waves; passive seismic interferometry; surface wave polarization; and passive techniques based on the analysis of the horizontal-to-vertical spectral ratio.

8.1 LOVE WAVES

The existence and properties of surface waves characterized by pure horizontal motion were first discussed by Love (1911). Therefore, these surface waves are called Love waves. They propagate as a pure shear disturbance; the particle displacement is horizontal, parallel to the free surface, and perpendicular to the direction of propagation, as shown in Figure 8.1. The Love wave propagation velocity and attenuation depend on the shear properties of the subsurface. Love waves can be acquired, processed, and inverted with methods that are similar to those used for Rayleigh waves.

In the next section, the basic properties of the Love waves are summarized, the experimental configuration to be used for the acquisition is discussed, and an example of joint Love and Rayleigh wave testing is presented.

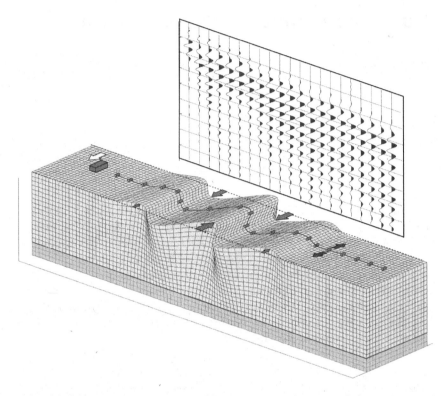

Figure 8.1 Schematic representation of Love wave propagation and their acquisition with horizontal sources and receivers on the ground surface.

8.1.1 The nature of Love waves

Love waves are surface waves containing only Shear, Horizontally polarised (SH) motion, while Rayleigh waves have coupled P and SV potentials (see Chapter 2). Although Rayleigh waves exist in a homogeneous half-space because of the interference between evanescent P-waves and phase shifted SV-waves, the Love waves do not exist in a homogeneous half-space.

The simplest model in which Love waves can exist is a low-velocity layer over a stiff half-space. The supercritical energy, impinging on the half-space with an angle larger than the critical angle, is totally reflected up and reaches the free surface where it is totally reflected down. Energy is then trapped in the low-velocity layer. In this case, the Love wave kinematics can be easily explained in terms of total internal reflections in the waveguide, interfering constructively and destructively as a function of the wavelength (Ewing et al. 1957).

Like Rayleigh waves, Love waves are modal and, in simple cases, they have simple modal shapes. In Figure 8.2, the displacement eigenfunctions

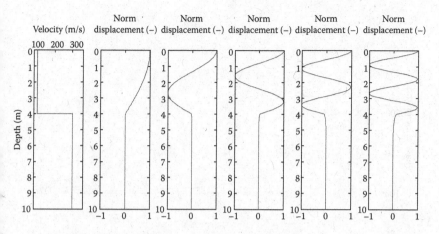

Figure 8.2 Displacement eigenfunctions of Love waves for the first five modes at 80 Hz for the shear wave velocity model reported in the panel on the left.

are plotted as a function of depth for the first five modes at the frequency of 80 Hz for a simple model of a single-layer waveguide (4 m with a shear wave velocity of 100 m/s) over a half-space (with a shear wave velocity of 300 m/s). The modal curves are plotted in Figure 8.3 and compared to the Rayleigh wave modal curves. Due to the high velocity contrast, the Rayleigh modes (Figure 8.3b) are more complex and show an osculation point between the fundamental mode and the first higher mode, indicated in the dashed box (Strobbia 2005).

The constructive interference of the SH-waves, within a limited range of incidence angles, creates resonance in the sedimentary layers. Usually, the SH resonances correspond to points of maximum curvature in the Love dispersion curves. The quarter wavelength rule often gives an acceptable estimate of the resonance frequency (i.e., the resonance frequency can be estimated as a quarter of the shear wave velocity of the top layer divided by its thickness).

This simple single-layer waveguide, however, is not the only case allowing the existence of Love waves. They exist in more complex models (e.g., in multilayer waveguides and with smoother velocity variations). In principle, Love waves also exist in the presence of a surface velocity inversion, with a low-velocity layer sandwiched between two stiffer layers. In practice, the generation and observation of Love waves can be difficult at sites with velocity inversions. The solution of the Love problem can be computed with similar approaches to the one used for Rayleigh waves (see Chapter 2).

The polarization of the shear component of the Rayleigh and of the Love waves is different—vertical for the Rayleigh wave, horizontal for the Love wave. In cases of material anisotropy, the Love wave and the Rayleigh wave are associated with the SH and the SV velocities, respectively.

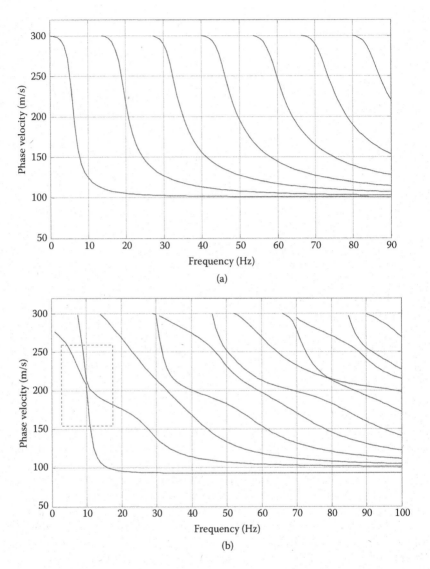

Figure 8.3 Modal curves for (a) Love and (b) Rayleigh modes for the simple model of Figure 8.2.

8.1.2 Experimental configurations

The particle displacement of the Rayleigh wave induced by a vertical force at the surface has a large vertical component at the free surface. Therefore, the acquisition of Rayleigh waves is straightforward with common engineering seismic equipment and with vertical source and receivers. On the contrary, the Love wave displacement is horizontal and parallel to the direction of propagation; moreover, the largest Love wave radiation is from a horizontal source, in the direction perpendicular to the source force. A horizontal force is needed to generate Love waves. The wave propagation mainly consists of horizontal crossline particle motion. The horizontal force also generates Rayleigh waves, even if mainly in the direction of application of the force.

Considering a horizontal force in the x direction at the origin of the reference system xyz, the Love wave has a maximum displacement in the x direction, u_x, along the y-axis, where the x component of the horizontal motion of a Rayleigh wave is minimal. Figure 8.4 shows the radiation of a source at the surface, with the maximum Love wave in the perpendicular direction. The wiggle graph (Figure 8.4b) represents a snapshot of the horizontal displacement u_x; along the x-axis, the contribution comes from the Rayleigh wave; along the y-axis, it comes from the Love wave.

The general principles for the acquisition of Love waves are similar to those for Rayleigh waves, as described in Chapter 3. The design of the measurement of the wave field requires the same considerations about the effects of coherent and incoherent noise, about the consequences of the sampling, and about the field equipment and procedures. However, to excite Love waves, a horizontal source has to be used and horizontal receivers are required to detect the induced particle motion.

Even a perfectly horizontal source induces a pressure component, which will generate multiple events such as refractions and reflections, affecting the records even in the direction of maximum shear radiation. The shear component, SH body waves and Love waves, will be superimposed onto these other events, and it can be difficult to separate them. This is well known in the acquisition of SH body waves, for reflection and refraction surveying (Deidda and Balia 2001), for which specific acquisition procedures have been developed. For example, reversing the polarity of a horizontal source reverses the radiated shear polarity but not that of the pressure component; the subtraction of the corresponding horizontal component of the particle motion, therefore, subtracts the pressure component and sums the shear component. An efficient alternative to this field procedure is the use of specially coupled receivers that perform the synchronous subtraction electrically (Sambuelli et al. 2001), increasing the accuracy of the procedure and the operational efficiency.

As mentioned before, Love waves can have marked resonances in very shallow layers, producing narrow band long traces. The resonance frequency is a function of the layer properties and can be used to estimate the characteristics of the waveguide. Data affected by such resonances have a sharply peaked amplitude spectrum. However, despite the apparently narrowband spectral content, usually the experimental data can be processed to extract a dispersion curve over a wide frequency range. Figure 8.5 shows an experimental shot gather with strong Love waves, acquired with a horizontal source. Traces show long-resonating signals with

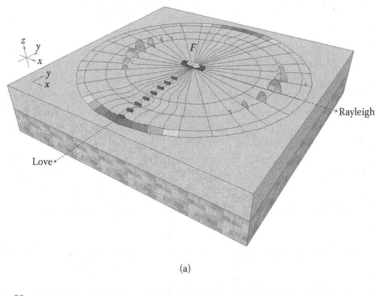

(a)

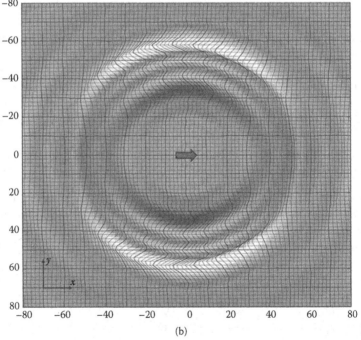

(b)

Figure 8.4 Radiation from a horizontal source (the arrows) with generation of Love waves and Rayleigh waves. (a) 3D view and (b) horizontal particle displacement in the x direction due to both Love and Rayleigh waves.

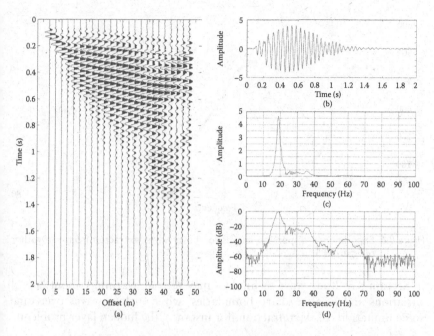

Figure 8.5 Example of Love wave experimental data: (a) shot gather; (b) example of a single trace (receiver offset equal to 30 m); (c) amplitude Fourier spectrum of the same trace in natural scale; (d) amplitude Fourier spectrum of the same trace in log scale.

a narrow spectrum at around 20 Hz. The traces at 30 m and its spectrum are plotted in Figures 8.5b and c. The representation of the spectrum in dB shows that at that offset the usable band is quite wide.

The data of Figure 8.5 are very dispersive as shown by the widening of the wave train, which reaches almost 2 s at 50 m offset. The experimental dispersion curve may be extracted with any of the methods reported in Chapter 4, and it is reported in Figure 8.6. The phase velocity varies from 50 to 500 m/s. The spectral peaks visible in Figure 8.5 (at about 19 Hz) correspond to the resonance of an extremely shallow layer, with a thickness of around 1 m and with a very large velocity contrast. The other peaks in the spectrum correspond to the resonance of the higher modes.

Rayleigh and Love waves' dispersion curves may be used for simultaneous inversion aimed at estimating the shear wave velocity profile under the assumption of isotropic behavior of the material.

Love waves can be acquired simultaneously to SH reflection and refraction data. The synergies between the techniques are similar to those between P-refraction and Rayleigh wave inversion, which will be discussed

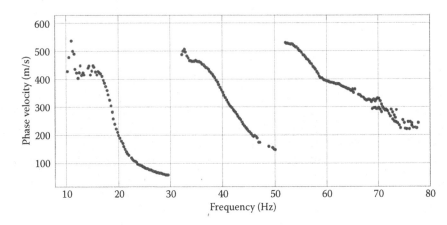

Figure 8.6 Experimental dispersion curves for Love waves extracted from the gather of Figure 8.5.

in Section 8.3. Body wave techniques may provide warnings about lateral variations and localize sharp boundaries, while surface waves overcome some limitations of the refraction (for instance, the hidden layer problem).

8.1.3 Real data example

A case history of acquisition of Love and Rayleigh waves is reported in the following. In particular, the availability of Love wave measurements allows some ambiguities in the Rayleigh wave dispersion extraction to be resolved.

The test site is located on Terceira Island in the Azores archipelago, in Sao Sebastiao, a village that suffered strong damage by an earthquake in 1980 due to local seismic site amplification. The village is located in a volcanic caldera, with basaltic bedrock and some lateral pyroclastic cones. As part of the site characterization, a geophysical campaign has been performed, acquiring surface waves (Rayleigh and Love) as well as refraction and reflection data (for P- and SH-waves) (Lopes et al. 2013).

Surface waves have been acquired with a multichannel seismograph with 24 active channels. For the Love wave acquisition, a wooden sleeper struck laterally using a sledgehammer has been used as source and horizontal swyphone geophones (Sambuelli et al. 2001) as receivers. Two persons were standing on the sleeper to increase the shear friction force on the ground. For the Rayleigh wave acquisition, a sledgehammer on steel plate as a seismic source has been used along with vertical 4.5 Hz geophones. The seismograms with the corresponding f–k spectra and dispersion curves are reported in Figures 8.7 and 8.8 for Rayleigh and Love waves, respectively.

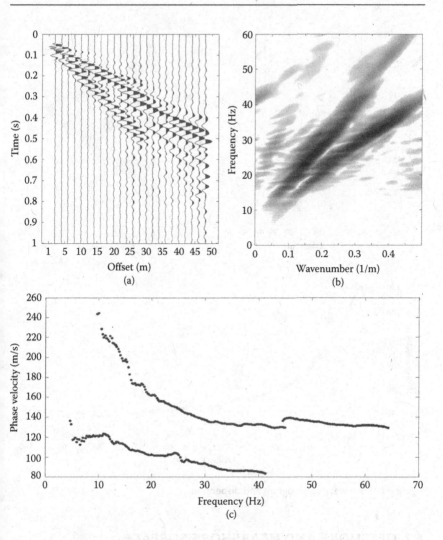

Figure 8.7 Rayleigh wave data: (a) seismogram; (b) frequency–wavenumber spectrum; (c) experimental curves.

The first Rayleigh mode suffers a strong attenuation and loses more than 1.3 dB/m (Figure 8.7). The higher mode, which is clearly visible in the *t–x* and *f–k* domains, is less attenuated because of the higher wavenumber. The Love wave gather shows a dominant fundamental mode with low velocity (Figure 8.8). The phase velocity of the Love wave first mode is better defined than the corresponding Rayleigh mode and confirms the velocity of the latter, resolving possible doubts about the Rayleigh modes.

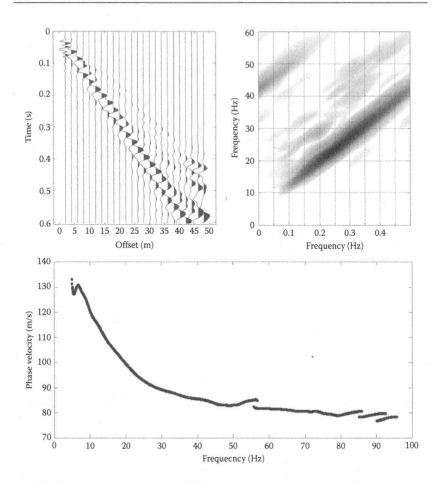

Figure 8.8 Love wave seismogram and dispersion.

8.2 OFFSHORE AND NEARSHORE SURFACE WAVE TESTING

The geophysical characterization of shallow soils and rocks under water is of great interest for a variety of applications, such as the design of port structures, offshore platforms, artificial islands, and sea bottom pipelines, and for environmental studies. The seismic parameters (velocity and attenuation) are needed for dynamic problems, for example, for the seismic site response assessment. They can also be useful in both offshore and nearshore applications for foundation design, slope stability evaluation, liquefaction potential assessment, and erosion potential studies. Moreover, the variations of properties in the shallow subseafloor are important in deep exploration applications to compensate for the near-surface distortions (for example, with static corrections). The geophysical underwater applications are often called *marine* even if they are applied to lakes, rivers, and water supply reservoirs.

Specific high-frequency acoustic reflection systems and techniques are used to image, with a high resolution, the morphology of the sediments below the water bottom. For instance, sub-bottom profilers use sources of high-frequency pulses that are propagated through the water layer and are reflected by the seafloor and by the sub-bottom layers. Frequency-modulated pulses are used to optimize the trade-off between penetration and resolution. The result is a detailed geometrical characterization.

On the contrary, the geophysical in situ measurement of the mechanical properties is not straightforward, with the conventional shallow seismic techniques that are effectively applied onshore.

The P-wave velocity of soft, saturated sediments is dominated by the pore water compressibility. The variations of P-wave velocity are low and difficult to detect with conventional refraction or reflection methods. The conventional shallow P-wave refraction can be used to detect major boundaries, such as the interface between a stiff bedrock and the sediments above. Yet, it offers limited potential to discriminate among loose sediments. Shear wave measurements play a major role in the characterization of saturated sediments. Indeed, the shear wave velocity is controlled by the sediment solid skeleton and has a stronger correlation with geotechnical parameters. Furthermore, the shear wave velocity can be much lower than the compression velocity in soft unlithified sediments. The V_S variation in the first tens of meters can be one order of magnitude greater than the V_P variation (Hamilton 1976). The lower velocity makes the wavelength shorter, providing a high resolution for body wave techniques.

However, the offshore acquisition of shear waves for refraction or reflection applications is challenging. The use of horizontal receivers and a horizontally polarized source presents difficulties even in very shallow water. Because the water does not propagate shear waves, sources and receivers have to be properly coupled to the seafloor. Deploying receivers at the bottom is a standard practice in marine and seabed exploration applications, with hydrophones and geophones either in cables (ocean bottom cable) or in independent nodes or stations (ocean bottom station). Converted waves from compression sources in the water can be used for the shear imaging and characterization. An alternative is the use of specific bottom dragged instrumentations able to generate and detect shear waves (Stoll and Bautista 1994; Winsborrow et al. 2003) and Love waves at the seafloor.

The analysis of Scholte wave propagation is an effective alternative for the estimation of the shear wave velocity of the shallow subseafloor. Scholte waves are. the surface waves propagating at the boundary between a solid and a liquid layer. They are similar to Rayleigh waves and can be analyzed for the underwater characterization with the same methods used for

Rayleigh waves. The shear wave velocity profile can be inferred inverting the dispersion relation obtained by processing the pressure wave field associated with interface waves. The generation and recording of Scholte waves in seabed applications can be performed with a simple experimental set-up, with no need for bottom sources.

8.2.1 Scholte waves

A system consisting of a liquid layer over a solid layered half-space allows two types of interface waves to be propagated: Scholte waves, the properties of which mainly depend on the shear wave velocity of the sub-bottom sediments, and P-guided waves, the properties of which mainly depend on the compression wave velocity of the sub-bottom sediments and of the water layer. Both these events are excited by sources in the water or at the seafloor and can be recorded as a pressure or particle velocity wave field at the seafloor. Schematically, the two events are represented in Figure 8.9, in which the qualitative shape of the displacement eigenfunctions for the fundamental mode of propagation is reported.

The Scholte waves (Scholte 1947) are the normal modes, that is, the eigenvalues related to real roots of the dispersion equation in the *soft bottom* case, when the shear velocity of the solid is lower than the compressional velocity in the liquid (Carcione and Helle 2004). They propagate parallel to the water–sediment interface, with most of the energy traveling at the solid–liquid interface. They are dispersive and modal and are similar to Rayleigh waves at the solid–air interface and to Stoneley waves at the solid–solid interface (Ewing et al. 1957).

The guided waves correspond to the leaky modes, with energy partially trapped in the waveguide. The SV reflections are supercritical, while the P is subcritical, and a fraction of the compression energy is radiated down out of the low-velocity layer. In most shallow sediments, with a high Poisson ratio, these waves degenerate toward the simple modes of acoustic guided waves. As Scholte waves, they are modal and dispersive, but even the fundamental

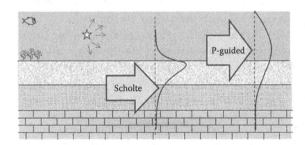

Figure 8.9 Schematic representation of interface waves propagating at the seafloor.

mode has a cutoff frequency that can be quite high compared to the usual frequency ranges of conventional seismic prospecting. The phase velocity is higher than the V_P of the water at any frequency, and it depends on the compressional wave velocity of the sediments.

The particle motion of Scholte waves is similar to that of Rayleigh waves. The energy is concentrated at the interface and decays rapidly with distance from the interface. For the case of two half-spaces in contact, the penetration is roughly equal to one wavelength in the solid and to a half wavelength in the liquid. In Figure 8.10a, the normalized displacement is plotted versus the normalized depth (fraction of the wavelength). The particle motion is elliptical retrograde, such as for the Rayleigh wave, and it changes direction at a depth of about a tenth of the wavelength (Figure 8.10a). High Scholte wave amplitudes can be expected at the seafloor, where both the pressure field and the vertical component of the particle motion have their maxima.

One of the differences with Rayleigh waves is that Scholte waves are dispersive even with a homogeneous solid half-space. Moreover, given a sediment column, the Scholte wave velocity is lower than the corresponding Rayleigh velocity (i.e., in the same model without a water layer on top). In the low-frequency range, the influence of the water layer is reduced because of the long wavelength; hence, Scholte wave velocity tends toward the Rayleigh wave velocity in the solid sub-bottom sediments. In the high-frequency range, Scholte wave velocity tends toward the Stoneley wave velocity at the solid–liquid interface. The dispersion curve of the Scholte wave for different water depths is plotted in Figure 8.10b. The velocity mainly depends on the shear wave velocity of the sediments, as for the Rayleigh wave.

Scholte waves can be generated by a source at the interface, as well as by a pressure source in the water layer. A large air gun array can excite strong Scholte waves in 100 m of water. The excitability of the modes acts as a high-cut filter; lower and lower frequencies are propagated as the water layer becomes thicker and thicker. The larger the distance of the source from the seafloor, the lower the frequency that is excited. This aspect has to be taken into account to plan the survey with respect to the targets because short wavelengths are required for increasing the resolution close to the sea bottom and long wavelengths for deep characterization.

Generally, Scholte waves are observed as low-frequency and low-velocity events. Indeed, the shallow unconsolidated sediments can have a very low shear wave velocity, and consequently, very low velocity Scholte waves are observed. In Figure 8.11, the properties of the solution are shown for a simple three-layer model (two layers of thickness 3 m each and shear velocity of 200 and 300 m/s and a half-space with a shear velocity of 400 m/s; the water layer is 3 m thick; the P-wave velocities are 1500, 1550, 1600, and 1700 m/s, respectively). The Rayleigh wave dispersion curves for the same model with no water layer are also reported for comparison in Figure 8.11.

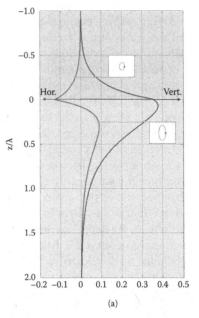

(a)

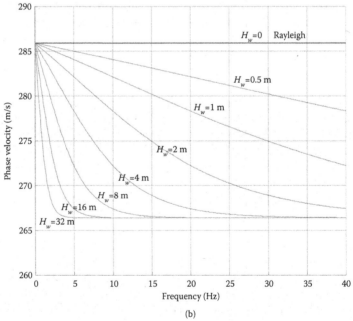

(b)

Figure 8.10 (a) Particle motion of the Scholte wave for the case of two homogeneous half-spaces in contact. (b) Dispersion curve for a bottom half-space
($V_S = 300$ m/s) with a top water layer of variable thickness.

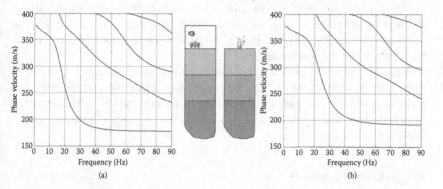

Figure 8.11 Example of (a) Scholte and (b) Rayleigh modes in a layered elastic medium with the same solid layers.

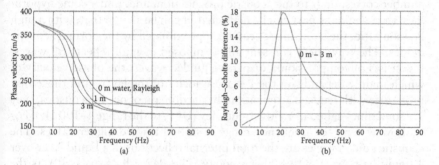

Figure 8.12 Example of Scholte and Rayleigh modes in a layered elastic medium with the same solid layers: (a) dispersion curves for different thicknesses of the water layer; (b) relative difference between Rayleigh and Scholte wave phase velocities for a 3 m thick water layer.

The fundamental mode and three higher modes exist in the considered frequency range for both types of waves. The effect of the water layer is a slight decrease of the phase velocity, particularly at high frequency. The Scholte fundamental mode does not have a cutoff frequency, and at a low frequency, its phase velocity tends to resemble that of Rayleigh waves.

A parametric analysis is shown in Figure 8.12 where the dispersion curves of the first normal mode for a layered system with different thickness of the top water layer are reported. The zero thickness case represents the Rayleigh wave solution. Even a rather shallow water layer affects the phase velocity. The analysis of marine data with a Rayleigh wave forward modeling is not correct, and it may lead to underestimation of the sediment velocity. The difference in phase velocity is a function of the frequency, as shown in Figure 8.12b for a 3 m thick water layer. Moreover, the thickness of the water layer may be variable along the survey line, especially in shallow waters. If this variability is not properly taken into account for the analysis, the results may be affected by artifact with false apparent variations of the sediment thicknesses.

A factor that is important to take into account in the Scholte wave inversion is the P-wave velocity of the sediments. In the inversion, if the Poisson's ratio is assumed a priori, its value should be compatible with saturated materials (see also the discussion in Section 7.1) unless trapped gas is expected at the specific site.

8.2.2 Guided waves

The complex roots of the dispersion equation of the Scholte problem correspond to leaky modes of propagation, with a phase velocity higher than the highest shear velocity in the system. The real part of the complex wavenumber corresponds to the acoustic propagation, and, under some hypotheses, the imaginary part can be neglected. It can be shown that, with a high Poisson ratio, the leaky modes can be approximated by the normal acoustic modes. The wave field is essentially composed of multireflected P-waves. It consists of compressional dispersive guided waves, the phase velocity of which mainly depends on the P-wave velocity of the layers and those that propagate with reduced attenuation. In the domain of ocean acoustics, they are sometimes called ultra low frequency (ULF, in the range 1–100 Hz). For shallow waters, the frequency range is shifted toward higher values. The acoustic guided waves are the total internal reflections in a liquid layer over a liquid half-space with higher velocity. The elastic half-space affects the dispersion with a low Poisson's ratio.

In the following example, a 20 m thick liquid layer with a velocity of 1500 m/s rests above a half-space with a P-wave velocity of 1700 m/s. The dispersion curves for the acoustic case (liquid half-space, i.e., $V_S = 0$ m/s) and for a solid elastic half-space (of variable V_S) are plotted in Figure 8.13. With a high Poisson's ratio (corresponding to $V_S = 300$ m/s), which is typical for soft uncemented sediments, the behavior is similar to that of the acoustic case.

The sub-bottom layering has to be considered when an accurate representation of the guided wave dispersion is required (e.g., if the experimental dispersion curves are inverted to obtain a velocity model). As an example, let us consider a simple model with 3 m of water ($V_P = 1500$ m/s, $V_S = 0$) and a 3 m layer of sediments ($V_P = 1560$ m/s, $V_S = 60$ m/s) above a bedrock ($V_P = 2000$ m/s, $V_S = 1000$ m/s). The acoustic approximation and the elastic solutions are plotted together in Figure 8.14. The acoustic approximation shows the general trend, the upper and lower velocity limits, but it fails to reproduce the shape of the curves. Considering the fundamental mode, it is evident that the frequency range needed to estimate the velocities of the two solid layers is wide (from 100 to 1000 Hz).

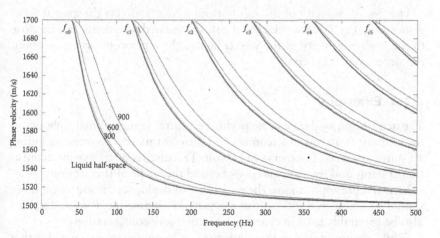

Figure 8.13 Dispersion modal curves for a 20 m thick water layer above a half-space. The acoustic approximation is valid for a high Poisson's ratio. For lower Poisson's ratios, or higher shear velocities in the half-space, differences cannot be neglected.

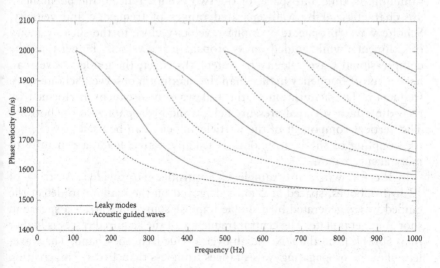

Figure 8.14 Modal curves for the guided P-waves for a two-layer system.

For wavelengths greater than the layer thickness, the fundamental mode is dominating, although higher modes become important for shorter wavelengths. It can be noted that the acoustic waves exist only above the cutoff frequency of the first mode. It can be computed with a simple relation for a single-layer waveguide. In this example, the cutoff frequency of the fundamental mode is about 100 Hz.

The P-wave velocity of the sub-bottom strongly affects the guided wave dispersion. The P-wave velocity of unconsolidated sediments can be close to the P-wave velocity of the water; hence, the a priori information about the water layer is crucial.

8.2.3 Example

The acquisition of data can be performed using strings of hydrophones at the seafloor or in towed streamers and different marine sources, including air guns, sparkers, boomers, and so on. The discussion of marine acquisition systems and technologies goes beyond the scope of this section.

In nearshore applications, the possibility of deploying the source on shore should be considered if the offset range makes it possible. Passive data can also be profitably used in nearshore and offshore configurations.

While designing the data acquisition, it is important to consider that Scholte waves and guided waves can be used in a joint inversion to estimate the shear and compressional wave velocities, better constraining the geometrical parameters. A joint acquisition requires a proper sampling, in time and space, of the two events. Indeed, the phenomena are characterized by well-separated ranges of frequency and velocity. Scholte waves propagate at a phase velocity close to the shear velocity of sediments, while guided waves propagate at a velocity higher than the compressional wave velocity in water. Moreover, the guided waves can have a cutoff frequency higher than the maximum observed Scholte wave frequency. The acquisition of the full wave field can be performed by recording the associated pressure field, using hydrophones at the bottom. The vertical component of the particle velocity can be used as well, but the coupling of the hydrophones is usually better, leading generally to more accurate data.

In the following, an example of acquisition is provided. A string of 12 hydrophones, spaced at 2 m, is deployed on the seafloor to detect the guided waves generated by a simple impulse source in water—a speargun shot at a metal plate. The central frequency of the acquired traces is higher than 500 Hz, and the S/N is high over a wide frequency band. The wavelength of the propagating waves ranges from less than 1 to 1.5 m, creating spatial aliasing. The dispersion curves extracted from the analysis of the fast event, in the first 25 ms, are shown in Figure 8.15.

The shot gather is dominated by the high-frequency components related to guided waves, with a phase velocity greater than 1500 m/s, but it also contains Scholte waves. This event, propagating with a velocity of less than 50 m/s, is visible in the first traces and can be enhanced by a simple low-pass filtering. The first two traces are shown in Figure 8.16.

The presence of the two phenomena is much more evident in a raw record gathered across the shoreline (Figure 8.17). The first 12 traces are gathered

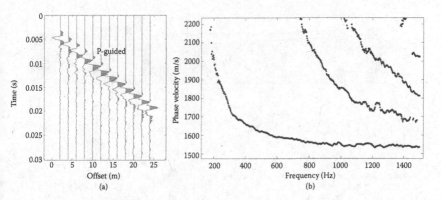

Figure 8.15 Example of marine data: (a) shot gather; (b) experimental dispersion curves for guided waves.

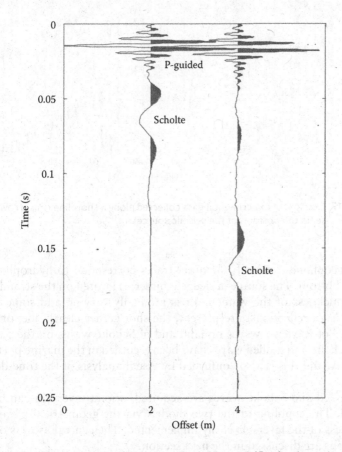

Figure 8.16 The first two traces of the shot gather of Figure 8.15a after low-pass filtering.

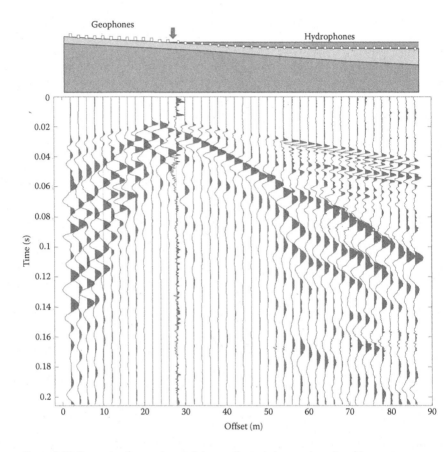

Figure 8.17 Example of experimental data collected along a shoreline (the arrow represents the position of the seismic source).

by 12 geophones on land; the other traces correspond to hydrophones at the sea bottom. The source (a sledgehammer) is located on the strand. Even if the thickness of the water layer is gradually varying and some lateral effects (e.g., reflections) are present, the shot gather clearly identifies the presence of Rayleigh waves on land and of Scholte waves on the seafloor. Both Scholte and guided waves have been recorded in the marine portion of the array, and this is also confirmed by visual analysis of the time-domain traces.

If guided and Scholte waves are acquired, a joint inversion can be performed. The coupling of the two models via the geometrical parameters (thickness of the layers) is easily implemented. The general aspects of joint inversions are discussed in the next section.

8.3 JOINT INVERSION WITH OTHER GEOPHYSICAL DATA

8.3.1 Joint inversion

The integration of multiple datasets and measurements is a crucial step in geophysical exploration. Multiple physical properties are estimated and used to quantify petrophysical parameters and for the discrimination of lithologies. Often even the imaging and identification of some targets require multiphysical surveying approaches. Gravity, magnetic, electromagnetic, and seismic methods are commonly used together in oil and gas exploration. Moreover, multiple datasets are often extracted from the same data acquisition or measurement; an example is the extraction and analysis of reflected and refracted body waves and surface waves from the same seismic dataset (see Section 7.5).

The integration and homogenization of multiphysics data into a consistent model can also improve the resolution, joining methods with different sensitivities to different subsurface features and to different portions of the investigated domain. Moreover, it can mitigate the nonuniqueness in the single-domain inversion of the individual measurements. With multiple physical domains, including a direct or indirect relationship between the parameters, it is possible to reduce the multidimensional model space to a smaller subspace.

The simplest integration of multiple datasets is based on the superposition and comparison of individual models for cross-validation. Still, the reconciliation of multiple models may be challenging and ambiguous. The synergies between methods and datasets can be exploited using mutual constraints, extracting from each the most reliable and trusted features. For example, a structural geometrical model can be built from one measurement and imposed on the other single-domain inversion. These approaches are subjective and rely on the operators' decisions, and hence several subjective solutions are possible. A more rigorous and robust approach for the integration of multiple measurements is based on joint inversions of the different datasets.

The term joint inversion indicates the process of estimating the unknown model parameters from multiple sets of geophysical data, building a unique objective function. The multiple physical domains of an Earth model are jointly estimated fitting multiple datasets. A *single-domain joint inversion* uses multiple datasets to estimate a model with a single physical domain or a type of unknown. For example, a single-domain joint inversion can be used to estimate a V_S model from the measured propagation properties of Rayleigh and Love waves. A *multiple-domain joint inversion* aims at the estimation of an Earth model with multiple properties; for example, to estimate a unique model with V_P and V_S from Rayleigh waves and P-wave refractions.

Different approaches to the joint inversion have been proposed. The *cooperative joint inversion* (Lines et al. 1988) essentially iterates inversion steps between the two domains. The result of one inversion step in one domain is converted and used as an initial model for the other domain. The process is repeated until convergence.

The *simultaneous joint inversion* inverts the datasets jointly with a link between the model domains. The links can be essentially of two kinds: geometrical or structural on one hand, intrinsic or petrophysical on the other hand.

The geometrical simultaneous joint inversion links the domains in terms of similarities among the geometrical distributions of model parameters. This can be achieved by minimizing the differences among the property distributions (for example, computed on the Laplacian of the model fields) (Haber and Oldenburg 1997). The cross-gradient links are also used to impose structural similarity across domains (Gallardo and Meju 2004).

The petrophysical link assumes a local intrinsic function among the domains. Analytical or empirical relationships may be used to link the physical parameters. These links may be either used as a priori constraints with fixed values or may be included in a global misfit function. In the latter case, the parameters of the petrophysical relationship become additional unknowns in the solution of the inversion problem (Jegen et al. 2009; De Stefano et al. 2011).

A schematic representation of the concept of joint inversion is provided in Figure 8.18. In Figure 8.18a, two datasets are inverted separately, and the two models are made compatible considering a possible link between them and the overlap between the two models. In a simultaneous joint inversion approach, the link is an input, together with the

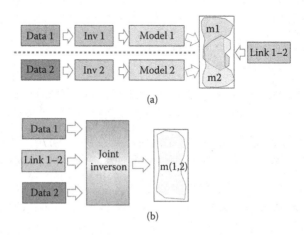

Figure 8.18 Schematic representation of (a) single-domain inversion and (b) joint inversion.

two experimental datasets, for the solution of the inversion process that produces a single model.

In the following section, we discuss different approaches to the joint inversion with examples of surface wave inversion with other geophysical data.

8.3.1.1 Geometrical joint inversion

The simplest link between the domains is given by the geometry of the models. Indeed, the spatial distribution of geophysical parameters is often controlled by common geological features. Parameters that appear as apparently uncorrelated can show similar spatial structure.

The geometrical resemblance can be included in the joint inversion, introducing structural links that encourage the similarity between the two models, without imposing or assuming any local relationship between the parameters. The resemblance between the two models can be defined, for example, using the cross-gradient approach (Gallardo and Meju 2003). The cross-gradient function is defined as the cross product of the gradients of the physical properties m_1 and m_2 in the two domains

$$\vec{\tau} = \nabla m_1(x,y,z) \times \nabla m_1(x,y,z) \tag{8.1}$$

The cross-gradient function measures the similarity between two images. It is minimum with the collinearity between the two domains. Figure 8.19 shows an example of two fields and the corresponding cross-gradients. The two fields m_1 and m_2 are plotted in Figure 8.19a and b, and the cross-gradient is shown in Figure 8.19c. The cross-gradient has low values in the portion where the dominant direction is similar (bottom part), even if the values are not directly positively correlated, and it has high values where the two fields show different geometrical distributions.

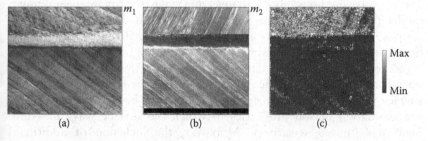

Figure 8.19 The similarity of the spatial distribution of two physical parameters can be evaluated using the cross-gradient: (a) model m_1; (b) model m_2; (c) cross-gradient m_1–m_2.

The joint inversion of the two domains can be performed, defining an objective function based on the optimization of the two datasets and of the geometrical similarity (for example, the residuals of the different datasets and the cross-gradient field). The structural link can be part of the objective function (Tryggvason and Linde 2006), or it can be imposed as an equality constraint, limiting the model search in the iterative process to the subspace of models with geometrical similarity.

In a 1D layered model, the geometrical joint inversion implies similar layer thicknesses, regardless of the fact that the geophysical parameters of each layer can be completely uncorrelated.

8.3.1.2 Petrophysical joint inversion

When the different domain parameters have an intrinsic petrophysical link, the domains can be linked using a local parameter relationship. In the petrophysical joint inversion, a local relationship among the multiple-domain model parameters is used. These links can be model-based physical laws or empirical relationships among geophysical parameters. Widely used relationships include Archie (1942), Wyllie et al. (1956), and Gardner et al. (1974).

In the petrophysical joint inversion, the objective function can be built from three different elements: the residuals of the different domains, the single-domain constraint and regularizations, and the cross-domain links. The petrophysical link can be imposed as an equality constraint or encouraged minimizing the local residual between an a-priori relationship and the model relationship. The three elements of the objective function can be combined using different weights. The different domains can be preconditioned and regularized independently, and multiple scales can be solved simultaneously.

8.3.2 Surface wave joint inversion

Surface wave methods rely on an inversion stage to estimate the unknown model parameters (i.e., shear wave velocity and damping ratio). The coupled inversion of velocity and attenuation described in Chapter 6 can be considered a joint inversion. In this paragraph, we discuss the joint inversion of surface wave data with other geophysical measurements, seismic and nonseismic. The joint inversion can be used to integrate the surface wave data with other geophysical data. This approach improves the estimation of soil properties to which the surface wave measurement has limited sensitivity. Moreover, the inclusion of additional information allows solution nonuniqueness of the inverse problem to be mitigated.

In the following section, joint inversions with electrical, electromagnetic, and seismic data are discussed. Other joint inversions have been proposed in the literature—for example, with microgravity data (Hayashi et al. 2005).

8.3.2.1 Joint inversion with electrical and electromagnetic measurements

The joint inversion of the surface wave dispersion with 1D electrical and electromagnetic data can be used with, for example, geometrical links to impose common boundaries. Lithological boundaries in the shallow sub-surface affect the shear wave velocity and the resistivity of shallow layers, and a geometrically coupled inversion is possible (Hering et al. 1995). The measured data have similar properties. The apparent resistivity curves from vertical electric sounding (VES) and time-domain electromagnetic (TDEM), similar to the dispersion curves, provide integral information about the corresponding physical parameters. Moreover, both forward models are based on a global transfer matrix built from layer matrices for the 1D solution.

With reference to the integration of VES and surface waves, the two sets of measured data can be the apparent resistivity versus the electrode array geometry and the Rayleigh wave phase velocity as a function of frequency. The model parameters can be the thickness, shear velocity, and resistivity of each layer.

The choice of geometrical coupling implies that interfaces (layer boundaries) are similar or coincident. However, this does not force to have a variation of one parameter when the other varies. For instance, the water table is a major interface for resistivity, but it is not for the shear wave velocity.

The model parameters can be joined in a single vector, creating a single Jacobian matrix with the partial derivatives for both set of parameters. The rows correspond to the p values of apparent resistivity ρ_{app} plus the m values of phase velocity v_r. The columns contain the derivatives with respect to the n resistivity values ρ, the $n-1$ thickness values h and the n shear wave velocity values V_S (for a model of n layers including the half-space). The absence of a petrophysical link leads to the zero values of the cross-domain derivatives.

An example is reported in Figure 8.20, in which a comparison between results from independent and joint inversions is shown. In this case, the tests were aimed at the identification of local stratigraphy in a geological context with alternate layers of sandy and clayey layers. In particular, the expected thickness of a shallow clayey layer was of interest for the assessment of the risk of seepage below a river embankment (Comina et al. 2004). The clayey layer is identified on the basis of its lower resistivity at a depth between 8 and 13 m. The solution of the joint inversion is deemed to

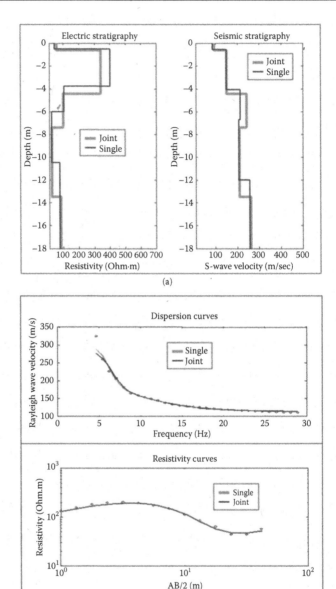

Figure 8.20 Joint inversion of surface wave and electrical resistivity data: (a) electrical
and seismic profiles from joint and individual inversions; (b) comparison
between experimental and numerical dispersion and resistivity data for
joint and individual inversions. (From Comina, C. et al., *Proceedings of ISC-2
on Geotechnical and Geophysical Site Characterization*, Millpress, Rotterdam,
pp. 451–458, 2004.)

be more robust and more reliable because it honors simultaneously the two experimental datasets.

A similar approach can be used for time-domain electromagnetic (TDEM) data and for magnetotelluric data with long-period surface waves.

8.3.2.2 Joint inversion with other seismic data

Considering seismic data, the joint inversion of Love and Rayleigh waves can be considered a single-domain joint inversion. The same layer model is constrained by two different datasets, and the unknowns can be the layer thickness and the shear wave velocity, under the assumption of isotropic behavior.

Particularly interesting is the case of the joint inversion of Rayleigh and guided waves. A joint inversion can be obtained with a global misfit function, where the Rayleigh and guided wave dispersion curves are used together. The distance between the picked curves and the determinant of the Haskell–Thomson matrix can be used, as discussed in Section 7.4. Alternatives include the use of the distance between curves modeled independently from the same stack of layers. An example is provided by Roth et al. (1998).

The joint inversion of surface waves and refracted P-waves is particularly interesting for several reasons: the two corresponding dataset are often available together, the two methods offer many synergies, and the intrinsic link between the two domains is often rather strong and dependent upon features of interest.

Actually, the two datasets can be extracted from the same seismic data, so the measurements can be performed simultaneously. They are normally acquired together in exploration seismic, and they can be acquired together in near-surface engineering testing, without significant additional operational effort. Because of the many basic differences between the two components of the wave field that are analyzed, the two methods complement each other in many respects (Foti et al. 2003).

Surface wave methods overcome some inherent limitations of the refraction technique caused by refraction equivalence such as the ones due to hidden layers, gradual velocity variations, and velocity inversions (Reynolds 1997). Also, the sensitivity to the shear wave velocity is particularly important in saturated sediments where the P-wave velocity is mainly controlled by the fluid compressibility.

On the contrary, refraction results provide very useful information for surface wave interpretation: constraints for surface wave inversion (e.g., the bedrock position), indication of the water table depth (which allows a good estimation of Poisson's ratio in surface wave modeling), and a warning in the case of lateral variation of velocities (e.g., topography of refractors).

When the two techniques give independently reliable results that are mutually confirmed, the information about compression and shear wave velocities allows the estimation of mechanical properties and porosity of the soil deposit. At the same time, the relationship between the elastic parameters can be used as a petrophysical link, or the similarity between the two spatial distributions can be used as a geometrical link.

8.3.2.3 Joint inversion of refracted and surface wave

Even if some benefits can be obtained just with the simple combined interpretation of the two results, the joint inversion provides a robust framework to exploit the synergies between the two methods.

In this section, we present an example of 1D joint inversion, where the refraction data suffer from equivalence due to the presence of a hidden layer. The seismic survey was carried out to determine the soil parameters for seismic site response assessment. The expected stratigraphy of the test site consists of a soft soil over a dense gravel layer and the presence of hard sandstone bedrock at a depth of less than 10 m.

The surface wave data show large dispersion and allow the extraction of a broadband dispersion curve, which is inverted for a layered model—confirming the presence of three main geological layers. The seismogram acquired with an unevenly spaced array, the extracted dispersion curve, and the inverted velocity profile are represented in Figure 8.21.

Refraction data acquired at the same location show a clear refraction in the first arrivals, with a large variation of velocity with offset. The picked travel times essentially show two velocities, with a large difference in velocity in the forward and backward shots, indicating the presence of a dipping layer.

The raw seismograms for the two end-off shots are shown in Figure 8.22. The interpretation of the travel time curves (Figure 8.22b) suggested a two-layer model, which is shown in Figure 8.22c. This result would locate a slightly dipping bedrock, of fairly high velocity, at a depth varying between 3.2 and 4 m, but the high velocity contrast (330 vs. 2600 m/s) between the two layers allows the presence of a hidden layer of significant thickness.

The "hidden layer" problem is one of the possible ambiguities of refraction methods. An intermediate layer might not be identified because its refracted arrivals reach the surface after the critically refracted waves from a deeper faster layer. In practice, the equivalence is due to the fact that multiple velocity profiles can generate exactly the same travel time curves. The example of Figure 8.23 shows how two different velocity profiles can generate exactly the same travel time curves.

The large difference in the bedrock depth (Figures 8.21 and 8.22) raises a further warning about the risk of a hidden layer. The evidence of dipping bedrock in forward and backward shots poses some questions

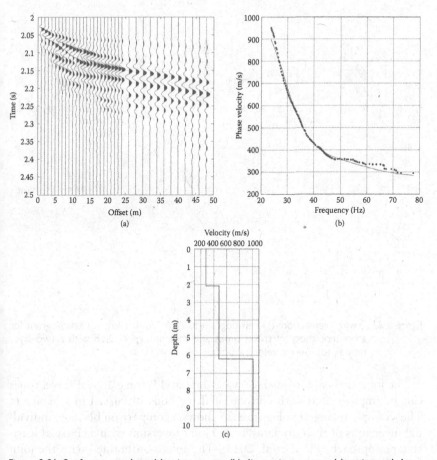

Figure 8.21 Surface wave data: (a) seismogram; (b) dispersion curves; (c) estimated shear wave velocity profile.

on the acceptability of the 1D hypothesis in surface wave interpretation. Essentially, the joint interpretation of the two measurements leads to a difficult reconciliation of the velocity profiles.

The two datasets (the picked travel times and the dispersion curve) are then inverted together, imposing common layer thicknesses for V_P and V_S models. A small dip is included among the model parameters to fit the travel time curves. The inversion is performed using a global Monte Carlo optimization, defining a joint misfit function as the weighted average of the two least-square misfits. The results are reported in Figure 8.24. The large contrast in shear wave velocity is confirmed, and the small dip is acceptable in terms of validity of the 1D assumption. The best 20 models are plotted in Figure 8.24c.

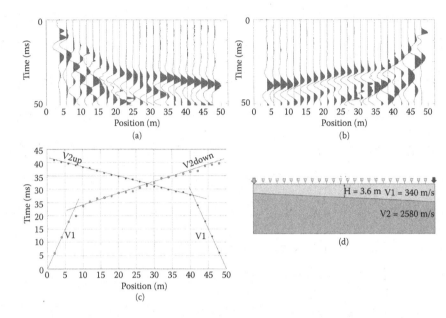

Figure 8.22 P-wave refraction: (a) seismogram for a forward shot; (b) seismogram for a reverse shot; (c) travel times and their interpretation with a two-layer model; (d) P-wave velocity model.

The joint inversion of surface wave data and P-wave arrival travel times can be implemented with conventional methods discussed in Chapter 6. The example reported in Figure 8.25 shows a comparison between individual inversions of the two datasets and joint inversion with a classical least-square approach (Piatti et al. 2013). The solution obtained with the joint inversion allows for an higher resolution and identifies the presence of a soft silt and clay layer at a depth around 20 m, which is embedded in between the top gravels and the bedrock below as confirmed by stratigraphic information obtained from borehole logs (Piatti et al. 2013).

8.3.2.3.1 Large-scale joint inversion of refracted and surface waves

In deep exploration, refracted body waves and surface waves are acquired over a large area using 3D geometries. The acquired traces can be processed to extract the two datasets—the surface wave propagation properties (typically, the dispersion volumes) and the refracted first arrival travel times. The two datasets have different spatial density, and the methods have different spatial resolution. The investigation depths can be different, resulting in two models with different resolutions and a partial overlap. A schematic representation of the sensitivity is provided in Figure 8.26. The travel times provide information on the P-wave velocity structure,

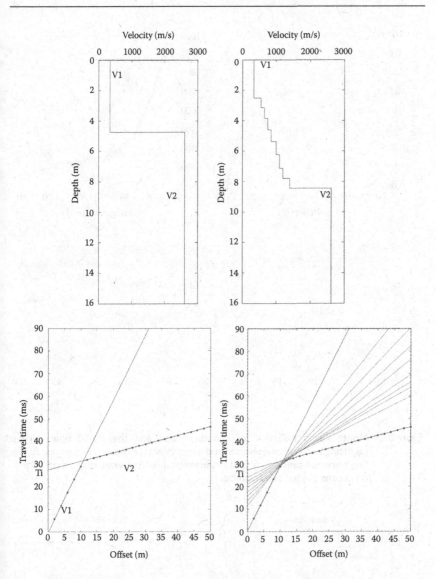

Figure 8.23 Equivalence of travel time curves for different profiles.

and the sensitivity is computed via ray-tracing in a 2D or 3D medium. The dispersion constrains the S-wave velocity structure, and the sensitivity is controlled by the eigenfunctions.

Both geometrical constraints and local intrinsic links can be used. A Poisson's ratio distribution can be used as petrophysical link. The example of Figure 8.27 shows a 2D application using a flexible joint inversion

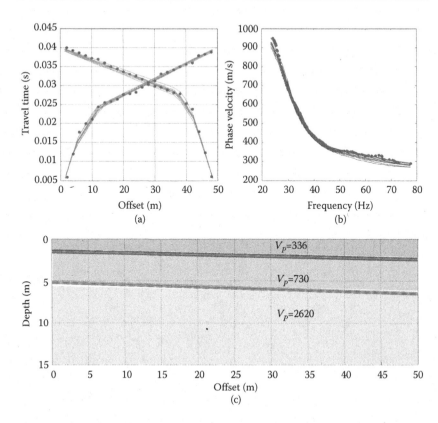

Figure 8.24 Monte Carlo inversion of the dispersion and the travel time curves: (a) fitting between theoretical and experimental P-wave travel times; (b) fitting between theoretical and experimental surface wave dispersion curves; (c) reconstructed seismic model.

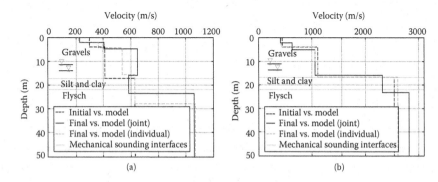

Figure 8.25 Joint inversion of P-wave seismic refraction and surface wave data: (a) S-wave model; (b) P-wave model.

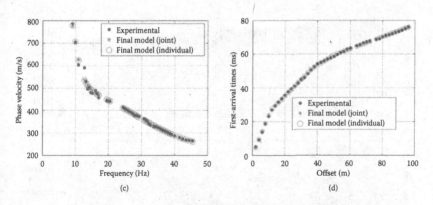

(c)   (d)

Figure 8.25 (Continued) Joint inversion of P-wave seismic refraction and surface wave data: (c) comparison between experimental and numerical dispersion curves; (d) comparison between experimental and numerical travel times. (From Piatti, C. et al., *Geophys Prospect*, 61(Suppl. 1), 77–93, 2013.)

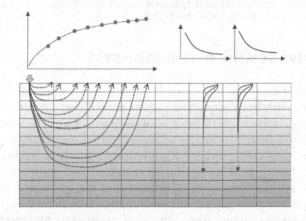

Figure 8.26 Schematic representation of the model sampling by refracted body waves and surface waves.

platform able to solve simultaneously for different scales and to handle separately the denser V_S unknowns to define the very shallow portions of the models and the coarser V_P unknowns to invert down to the main refractors of the model. The line, 10 km long, is located in Egypt along the Gulf of Suez and is discussed in Strobbia et al. (2010). The P-wave velocity model for the shallowest 400 m is plotted, together with a shallow stack section, in Figure 8.27. The correlation of the velocity distribution with the reflection stack section confirms the resolution of the obtained velocity distribution.

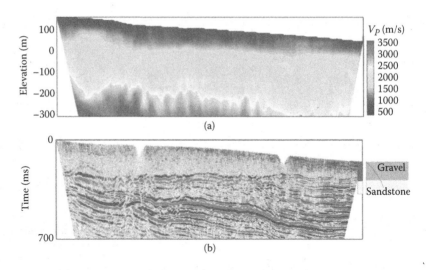

Figure 8.27 Joint inversion of surface wave and P-wave data: (a) velocity distribution from joint inversion; (b) stack section from P-wave reflection. (From Strobbia, C. et al., *First Break*, 28(8), 85–91, 2010.)

8.4 PASSIVE SEISMIC INTERFEROMETRY

Seismic interferometry has recently gained large popularity in many applications with both active and passive data. Generally, interferometry indicates the study of interference between pairs of signals. Seismic interferometry indicates the principle of generating seismic traces from virtual sources via cross-correlation or convolution of the observations at different receiver locations. The applications can have two main objectives: for the estimation or isolation of a component of the wave field and for the estimation of the properties of the medium in which the waves propagate. In this section, we discuss some aspects of passive seismic interferometry as a tool to analyze surface waves in ambient noise or microtremor data.

Recent theoretical studies have demonstrated that the diffuse wave field of the ambient noise can be processed and transformed, via cross-correlation, into a deterministic seismic impulse response, estimating the Green's function between receivers (Weaver and Lobkis 2001, 2002; Snieder 2004; Wapenaar 2004; Wapenaar and Fokkema 2006). Applications at the global seismology scale, with coda waves and long noise records, showed that the surface wave component of the Green's function between two receivers can be retrieved from the cross-correlation of observations at the receivers (Campillo and Paul 2003; Shapiro and Campillo 2004). The potential of this approach in the higher frequency range also has been proven at the engineering seismology scale. Application with active seismic data, in the exploration seismology, confirmed the possibility of estimating the direct and the diffracted surface wave component of the wave field using interferometry (Halliday et al. 2010a).

The principle of seismic interferometry is rather simple. The summation of the cross-correlation between two receivers for multiple sources creates a signal that can be interpreted as the response that would be measured at a receiver location if a source was active at the other. The principle of interferometry is illustrated here with a simple example, following the review of Wapenaar et al. (2010a).

Let us consider two receivers, R_1 and R_2, observing the surface wave signals from different sources S_i surrounding them. For each of the sources, the recorded signals $T(s_j, R_1)$ and $T(s_j, R_2)$ are cross-correlated, thus generating a trace $T_{12}(s_j)$. Such correlated trace depends on the position of the sources and on the properties of the subsurface around the receivers (Figure 8.28).

The process is repeated for a number of sources surrounding the two receivers, generating a multiplicity of correlated traces, one from each of the source points. These traces are collected in a correlogram gather (Figure 8.29); in the example, they are sorted according to the azimuth of the source. This gather shows the variation of the correlation according to the position of the source. The behavior is regular and smooth despite the difference in distance of the individual sources. Indeed, the correlation removes the effects of the propagation from the source to the receivers, and the correlogram gather is more regular than the two individual gathers of the traces at the two receivers because it mainly depends on the azimuth.

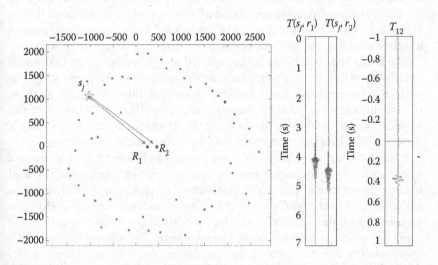

Figure 8.28 Two receivers surrounded by sources. For each source, the two traces are cross-correlated, creating a signal.

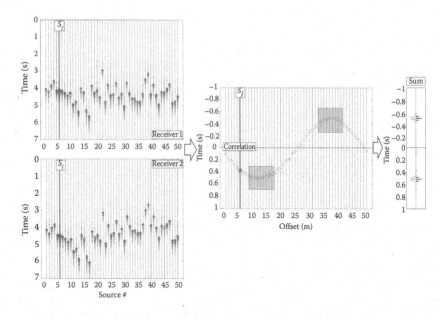

Figure 8.29 Seismic interferometry: signals received by the two detectors—their correlation and the sum.

In the summation of the correlated traces, contributions from opposite points tend to cancel out. The summation of the traces from the phase stationary points constructively produces a trace with the causal and the anticausal response from the first receiver to the second receiver. In other words, the contributions from sources almost aligned with the two receivers do not cancel out. The resulting trace can be interpreted as the impulse response of a source placed at the first receiver, which is considered the virtual source.

The extracted signals can be interpreted and processed as virtual active data. One of the main applications of this process is the retrieval of the surface wave response from ambient noise and the subsequent analysis of the surface wave velocity (Shapiro and Campillo 2004). Most of the applications are based on the analysis of 2D sparse arrays and lead to a tomographic determination of the surface wave velocity distribution.

It has to be stressed that in case of a nonisotropic wave field, the use of a linear array does not allow correcting for the apparent velocity and can lead to largely overestimated velocities. But under the hypothesis of a diffuse ambient noise wave field, approaches similar to refraction microtremor (ReMi) can be used. The Green's function can be built from passive data with a linear array by using one detector as the virtual source. The advantage over ReMi approaches is that the seismogram can be processed as an active dataset. The validation of the hypotheses will be easier, and the corresponding f–k spectrum will have features more similar to the usual spectra from active data.

To compare the two approaches, simple synthetics are created using 1000 sources around a linear array of receivers in a layered medium. The interferogram is plotted in Figure 8.30. Its symmetry around $t = 0$ is due to the basic properties of the cross-correlation and confirms the uniform distribution of the sources.

The f–k spectrum of the interferogram is plotted in Figure 8.31. The maximum correctly indicate the position of the true wavenumber. The interferometric process cancels out opposite contributions, and the

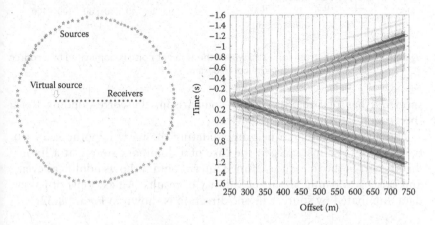

Figure 8.30 A linear array surrounded by sources, and the correlogram obtained from interferometric processing of 1000 sources.

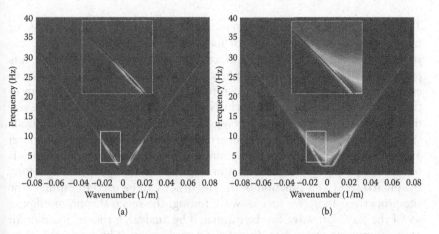

Figure 8.31 (a) The f–k spectrum of the correlogram gather of Figure 8.28. The maximum (black dots in the zoomed window) is located on the true theoretical wavenumber. (b) The ReMi spectrum obtained from the same synthetic data.

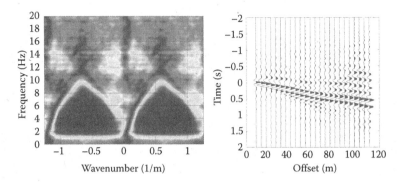

Figure 8.32 ReMi spectrum for a site where the ambient noise is dominated by a source in one direction and the correlogram.

spectrum has real maxima. As a comparison, the corresponding ReMi spectrum is plotted on Figure 8.31b.

The interferogram can help in validating the basic hypotheses of isotropic noise. For instance, in the case of a dominant source at a limited distance from the line, the obtained interferogram shows nonlinear events and raises a warning about the validity of results. An example of passive data dominated by sources in one direction is shown in Figure 8.32.

8.5 MULTICOMPONENT SURFACE WAVE ANALYSIS, POLARIZATION STUDIES, AND HORIZONTAL-TO-VERTICAL SPECTRAL RATIO

The amplitude of surface wave modal contributions as a function of frequency depends on the source location and spectrum, but it is also strongly affected by the site subsurface velocity structure. In principle, it is possible to use the measured amplitude spectrum as experimental data to estimate the site properties, but controlling or measuring the spectrum of the source radiation is not straightforward. However, the radial and the vertical components of the surface wave modal displacements have amplitude spectra that depend on the site properties. The ratio between the horizontal and the vertical components of the surface wave displacement only depends on the site properties for a given source location and direction. Therefore, the polarization as a function of the frequency can be used to estimate the site properties. In active surface wave testing, the spectral ratio or ellipticity of the Rayleigh wave can be estimated by analyzing the polarization of multicomponent active data. In passive data dominated by Rayleigh waves, the average ratio between the horizontal and the vertical components can be estimated over long records and used to evaluate the local site properties.

In a homogeneous medium, the ratio between the vertical and the horizontal components is constant, and it is a function of the Poisson's ratio. In layered media, the ellipticity is frequency dependent. The effect of the layering on the ellipticity is illustrated in Figure 8.33 (Boaga et al. 2013). A simple model, with a 25 m thick layer over half-space is considered; the velocity of the half-space is kept constant at 500 m/s, and the velocity of the layer varies from 500 m/s (thus giving a homogeneous half-space) to 200 m/s. The modal curves are plotted in Figure 8.33a, and the vertical ellipticity curves (V/H) for the fundamental mode are plotted in Figure 8.33b. The ratio between the vertical and horizontal components assumes a wider range for larger velocity contrasts; the polarization is more vertical in the high-frequency range and less in the low-frequency range. In the example, for a contrast larger than two, the horizontal component becomes larger than the vertical.

When the contrast is large, the amplitude spectrum of the vertical component has a strong peak, but it also has a zero (i.e., a frequency at which the vertical component vanishes and the motion is purely horizontal). The elliptical orbit changes direction across this point. Close to the maximum

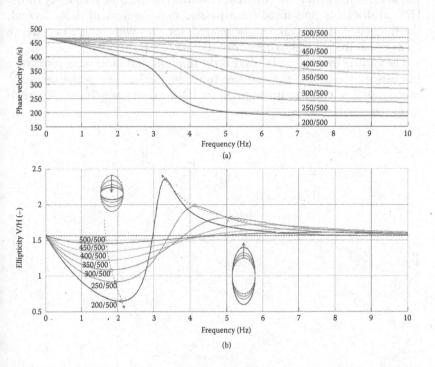

Figure 8.33 (a) Modal curves and (b) ellipticity of the fundamental mode for a single-layer model. The seven models have a 25 m thick layer with a V_S varying from 200 to 500 m/s, over a half-space with velocity 500 m/s.

of amplitude of the vertical component, the horizontal component vanishes and the motion is purely vertical.

Most applications use the horizontal ellipticity (i.e., the ratio between the horizontal and the vertical components, also called the H/V spectral ratio). The ratio between the horizontal and the vertical components for a simple model with a shear wave velocity ratio of four is plotted in a logarithmic scale in Figure 8.34: a low-velocity layer, 25 m thick, with 200 m/s shear wave velocity, over a 800 m/s half-space (V_P of 1500 m/s, 2000 m/s). In the very low-frequency range (A), the particle motion is elliptical and counterclockwise. At the first singularity, f_H, the motion becomes purely horizontal (B); in the intermediate frequency range, the motion is clockwise (C) and becomes more and more vertically polarized until the second singularity f_V where the motion is purely vertical (D) and the ratio H/V goes to zero. In the high-frequency range, the polarization tends toward the value of the first layer.

In the case of large impedance contrasts, the singularities of the spectral ratio are a strong signature of the site properties and have a very direct dependence on the thickness and velocity of the low-velocity sediments.

The peak of the spectral ratio H/V, where the vertical component vanishes, is important for different reasons. For one, the frequency of the H/V peak can be measured from passive multicomponent data. Second, the frequency of the peak can be related, for sites with a high impedance contrast, to the natural frequency of the site.

The natural frequency of the site is the fundamental shear wave resonance frequency. It refers to site transfer function for incident plane vertically propagating S-waves routinely used in 1D local site amplification assessment (Kramer 1996).

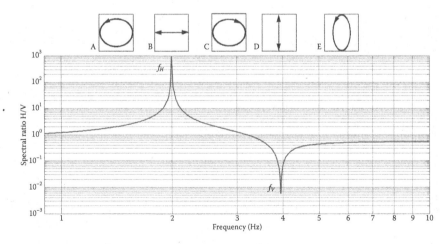

Figure 8.34 Horizontal ellipticity (H/V) of the fundamental mode for a simple model with 25 m at 200 m/s and a half-space at 800 m/s.

With a homogeneous low-velocity layer with a shear velocity V_S and a thickness H, the natural frequency is equal to the $V_S/4H$, and the peak of the ellipticity of Rayleigh waves tends to be very close to this value. In case of gradient or layering in the low-velocity layer, the zero of the vertical component is at a slightly higher frequency than the fundamental frequency. The Poisson's ratio and the mass density give second-order effects. In Figure 8.35a and b, the transfer function and the ellipticity for a single layer are plotted. The peak of the horizontal ellipticity is close to the natural frequency. The cross-plot (Figure 8.35c) shows the relationship between the natural frequency and the peak ellipticity for a Monte Carlo sample of 2000 cases with a sublayered low-velocity layer over bedrock, with an impedance contrast larger than 3.5.

The ellipticity can be used to estimate relevant site properties, but measuring the ellipticity with active data requires multicomponent recording, which is not a standard practice. On the contrary, with passive data, some of the single-station multicomponent techniques are based on the estimation of the average spectral ratio.

8.5.1 Mode identification in the case of high velocity contrasts

One application of multicomponent data is the identification of modes, particularly for ambiguous cases. The mode osculation is the phenomenon where two modes get very close to each other, and their amplitude is also creating a smooth transition. It generally happens with large impedance contrasts. In Figure 8.36, the transition between the fundamental mode and the first higher mode is seamless and smooth for the velocity and the amplitude.

In this case, the energy jumps from the fundamental mode to the higher mode, creating an apparent continuous curve. Even in less extreme cases, the transition can be continuous. With the limited spectral resolution of real data, the identification of the mode superposition and transition may be impossible (see also Section 7.4). In Figure 8.37, the limited spatial resolution leads to an apparent dispersion curve that is associated with the fundamental mode in the high-frequency range and to the first higher mode in the low-frequency range (Boaga et al. 2013).

Misidentification can be avoided if the horizontal component is recorded. Indeed, the horizontal in-line component allows for identifying the fundamental mode below the osculation frequency as shown in Figure 8.38.

8.5.2 Passive H/V

The single-station passive methods, based on the analysis of the spectral ratio between the horizontal-to-vertical components, have been extensively debated and investigated. A general agreement on the important role played

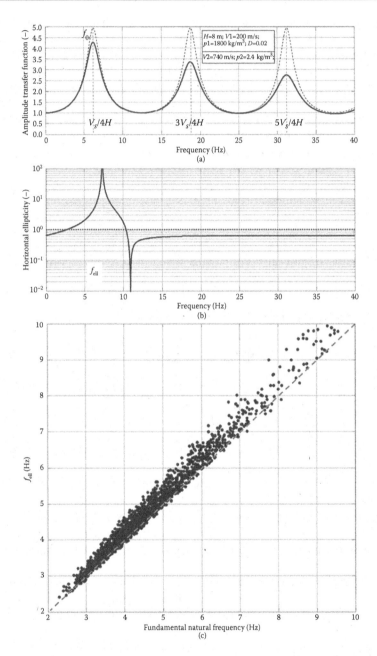

Figure 8.35 Horizontal ellipticity (H/V) of the fundamental mode: (a) amplification function for vertically propagating shear waves; (b) Rayleigh wave ellipticity; (c) comparison of the natural frequency with the frequency at which the maximum ellipticity is obtained.

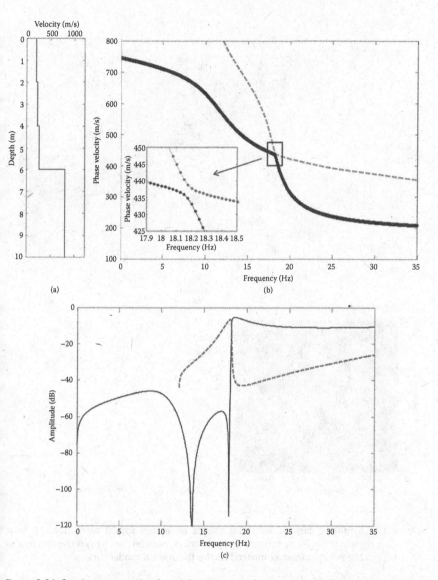

Figure 8.36 Synthetic example of model with an osculation point: (a) shear wave velocity model; (b) Rayleigh wave dispersion curve; (c) modal amplitudes for the first two higher modes.

by the Rayleigh wave ellipticity has been reached. Even if, in some cases, the role of body waves cannot be neglected, passive H/V measurements are related to surface wave propagation in most cases.

Ambient vibration techniques with single multicomponent stations and the H/V method can contribute effectively to site effect evaluation,

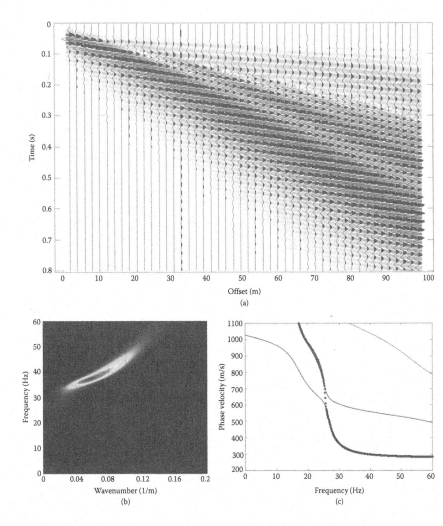

Figure 8.37 Synthetic data showing the effects of limited spatial resolution: (a) shot gather; (b) *f–k* spectrum; (c) apparent dispersion curve from the maxima of the *f–k* spectrum compared to the theoretical modal curves.

especially in seismic hazard microzonation, and particularly in urban areas. These low-cost techniques are also of great interest in moderate seismicity and in developing countries.

The H/V spectral ratio technique, based on the interpretation or inversion of the ratio between the Fourier amplitude of the horizontal and vertical components of microtremors, was introduced by Nogoshi and Igarashi in 1971 and became popular with the work by Nakamura (1989, 1996, 2000). Nakamura's theoretical explanation assumes an ambient noise wave field

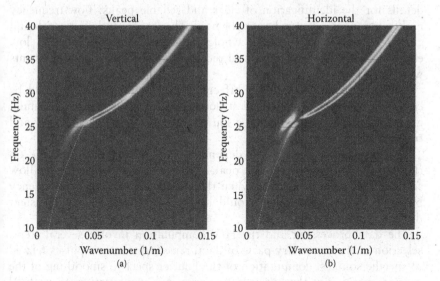

Figure 8.38 The *f–k* spectra for the same synthetic dataset of Figure 8.37: (a) vertical component; (b) horizontal component.

dominated by body waves, with deep sources, and explains the shape of the curve with the shear wave resonance in the sediments above bedrock. If the H/V curves are controlled by the shear wave resonance in the sediments, peak frequency and the amplitude depend on the soil transfer function and can be interpreted as the fundamental resonance frequency and the site amplification factor for low strain.

A large number of experiments and theoretical and numerical studies have demonstrated that, in most situations, the noise wave field is dominated by surface waves (Horike 1985; Arai and Tokimatsu 1998) and that the peak observed in the spectral ratio curve is related to the ellipticity of the Rayleigh waves.

When data are acquired and processed properly, the H/V ratio depends on the site properties. When the sources of the ambient noise are inside of the sedimentary layer or at the surface and close to the recording station, the peak is related to the horizontal ellipticity of the fundamental mode of the Rayleigh wave. A second peak related to the head shear wave can appear with far sources. With deep sources, in the bedrock, the peaks are due to the shear wave resonance of the head wave. The relative importance of the Rayleigh and Love waves can affect the shape of the curves, and this can vary from site to site.

8.5.3 How to compute the H/V

The acquisitions and processing of H/V data are discussed extensively in the documentation of the SESAME project (SESAME 2004), as well as details for the identification of clear and reliable peaks. Low-frequency multicomponent receivers have to be used. The sensors and digitizer should guarantee sensitivity over the whole frequency range of interest, low electronic noise, and synchronized recording of the different components with the same system response.

The experimental conditions affect the measurement. Data should be collected avoiding underground structures, soft and irregular soil (mud, grass, ice, gravel, etc.), and some sources of noise such as machinery and traffic loads.

The duration of the records depends on the noise conditions and on the expected frequency of the peak—from a few minutes for shallow soil deposits to half an hour when the expected peak is at a frequency below 1 Hz. The duration should be increased if there are transient disturbances.

The data processing and the H/V computation involve several steps: selection of the stationary parts of data, rejecting transient pulses related to specific sources, computation of the Fourier spectra, smoothing of the spectra, merging of the horizontal components, computation of the H/V ratio, averaging of acquisition over different time windows, and estimation of average values and associated standard deviation to identify the reliable peaks.

8.5.4 Interpretation of H/V

In the original interpretation proposed by Nakamura (1989), the spectral ratio is directly related to the S-wave site response. However, the agreement in the seismological and engineering community is that the surface waves dominate the H/V response. With large velocity contrasts and with shallow sources close to the recording station, the role of the surface waves dominates. In some cases, with a noise wave field dominated by fundamental mode Rayleigh waves, not only can a clear large maximum be identified but a minimum may also be visible (related to the zero of the H/V ellipticity) such as in Figure 8.39.

In some cases, the shape of the curve indicates quite clearly which types of events have been recorded and analyzed; the curve can be interpreted as the response of the site to a certain phenomenon. The frequency of the H/V peak can be interpreted as the peak of the fundamental mode ellipticity and can be used to estimate the natural frequency of the site. In this case, the value of the peak is not related to the site amplification factor.

However, in general, there are different factors that can affect the shape of the H/V curve. The unknown nature and position of the seismic sources play a dominant role in determining which type of seismic waves compose the passive data, and consequently, this largely affects the H/V response.

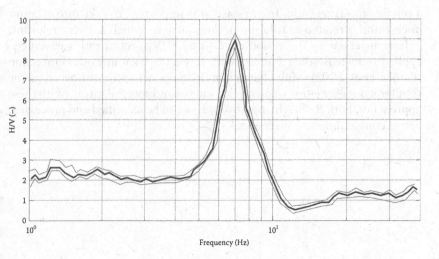

Figure 8.39 Example of H/V spectral ratio from experimental data.

The possible ambiguity and nonuniqueness of the computed curve should not be neglected. Using a single curve for the characterization of a site can lead to severe uncertainties.

The interpretation of the H/V curves requires a preliminary careful evaluation of their characteristics. As first criterion, the clarity of the peaks can be used. When the peak is clear (and does not have an industrial origin), it indicates the presence of a large impedance contrast at some depth, and the peak frequency is close to the fundamental frequency of the site. This is quite independent from the nature of the wave field. If the thickness is known, an estimation of the shear wave velocity can be obtained. If, on the contrary, the curve is flat (typically, values between 0.5 and 2), it is likely that the local velocity structure has no impedance contrasts. Different situations are possible, and detailed interpretation guidelines have been produced (SESAME 2004).

More advanced processing and interpretation techniques involve the inversion of the curve for the identification of model parameters as in surface wave dispersion inversion (Fäh et al. 2001). The curves are first corrected for the contamination by SH and Love waves and then inverted. Arai and Tokimatsu (2000) showed that H/V spectral ratios can be better reproduced if the contribution of higher modes of Rayleigh waves and Love waves is also taken into account. For the inversion, the strong non-linearity of the curves makes global search preferable, and the limited amount of information in the ellipticity curves requires the introduction of a priori information (e.g., on bedrock depth) or further constraints for a reliable interpretation. In any case, it is important to take into account uncertainties related to solution nonuniqueness, which are very relevant.

Joint inversion of dispersion curves and H/V spectral ratios have been applied to increase the reliability with respect to each individual inversion (Parolai et al. 2005). Multichannel, multicomponent acquisitions allow the simultaneous extraction of dispersion curves and spectral ratios. With these approaches, both Rayleigh and Love waves can be identified and processed, better constraining the inversion.

References

Abo-Zena, A. M. (1979). Dispersion function computations for unlimited frequency values. *Geophys J Roy Astron Soc* 58, 91–105.

Achenbach, J. D. (1984). *Wave Propagation in Elastic Solids*. Amsterdam, The Netherlands: North-Holland Publishing.

Aki, K. (1957). Space and time spectra of stationary stochastic waves, with special reference to microtremors. *Bull Earthquake Res Inst* 35, 416–456.

Aki, K. (1965). A note on the use of microseisms in determining the shallow structure of the Earth's crust. *Geophysics* 30, 665–666.

Aki, K., and P. G. Richards. (2002). *Quantitative Seismology*. Sausalito, CA: University Science Books.

Al-Hunaidi, M. O. (1992). Difficulties with phase spectrum unwrapping in spectral analysis of surface waves nondestructive testing of pavements. *Can Geotech J* 29, 506–511.

Al-Hunaidi, M. O. (1994). Analysis of dispersed multi-mode signals of the SASW method using the multiple filter/crosscorrelation technique. *Soil Dynam Earthquake Eng* 13, 13–24.

Al-Hunaidi, O. (1998). Evolution based genetic algorithm for the analysis of nondestructive surface wave test on pavements. *Nondestr Test Eval* 31, 273–280.

Alkire, B. D. (1992). Seasonal soil strength by spectral analysis of surface waves. *J Cold Regions Eng* 6(1), 22–38.

Andrus, R. D., R. M. Chung, K. H. Stokoe, and J. A. Bay. (1998). Delineation of densified sand at Treasure Island by SASW testing. In: *Geotechnical Site Characterization* (P. K. Robertson and P. W. Mayne, eds.). Vol. 1. Balkema, Rotterdam, The Netherlands; pp. 459–464.

Arai, H., and K. Tokimatsu. (1998). The Effects of surface geology on Seismic motion: recent progress and new horizon on ESG study. In: Proceeding of the Second International Symposium on the Effects of Surface Geology on Seismic Motion, Japan, December 1–3; pp. 673–680.

Arai, H., and K. Tokimatsu. (2000). Effects of Rayleigh and Love waves on microtremor H/V spectra. Paper presented at 12th World Conference on Earthquake Engineering, New Zealand Society for Earthquake Engineering, Auckland.

Archie, G. E. (1942). The electrical resistivity log as an aid in determining some reservoir characteristics. *Petrol Trans AIME* 146, 54–62.

Asten, M. W., and J. D. Henstridge. (1984). Array estimators and the use of microseisms for reconnaissance of sedimentary basins. *Geophysics* 49(11), 1828–1837.

Aster, R. C., B. Borchers, and C. H. Thurber. (2005). *Parameter Estimation and Inverse Problems*. Amsterdam: Elsevier Academic Press.

Atkinson, J. H. (2000). Non-linear soil stiffness in routine design. *Gèotechnique* 50(5), 487–508.

Backus, G. E., and F. Gilbert. (1970). Uniqueness in the inversion of inaccurate gross earth data. *Phil Trans Roy Soc Lond A* 266, 123–192.

Badsar, S. A., M. Schevenels, W. Haegeman, and G. Degrande. (2010). Determination of the material damping ratio in the soil from SASW tests using the half-power bandwidth method. *Geophys J Int* 182(3), 1493–1508.

Badsar, S. A., M. Schevenels, W. Haegeman, and G. Degrande. (2011). Determination of the material damping ratio in the soil from SASW test using the half-power bandwidth method and the Arias intensity. In: Proceedings of the 8th International Conference on Structural Dynamics (G. De Roeck, G. Degrande, G. Lombaert, and G. Müller, eds.), Department of Civil Engineering, Ghent University, Ghent, Belgium, July 4–6.

Baecher, G. B., and J. T. Christian. (2003). *Reliability and Statistics in Geotechnical Engineering*. Chichester: Wiley.

Bain, L. J., and M. Engelhardt. (1992). *Introduction to Probability and Mathematical Statistics*. Boston: PWS-KENT.

Ballard, R. F. (1964). *Determination of Soil Shear Moduli at Depth by In Situ Vibratory Techniques*. Miscellaneous paper No. 4-691. Waterways Experiment Station, Vicksburg, Mississippi.

Barrowdale, I., and F. D. K. Roberts. (1974). Solution of an overdetermined system of equations in the L1 norm. *Commun ACM* 17(6), 319–326.

Båth, M., and A. J. Berkhout. (1984). *Mathematical Aspects of Seismology*. Amsterdam, The Netherlands: Elsevier.

Beaty, K. S., D. R. Schmitt, and M. Sacchi. (2002). Simulated annealing inversion of multimode Rayleigh-wave dispersion curves for geological structure. *Geophys J Int* 151, 622–631.

Bendat, J. S., and A. G. Piersol. (2010). *Random Data: Analysis and Measurement Procedures*. 4th ed. Hoboken, NJ: Wiley.

Ben-Menahem, A. (1995). A concise history of mainstream seismology: Origins, legacy and perspectives. *Bull Seismol Soc Am* 85(4), 1202–1225.

Ben-Menahem, A., and S. J. Singh. (2000). *Seismic Waves and Sources*. Mineola, NY: Dover Publications.

Bergamo, P., C. Comina, S. Foti, and M. Maraschini. (2011). Seismic characterization of shallow bedrock sites with multimodal Monte Carlo inversion of surface wave data. *Soil Dynam Earthquake Eng* 31(3), 530–534.

Biot, M. A. (1956a). Theory of propagation of elastic waves in a fluid saturated porous solid. I. Low-frequency range. *J Acoust Soc Am* 28, 168–178.

Biot, M. A. (1956b). Theory of propagation of elastic waves in a fluid saturated porous solid. II. Higher frequency range. *J Acoust Soc Am* 28, 179–191.

Bleistein, N. (1984). *Mathematical Methods for Wave Phenomena*. Academic Press, London, UK.

Boaga, J., G. Cassiani, C. L. Strobbia, and G. Vignoli. (2013). Mode misidentification in Rayleigh waves: Ellipticity as a cause and a cure. *Geophysics* 78(4), EN17–EN28.

Bodin, T., M. Sambridge, H. Tkalčić, P. Arroucau, K. Gallagher, and N. Rawlinson. (2012). Transdimensional inversion of receiver functions and surface wave dispersion. *J Geophys Res Solid Earth* 117(B2), 1–24.

Bonnefoy-Claudet, S., F. Cotton, and P.-Y. Bard. (2006). The nature of noise wavefield and its applications for site effects studies: A literature review. *Earth Sci Rev* 79(3), 205–227.

Boore, D. M. (1972). *Finite Difference Methods for Seismic Wave Propagation in Heterogeneous Materials*. New York: Academic Press.

Boore, D. M., and M. W. Asten. (2008). Comparisons of shear-wave slowness in the Santa Clara Valley, California, using blind interpretations of data from invasive and non-invasive methods. *Bull Seismol Soc Am* 98, 1983–2003.

Borcherdt, R. (1973a). Energy and plane waves in linear viscoelastic media. *J Geophys Res* 78(14), 2442–2453.

Borcherdt, R. (1973b). Rayleigh-type surface wave on a linear viscoelastic half-space. *J Acoust Soc Am* 54(6), 1651–1653.

Borcherdt, R., and L. Wennerberg. (1985). General P, type-I S, and type-II S waves in anelastic solids: Inhomogeneous wave fields in low-loss solids. *Bull Seismol Soc Am* 75(6), 1729–1763.

Borcherdt, R. D. (1971). Inhomogeneous body & surface plane waves in a generalised viscoelastic half-space. PhD Thesis, University of California.

Borcherdt, R. D. (1994). Estimates of site-dependent response spectra for design (methodology and justification). *Earthquake Spectra* 10, 617–654.

Borcherdt, R. D. (2009). *Viscoelastic Waves in Layered Media*. Cambridge: Cambridge University Press.

Boyd, S., and L. Vandenberghe. (2004). *Convex Optimization*. Cambridge University Press, Cambridge, UK.

Bracewell, R. N. (1978). *The Fourier Transform and Its Applications*. 2nd ed. New York: McGraw-Hill.

Brown, L. T., J. G. Diehl, and R. L. Nigbor. (2000). A simplified procedure to measure average shear-wave velocity to a depth of 30 meters (VS30). In: 12th World Conference on Earthquake Engineering, Aucukland, NZ, January 30–February 4.

Buchen, P. (1971). Plane waves in linear viscoelastic media. *Geophys J Roy Astron Soc* 23, 531–542.

Burland, J. B. (1989). "Small is beautiful": The stiffness of soils at small strains. *Can Geotech J* 26(4), 499–516.

Buttkus, B. (2000). *Spectral Analysis and Filter Theory in Applied Geophysics*. Berlin: Springer-Verlag.

Campillo, M., and A. Paul. (2003). Long-range correlations in the diffuse seismic coda. *Science* 299, 547–549.

Carcione, J., and H. Helle. (2004). The physics and simulation of wave propagation at the ocean bottom. *Geophysics* 69(3), 825–839.

Chapman, C. H. (2003). Yet another elastic plane-wave, layer-matrix algorithm. *Geophys J Int* 154, 212–223.

Chen, X. (1993). A systematic and efficient method of computing normal modes for multilayered half space. *Geophys J Int* 115, 391–409.

Christensen, R. M. (1971). *Theory of Viscoelasticity: An Introduction*. Academic Press, New York, London.

Claerbout, J. F., and F. Muir. (1973). Robust modelling with erratic data. *Geophysics* 38, 826–844.

Clough, R. W., and J. Penzien. (1993). *Dynamics of Structures*. 2nd ed. New York: McGraw-Hill.

Cole, K. S., and R. H. Cole. (1941). Dispersion and absorption in dielectrics. I. Alternating current characteristics. *J Chem Phys* 9, 341–351.

Comina, C., S. Foti. (2007). Surface wave tests for vibration mitigation studies, Technical Note, *Journal of Geotechn. and Geoenv. Eng.*, ASCE, 133(10), 1320–1324.

Comina, C., S. Foti, D. Boiero, and L. V. Socco. (2011). Reliability of $V_{S,30}$ evaluation from surface waves tests. *J Geotech Geoenviron Eng* 137(6), 579–586.

Comina, C., S. Foti, L. V. Socco, and C. Strobbia. (2004). Geophysical characterization for seepage potential assessment along the embankments of the Po River. In: Proceedings of ISC-2 on Geotechnical and Geophysical Site Characterization (Viana da Fonseca A. and Mayne, P. W., eds.). Millpress, Rotterdam, pp. 451–458.

Comina, C., S. Foti, L. Sambuelli, L. V. Socco, and C. Strobbia. (2002). Joint inversion of VES and surface wave data. In: Proceeding of Symposium on the Application of Geophysics to Engineering and Environmental Problems, SAGEEP 2002, Las Vegas, USA, February 10–14.

Constable, S. C., R. L. Parker, and C. Constable. (1987). Occam's inversion: A practical algorithm for generating smooth models from electromagnetic sounding data. *Geophysics* 52(3), 289–300.

Cornou, C., M. Ohrnberger, D. M. Boore, K. Kudo, and P.-Y. Bard. (2007). Derivation of structural models from ambient vibration array recordings: Results from an international blind test. In: 3rd International Symposium on the Effects of Surface Geology on Seismic Motion, Paper number: NBT, Grenoble, France, August 30–1 September.

Courant, R., and D. Hilbert. (2004). *Methods of Mathematical Physics*. John Wiley, Weinheim, Germany.

Cox, B. R., and C. M. Wood. (2011). *Surface Wave Benchmarking Exercise: Methodologies, Results, and Uncertainties*. Georisk 2011. Atlanta, GA: American Society of Civil Engineers.

Dal Moro, G., M. Pipan, and P. Gabrielli. (2007). Rayleigh wave dispersion curve inversion via genetic algorithms and marginal posterior probability density estimation. *J Appl Geophys* 61, 39–55.

Deidda, G. P., and R. Balia. (2001). An ultrashallow SH-wave seismic reflection experiment on a subsurface ground model. *Geophysics* 66(4), 1097–1104.

De Stefano, M., F. Golfré Andreasi, S. Re, M. Virgilio, and F. Snyder. (2011). Multiple-domain, simultaneous joint inversion of geophysical data with application to subsalt imaging. *Geophysics* 76(3), R69–R80.

Dobry, R. (1970). *Damping in Soils: Its Hysteretic Nature and the Linear Approximation*. Research Report R70-14. Massachusetts Institute of Technology, Cambridge, MA; pp. 82.

Doyle, H. (1995). *Seismology*. Chichester: John Wiley.

Dunkin, J. W. (1965). Computation of modal solutions in layered, elastic media at high frequencies. *Bull Seismol Soc Am* 55(2), 335–358.

Dziewonski, A., S. Bloch, and M. Landisman. (1969). A technique for the analysis of transient seismic signals. *Bull Seismol Soc Am* 59, 427–444.

Dziewonski, A. M., and A. L. Hales. (1972). Numerical analysis of dispersed seismic waves. In: *Methods in Computational Physics Vol. 11 Seismology: Surface Waves and Earth Oscillations* (B. A. Bolt, ed.). New York: Academic Press, pp. 39–85.

Dziewonski, A. M., and J. M. Steim. (1982). Dispersion and attenuation of mantle waves through waveform inversion. *Geophys J Roy Astron Soc* 70(2), 503–527.

Engl, H. W. (1993). Regularization methods for the stable solution of inverse problems. *Surv Math Ind* 3, 71–143.

Engl, H. W., M. Hanke, and A. Neubauer. (1996). *Regularization of Inverse Problems*. Dordrecht: Kluwer Academic.

Eringen, A. C., and E. S. Suhubi. (1975). *Elastodynamics*. Academic Press, New York, London.

Ewing, W. M., W. S. Jardetzky, and F. Press. (1957). *Elastic Waves in Layered Media*. New York: McGraw-Hill.

Faccioli, E., F. Maggio, A. Quarteroni, and A. Taghan. (1996). Spectral-domain decomposition methods for the solution of acoustic and elastic wave equations. *Geophysics* 61(4), 1160–1174.

Fäh, D., F. Kind, and D. Giardini. (2001). A theoretical investigation of average H/V ratios. *Geophysical Journal International* 145(2), 535–549.

Fäh, D., F. Kind, and D. Giardini. (2003). Inversion of local S-wave velocity structures from average H/V ratios, and their use for the estimation of site-effects. *J Seismol* 7, 449–467.

Foti, S. (2000). Multi-station methods for geotechnical characterisation using surface waves. PhD Dissertation, Politecnico di Torino.

Foti, S. (2002). Numerical and experimental comparison between 2-station and multistation methods for spectral analysis of surface waves. *Riv Ital Geotecnica* 36(1), 11–22.

Foti, S. (2003). Small strain stiffness and damping ratio of Pisa clay from surface wave tests. *Geotechnique* 53, 455–461.

Foti, S., C. Comina, and D. Boiero. (2007). Reliability of combined active and passive surface wave methods. *Riv Ital Geotecnica* 41(2), 39–47.

Foti, S., C. Comina, D. Boiero, and L. V. Socco. (2009). Non-uniqueness in surface wave inversion and consequences on seismic site response analyses. *Soil Dynam Earthquake Eng* 29(6), 982–993.

Foti, S., C. G. Lai, and R. Lancellotta. (2002). Porosity of fluid-saturated porous media from measured seismic wave velocities. *Geotechnique* 52(5), 359–373.

Foti, S., L. Sambuelli, L. V. Socco, and C. Strobbia. (2003). Experiments of joint acquisition of seismic refraction and surface wave data. *Near Surf Geophys* 1(3), 119–129.

Foti, S., and C. Strobbia. (2002). Some notes on model parameters for surface wave data inversion. In: Proceedings of SAGEEP 2002, Las Vegas, USA, February 10–14, CD-Rom.

Fung, Y. C. (1965). *Foundations of Solid Mechanics*. Englewood Cliffs, NJ: Prentice Hall.

Gabriels, P., R. Snieder, and G. Nolet. (1987). *In situ* measurement of shear wave velocity in sediments with higher-mode Rayleigh waves. *Geophys Prospect* 35, 187–196.

Gallardo, L. A., and M. A. Meju. (2003). Characterization of heterogeneous near surface materials by joint 2D inversion of dc resistivity and seismic data. *Geophysical Research Letters*, 30(13).

Gallardo, L. A., and M. A. Meju. (2004). Joint two-dimensional DC resistivity and seismic travel time inversion with cross-gradients constraints. *J Geophys Res* 109(B3), B03311.

Ganji, V., N. Gucunski, and A. Maher. (1997). Detection of underground obstacles by SASW method: Numerical aspects. *J Geotech Geoenviron Eng* 123(3), 212–219.

Ganji, V., N. Gucunski, and S. Nazarian. (1998). Automated inversion procedure for spectral analysis of surface waves. *J Geotech Geoenviron Eng* 124, 757–770.

Ganpan Ke, H. D., Å. Kristensen, and M. Thompson. (2011). Modified Thomson–Haskell matrix methods for surface-wave dispersion-curve calculation and their accelerated root-searching schemes. *Bull Seismol Soc Am* 101(4), 1692–1703.

Gardner, G. H. F., L. W. Gardner, and A. R. Gregory. (1974). Formation velocity and density-the diagnostic basics for stratigraphic traps. *Geophysics* 39, 770–780.

Gilbert, F., and G. E. Backus. (1966). Propagator matrices in elastic wave and vibration problems. *Geophysics* 31, 326–332.

Golub, G. H., and C. F. Van Loan. (1996). *Matrix Computations*. Baltimore: Johns Hopkins University Press.

Graff, K. F. (1975). *Wave Motion in Elastic Solids*. Columbus: Ohio State University Press.

Groetsch, C. W. (1993). *Inverse Problems in the Mathematical Sciences*. Braunschweig: Vieweg.

Gucunski, N., V. Ganji, and M. H. Maher. (1996). Effects of obstacles on Rayleigh wave dispersion obtained from the SASW test. *Soil Dynam Earthquake Eng* 15, 223–231.

Gucunski, N., V. Krstic, and M. H. Maher. (1998). Experimental procedures for detection of underground objects. In: *Geotechnical Site Characterization* (Robertson, P. R., and Mayne, P. W. eds.). Vol. 1. Balkema, Rotterdam, The Netherlands; pp. 469–472.

Gucunski, N., and R. D. Woods. (1991). Use of Rayleigh modes in interpretation of SASW tests. In: Proceedings of the Second International Conference on Recent Advances in Geotechnical Earthquake Engineering and Soil Dynamics, Vol. 2, St. Louis, Missouri, March 11–15; pp. 1399–1408.

Haber, E., and D. Oldenburg. (1997). Joint inversion: A structural approach. *Inverse Probl* 13(1), 63–77.

Hadamard, J. (1923). *Lectures on the Cauchy Problem in Linear Partial Differential Equations*. New Haven, CT: Yale University Press.

Haegeman, W., and W. F. Van Impe. (1998). SASW control of a vacuum consolidation on a sludge disposal. In: *Geotechnical Site Characterization* (Robertson, P. K., and Mayne, P. W. eds.). Vol. 1. Balkema, Rotterdam, The Netherlands; pp. 473–477.

Halliday, D. F., A. Curtis, P. Vermeer, C. Strobbia, A. Glushchenko, D.-J. van Manen, and J. O. A. Robertsson. (2010). Interferometric ground-roll removal: Attenuation of scattered surface waves in single-sensor data. *Geophysics* 75(2), SA15–SA25.

Hamilton, E. L. (1976). Variations of Density and Porosity with Depth in Deep-sea Sediments. *Journal of Sedimentary Petrology* 46(2), 280–300.

Harr, M. E. (1996). *Reliability-Based Design in Civil Engineering*. Mineola, NY: Dover Publications.

Harvey, D. (1981). Seismogram synthesis using normal mode superposition: The locked mode approximation. *Geophys J Roy Astrom Soc* 66, 37–70.

Haskell, N. A. (1953). The dispersion of surface waves on multilayered media. *Bull Seismol Soc Am* 43(1), 17–34.

Hayashi, K.; T. Matsuoka, and H. Hatakeyama. (2005). Joint analysis of a surface-wave method and micro-gravity survey. *J Environ Eng Geophys* 10(2), 175–184.

Heisey, J. S., K. H. Stokoe, II, and A. H. Meyer. (1982). Moduli of pavement systems from spectral analysis of surface waves. *Transport Res Rec* 852, 22–31.

Hering, A., R. Misiek, A. Gyulai, T. Ormos, M. Dobroka, and L. Dresen. (1995). A joint inversion algorithm to process geoelectric and surface wave seismic data. Part I: Basic ideas. *Geophys Prospect* 43, 135–156.

Herrmann, R. B. (1994). *Computer Programs in Seismology*. User's Manual. MO: St. Louis University, St. Louis, Missouri.

Herrmann, R. B. (2007). *Computer Programs in Seismology*. Version 3.30. St. Louis, MO: Saint Louis University.

Hisada, Y. (1994). An efficient method for computing Green's functions for a layered half-space with sources and receivers at close depths. *Bull Seismol Soc Am* 84(5), 1456–1472.

Hisada, Y. (1995). An efficient method for computing Green's functions for a layered half-space with sources and receivers at close depths (part 2). *Bull Seismol Soc Am* 85(4), 1080–1093.

Holzlohner, U. (1980). Vibrations of the elastic half-space due to vertical surface loads. *Earthquake Eng Struct Dynam* 8, 405–414.

Horike, M. (1985). Inversion of phase velocity of long period microtremors to the S-wave velocity structure down to the basement in urbanized areas. *J Phys Earth* 33, 59–96.

Horst, R., P. M. Pardalos, and N. V. Thoai. (2000). *Introduction to Global Optimization*. Boston, MA: Kluwer Academic.

Hudson, J. A. (1980). *The Excitation and Propagation of Elastic Waves*. Cambridge: Cambridge University Press.

Ishihara, K. (1996). *Soil Behaviour in Earthquake Geotechnics*. Oxford, UK: Oxford Science.

Jamiolkowski, M. (2012). Role of geophysical tests in geotechnical site characterization, III De Mello lecture. *Soils Rocks* 35(2), 117–137.

Jegen, M. D., R. W. Hobbs, P. Tarits, and A. Chave. (2009). Joint inversion of marine magnetotelluric and gravity data incorporating seismic constraints. Preliminary results of sub-basalt imaging off the Faroe Shelf. *Earth Planet Sci Lett* 282(1–4), 47–55.

Jocker, J., D. Smeulders, G. Drijkoningen, C. Van Der Lee, and A. Kalfsbeek. (2004). Matrix propagator method for layered porous media: Analytical expressions and stability criteria. *Geophysics* 69(4), 1071–1081.

Johnson, D. H., and D. E. Dudgeon. (1993). *Array Signal Processing: Concepts and Techniques*. Upper Saddle River, NJ: PTR Prentice Hall.

Johnston, D. H., M. N. Toksöz, and A. Timur. (1979). Attenuation of seismic waves in dry and saturated rocks: II. Mechanisms. *Geophysics* 44(4), 691–711.

Jones, R. B. (1958). In-situ measurement of the dynamic properties of soil by vibration methods. *Geotechnique* 8(1), 1–21.

Jones, R. B. (1962). Surface wave technique for measuring the elastic properties and thickness of roads: Theoretical development. *Br J Appl Phys* 13, 21–29.

Kac, M. (1966). Can one hear the shape of the drum? *Am Math Mon* 73, 1–23.

Karray, M., and G. Lefebvre. (2009). Techniques for mode separation in Rayleigh wave testing. *Soil Dynam Earthquake Eng* 29, 607–619.

Kausel, E. (1981). *An Explicit Solution for the Green Functions for Dynamic Loads in Layered Media*. Massachusetts Institute of Technology, Cambridge, MA; p. 79.

Kausel, E., and J. M. Roesset. (1981). Stiffness matrices for layered soils. *Bull Seismol Soc Am* 71(6), 1743–1761.

Kavazanjian, E., N. Matasovic, K. H. Stokoe II, and J. D. Bray. (1996). In situ shear wave velocity of solid waste from surface wave measurements. In Environmental Geotechnics, Proc. 2nd Int. Congr. On Environm. Geotechnics, Osaka, Japan, November 5–8 1996; vol. 1, 97–102.

Keilis-Borok, V. I. (1989). *Seismic Surface Waves in a Laterally Inhomogeneous Earth*. Kluwer Academic Publishers, Dordrecht, The Netherlands.

Kennett, B. L. N. (1974). Reflections, rays, and reverberations. *Bull Seismol Soc Am* 64(6), 1685–1696.

Kennett, B. L. N. (1983). *Seismic Wave Propagation in Stratified Media*. UK: Cambridge University Press, Cambridge, UK.

Kennett, B. L. N., and N. J. Kerry. (1979). Seismic waves in a stratified half-space. *Geophys J Roy Astron Soc* 57, 557–583.

Kim, D. S., H. J. Park, and E. S. Bang. (2013). Round Robin test for comparative study of in-situ seismic tests. In: *Geotechnical and Geophysical Site Characterization 4* (R. Q. Coutinho and P. W. Mayne, eds.). Leiden: CRC Press, pp. 1427–1434.

Kim, D. S., M. K. Shin, and H. C. Park. (2001). Evaluation of density in layer compaction using SASW method. *Soil Dynam Earthquake Eng* 21, 39–46.

Kim, Y. R., and B. Xu. (2000). A new backcalculation procedure based on dispersion analysis of FWD time-history deflections and surface wave measurements using artificial neural network. In: *Nondestructive Testing of Pavements and Backcalculation of Moduli* (S. D. Tayabji and E. O. Lukanen, eds.). West Conshohocken, PA: American Society for Testing and Materials, pp. 297–312.

Kirsch, A. (2011). *An Introduction to the Mathematical Theory of Inverse Problems*. New York: Springer.

Kjartansson, E. (1979). Constant Q-wave propagation and attenuation. *J Geophys Res* 84, 4737–4748.

Knopoff, L. (1964). A matrix method for elastic wave problems. *Bull Seismol Soc Am* 54(1), 431–438.

Komatitsch, D., and J. P. Vilotte. (1998). The spectral element method: An efficient tool to simulate the seismic response of 2D and 3D geological structures. *Bull Seismol Soc Am* 88(2), 368–392.

Kramer, S. L. (1996). *Geotechnical Earthquake Engineering*. Upper Saddle River, NJ: Prentice Hall.

Laake, A., M. Sheneshen, C. Strobbia, L. Velasco, and A. Cutts. (2010). Surface-subsurface integration reveals faults in Gulf of Suez oilfields. 72nd EAGE meeting, Barcelona, Spain, Expanded Abstracts, F008.

Lacoss, R. T., E. J. Kelly, and M. N. Toksoz. (1969). Estimation of seismic noise structure using arrays. *Geophysics* 34, 21–38.

Lai, C. G. (1998). Simultaneous inversion of Rayleigh phase velocity and attenuation for near-surface site characterization. PhD Thesis, Georgia Institute of Technology.

Lai, C. G. (2005). Surface waves in dissipative media: Forward and inverse modelling. In: *Surface Waves in Geomechanics: Direct and Inverse Modelling for Soil and Rocks* (C. G. Lai and K. Wilmanski, eds.). New York: Springer-Verlag, p. 385.

Lai, C. G., S. Foti, and G. J. Rix. (2005). Propagation of data uncertainty in surface wave inversion. *J Environ Eng Geophys* 10(2), 219–228.

Lai, C. G., D. C. F. Lo Presti, O. Pallara, and G. J. Rix. (1999). *Misura simultanea del modulo di taglio e dello smorzamento intrinseco dei terreni a piccole deformazioni.* Torino: ANIDIS. (in Italian).

Lai, C. G., M.-D. Mangriotis, G. J. Rix. (2014). An explicit relation for the apparent phase velocity of Rayleigh-waves in a vertically-heterogeneous elastic half space. *Geophysical Journal International*, Under revision.

Lai, C. G., and G. J. Rix. (2002). Solution of the Rayleigh eigenproblem in viscoelastic media. *Bull Seismol Soc Am* 92(6), 2297–2309.

Lai, C. G., G. J. Rix, S. Foti, and V. Roma. (2002). Simultaneous measurement and inversion of surface wave dispersion and attenuation curves. *Soil Dynam Earthquake Eng* 22(9–12), 923–930.

Lamb, H. (1904). On the propagation of tremors over the surface of an elastic solid. *Philos Trans Roy Soc Lond Ser A* 203, 1–42.

Lanczos, C. (1961). *Linear Differential Operators.* New York: Van Nostrand, p. 132.

Landisman, M. A., A. Dziewonski, and Y. Sato. (1969). Recent improvements in the analysis of surface wave observations. *Geophys J Roy Astron Soc* 17, 369–463.

Lawson, C. L., and R. J. Hanson. (1974). *Solving Least Squares Problems.* Englewood Cliffs, NJ: Prentice Hall.

Lebedev, N. N. (1972). *Special Functions & Their Applications.* New York: Dover Publications.

Lee, W. B., and S. C. Solomon. (1979). Simultaneous inversion of surface-wave phase velocity and attenuation: Rayleigh and Love waves over continental and oceanic paths. *Bull Seismol Soc Am* 69(1), 65–95.

Leurer, K. C. (1997). Attenuation in fine-grained marine sediments: Extension of the Biot-stoll model by the effective grain model (EGM). *Geophysics* 62(5), 1465–1479.

Levshin, A., M. Ritzwoller, and L. Ratnikova. (1994). The nature and cause of polarization anomalies of surface waves crossing northern and central Eurasia. *Geophys J Int* 117, 577–591.

Lines, L., A. Schultz, and S. Treitel. (1988). Cooperative inversion of geophysical data. *Geophysics* 53(1), 8–20.

Linville, A. F., and S. J. Laster. (1966). Numerical experiments in the estimation of frequency-wavenumber spectra of seismic events using linear arrays. *Bull Seismol Soc Am* 56(6), 1337–1355.

Liu, H. P., D. L. Anderson, and H. Kanamori. (1976). Velocity dispersion due to anelasticity: Implications for seismology and mantle composition. *Geophys J Roy Astron Soc* 47, 41–58.

Liu, T. (2010). Efficient reformulation of the Thomson-Haskell method for computation of surface waves in layered half-space. *Bull Seismol Soc Am* 100(5A), 2310–2316.

Lockett, F. J. (1962). The reflection and refraction of waves at an interface between viscoelastic materials. *J Mech Phys Solids* 10(53), 53–64.

Logan, J. D. (2006). *Applied Mathematics*. Hoboken, NJ: Wiley-Interscience.

Lopes, I., G. P. Deidda, M. Mendes, C. Strobbia, and J. Santos. (2013). Contribution of *in situ* geophysical methods for the definition of the São Sebastião crater model (Azores). *J Appl Geophys* 98, 265–279.

Lo Presti, D. C. F., and O. Pallara. (1997). Damping ratio of soils from laboratory and in-situ tests. In: 14th International Conference on Soil Mechanics and Foundation Engineering, Hamburg, Germany, September 6–12.

Louie, J. N. (2001). Faster, better: Shear-wave velocity to 100 meters depth from refraction microtremor arrays. *Bull Seismol Soc Am* 91(2), 347–364.

Love, A. E. H. (1911). *Some Problems of Geodynamics*. Cambridge: Cambridge University Press.

Luco, J. E., and R. J. Apsel. (1983). On the Green's function for a layered half-space. Part I. *Bull Seismol Soc Am* 73, 909–929.

Luke, B. A. (1994). In situ measurements of stiffness profiles in the seafloor using the Spectral-Analysis-of-Surface-Waves (SASW) method. PhD Dissertation, University of Texas, Austin.

Luke, B. A., and K. H. Stokoe. (1998). Application of SASW method underwater. *J Geotech Geoenviron Eng* 124, 523–531.

Lysmer, J., and L. A. Drake. (1972). *A Finite Element Method for Seismology*. New York: Academic Press.

Lysmer, J., and G. Waas. (1972). Shear waves in plane infinite structures. *J Eng Mech Div* (18), 859–877.

Malischewsky, P. (1987). *Surface Waves and Discontinuities*. Elsevier, Amsterdam, The Netherlands.

Manesh, S. M. (1991). Theoretical investigation of the spectral-analysis-of-surface-waves (SASW) technique for application offshore. PhD Dissertation, University of Texas, Austin.

Manolis, G. D., and D. E. Beskos. (1988). *Boundary Element Methods in Elastodynamics*. London: Unwin Hyman.

Maraschini, M., D. Boiero, S. Foti, and V. Socco. (2011). Scale properties of the seismic wavefield: Perspectives for full waveform inversion. *Geophysics* 76(5), A37–A44.

Maraschini, M., F. Ernst, S. Foti, and V. Socco. (2010). A new misfit function for multimodal inversion of surface waves. *Geophysics* 75(4), 31–43.

Maraschini, M., and S. Foti. (2010). A Monte Carlo multimodal inversion of surface waves. *Geophys J Int* 182(3), 1557–1566.

Marosi, K. T., and D. R. Hiltunen. (2004a). Characterization of SASW phase angle and phase velocity measurement uncertainty. *Geotech Test J* 27(2), 205–213.

Marosi, K. T., and D. R. Hiltunen. (2004b). Characterization of spectral analysis of surface waves shear wave velocity measurement uncertainty. *J Geotech Geoenviron Eng* 130(10), 1034–1041.

Matthews, M. C., V. S. Hope, and C. R. I. Clayton. (1996). The use of surface waves in the determination of ground stiffness profiles. *Geotech Eng* 119, 84–95.

Mayne, P. (2000). *Results of Seismic Piezocone Penetration Tests Performed in Memphis, Tennessee.* GTRC Project Nos. E-20-F47/F34 Submitted to USGS/MAE Hazard Mapping Program Central Region. Atlanta, GA: Civil and Environmental Engineering, Geosystems Engineering Group, Georgia Tech Research Corporation.

McMechan, G. A., and M. J. Yedlin. (1981). Analysis of dispersive waves by wave field transformation. *Geophysics* 46, 869–874.

Meier, R. W., and G. J. Rix. (1993). An initial study of surface wave inversion using artificial neural networks. *Geotech Test J* 16(4), 425–431.

Menke, W. (1989). *Geophysical Data Analysis Discrete Inverse Theory.* San Diego, CA: Academic Press.

Menzies, B., and M. Matthews. (1996). The continuous surface-wave system: A modern technique for site investigation. In: Special Lecture: Indian Geotechnical Conference, Madras, December 11–14.

Meza-Fajardo, K. C., and C. G. Lai. (2007). Explicit causal relations between material damping ratio and phase velocity from exact solutions of the dispersion equations of linear viscoelasticity. *Geophys J Int* 171, 1247–1257.

Misbah, S., and S. Strobbia. (2014). Joint estimation of modal attenuation and velocity from multichannel surface wave data. *Geophysics* 79(3), 25–38.

Misiek, R., A. Liebig, A. Gyulai, T. Ormos, M. Dobroka, and L. Dresen. (1997). A joint inversion algorithm to process geoelectric and surface wave seismic data. Part II: Applications. *Geophys Prospect* 45, 65–85.

Moore, E. H. (1920). On the reciprocal of the general algebraic matrix. *Bull Am Math Soc* 26, 394–395.

Moss, R. E. S. (2008). Quantifying measurement uncertainty of thirty-meter shear-wave velocity. *Bull Seismol Soc Am* 98(3), 1399–1411.

Muyzert, E. (2007a). Seabed property estimation from ambient-noise recordings: Part I: Compliance and Scholte wave phase-velocity measurements. *Geophysics* 72(2), U21–U26.

Muyzert, E. (2007b). Seabed property estimation from ambient-noise recordings: Part 2: Scholte-wave spectral-ratio inversion. *Geophysics* 72(4), U47–U53.

Nakamura, Y. (1989). A method for dynamic characteristics estimation of subsurface using microtremor on the ground surface. *Q Rep Railway Tech Res Inst* 30, 25–33.

Nakamura, Y. (1996). Real time information systems for seismic hazards mitigation UrEDAS, HERAS and PIC. *Q Rep Railway Tech Res Inst* 37(3), 112–127.

Nakamura, Y. (2000). Clear identification of fundamental idea of Nakamura's technique and its applications. In: Proceedings of the 12th World Conference on Earthquake Engineering, Auckland, New Zealand, January 30–February 4.

Nasseri-Moghaddam, A., G. Cascante, and J. Hutchinson. (2005). A new quantitative procedure to determine the location and embedment depth of a void using surface waves. *J Environ Eng Geophys* 10, 51–64.

Nazarian, S. (1984). *In situ* determination of elastic moduli of soil deposits and pavement systems by spectra-analysis-of-surface-waves Method. PhD Dissertation, The University of Texas, Austin.

Nazarian, S., and K. H. Stokoe, II. (1984). *In situ* shear wave velocities from spectral analysis of surface waves. In: 8th Conference on Earthquake Engineering, San Francisco, CA, July 21–28; pp. 31–38.

Nogoshi, M., and T. Igarashi. (1971). On the amplitude characteristics of microtremor (part 2). *J Seismol Soc Jpn* 24, 26–40. (Japanese with English abstract)

Nolet, G., and G. F. Panza. (1976). Array analysis of seismic surface waves: Limits and possibilities. *Pure Appl Geophys* 114, 775–790.

O'Connell, R. J., and B. Budiansky. (1978). Measures of dissipation in viscoelastic media. *Geophys Res Lett* 5, 5–8.

Ohori, M., A. Nobata, and K. Wakamatsu. (2002). A comparison of ESAC and FK methods of estimating phase velocity using arbitrarily shaped microtremor analysis. *Bull Seismol Soc Am* 92, 2323–2332.

O'Neill, A. (2003). Full-waveform reflectivity for modelling, inversion and appraisal of surface wave dispersion in shallow site investigations. PhD Thesis, University of Western Australia, Perth.

O'Neill, A. (2004). Full waveform reflectivity for inversion of surface wave dispersion in shallow site investigations. In: *Geotechnical and Geophysical Site Characterization. A. Vianada Fonseca and P. Mayne* eds., Millpress, Rotterdam, 2, 547–554.

Oppenheim, A. V., and Willsky A. S. (1997). *Signals and Systems*. Upper Saddle River, NJ: Prentice Hall.

Park, C. B., R. D. Miller, and J. Xia. (1999). Multichannel analysis of surface waves. *Geophysics* 64, 800–808.

Parker, R. L. (1977). Understanding inverse theory. *Ann Rev Earth Planet Sci* 5, 35–64.

Parker, R. L. (1994). *Geophysical Inverse Theory*. Princeton: Princeton University Press.

Parolai, S., M. Picozzi, S. M. Richwalski, and C. Milkereit. (2005). Joint inversion of phase velocity dispersion and H/V ratio curves from seismic noise recordings using a genetic algorithm, considering higher modes. *Geophys Res Lett* 32. doi: 10.1029/2004GL 021115.

Pei, D. H., J. N. Louie, and S. K. Pullammanappallil. (2008). Improvements on computation of phase velocity of Rayleigh wave based on the generalized R/T coefficient method. *Bull Seismol Soc Am* 98, 280–287.

Pekeris, C. K. (1955). The seismic buried pulse. *Proc Natl Acad Sci U S A* 41, 629–639.

Penrose, R. (1955). A generalized inverse for matrices. *Proc Camb Philos Soc* 51, 406–413.

Pestel, E., and F. A. Leckie. (1963). *Matrix Methods in Elasto-Mechanics*. New York: McGraw-Hill.

Pezeshk, S., and M. Zarrabi. (2005). A new inversion procedure for spectral analysis of surface waves using a genetic algorithm. *Bull Seismol Soc Am* 95, 1801–1808.

Philippacopoulos, A. J. (1988). Lamb's problem for fluid-saturated, porous media. *Bull Seismol Soc Am* 78, 908–923.

Piatti, C., D. Boiero, S. Foti, and L. V. Socco. (2013). Constrained 1D joint inversion of seismic surface waves and P-wave refraction traveltimes. *Geophys Prospect* 61(Suppl. 1), 77–93.

Pilant, W. L., and L. Knopoff. (1964). Observations of multiple seismic events. *Bull Seismol Soc Am* 54, 19–39.

Pipkin, A. C. (1986). *Lectures on Viscoelasticity Theory*. Berlin: Springer-Verlag.

Poggiagliolmi, E., A. J. Berkhout, and M. M. Boone. (1982). Phase unwrapping, possibilities and limitations. *Geophys Prospect* 30, 281–291.

Press, W. H., S. A. Teukolsky, W. T. Vetterling, and B. P. Flannery. (1992). *Numerical Recipes in Fortran: The Art of Scientific Computing*. Cambridge University Press, Cambridge, UK.

Raptakis D. G. (2012). Pre-Loading effect on dynamic soil properties: Seismic methods and their efficiency in geotechnical aspects. *Soil Dynamics and Earthq. Eng.*, 34, 69–77.

Rayleigh, J. W. S. (1877). The theory of sound, Dover Publications, New York, 2nd edition (1945), New-York.

Rayleigh, J. W. S. (1885). On waves propagated along the plane surface of an elastic solid. *Proc Lond Math Soc* 17, 4–11.

Read, W. T. (1950). Stress analysis for compressible viscoelastic materials. *J Appl Phys* 21, 671.

Remmert, R. (1997). *Classical Topics in Complex Function Theory*. New York: Springer.

Reynolds, J. M. (1997). *An Introduction to Applied and Environmental Geophysics*. Wiley, Chichester, UK.

Richart, F. E., Jr., R. D. Woods, and J. R. Hall. (1970). *Vibrations of Soils and Foundations*. Englewood Cliffs, NJ: Prentice Hall.

Rietch, E. (1977). The maximum entropy approach to inverse problems. *J Geophys* 42, 489–506.

Rix, G. J., and C. G. Lai. (2014). Surface wave inversion using a continuous subsurface model. *J Geotech Geoenviron Eng*. In preparation.

Rix, G. J., C. G. Lai, and S. Foti. (2001). Simultaneous measurement of surface wave dispersion and attenuation curves. *Geotech Test J* 24, 350–358.

Rix, G. J., C. G. Lai, S. Foti, and D. Zywicki. (1998). Surface wave tests in landfills and embankments. In: *Geotechnical Earthquake Engineering and Soil Dynamics III* (P. Dakoulas, M. Yegian, and R. D. Holtz, eds.). ASCE Geotechnical Special Publication No. 75, ASCE, Reston, Virginia. pp. 1008–1019.

Rix, G. J., C. G. Lai, and A. W. Spang. (2000). *In situ* measurements of damping ratio using surface waves. *J Geotech Geoenviron Eng* 126, 472–480.

Rodríguez-Castellanos, A., R. Ávila-Carrera, and F. J. Sánchez-Sesma. (2007). Scattering of Rayleigh-waves by surface-breaking cracks: An integral formulation. *Geofís Int* 46(4), 241–248.

Rokhlin, S. I., and L. Wang. (2002). Stable reformulation of transfer matrix method for wave propagation in layered anisotropic media. *J Acoust Soc Am* 112, 822–834.

Romanowicz, B. (2002). Inversion of surface waves: A review. In: *International Handbook of Earthquake and Engineering Seismology*. Academic Press, Waltham, MA.

Roth, M., K. Holliger, and A. G. Green. (1998). Guided waves in near-surface seismic surveys. *Geophys Res Lett* 25(7), 1071–1074.

Russell, B. (1946). *A History of Western Philosophy*. London: George Allen & Unwin, pp. 494–495.

Ryden, N., and C. Park. (2006). Fast simulated annealing inversion of surface waves on pavement using phase-velocity spectra. *Geophysics* 71(4), R49–R58.

Ryden, N., C. Park, P. Ulriksen, and R. Miller. (2004). Multimodal approach to seismic pavement testing. *J Geotech Geoenviron Eng* 130(6), 636–645.

Sambuelli, L., G. P. Deidda, G. Albis, E. Giorcelli, and G. Tristano. (2001). Comparison of standard horizontal geophones and newly designed horizontal detectors. *Geophysics* 66, 1827–1837.

Sànchez-Salinero, I. (1987). Analytical investigation of seismic methods used for engineering applications. PhD Dissertation, University of Texas, Austin.

Santamarina, J. C., and D. Fratta. (2010). *Discrete Signals and Inverse Problems: An Introduction for Engineers and Scientists*. Wiley, Chichester, UK.

Santamarina, J. C., K. A. Klein, and M. A. Fam. (2001). *Soils and Waves*. Chichester: Wiley.

Sato, H., and M. C. Fehler. (1997). *Seismic Wave Propagation and Scattering in the Heterogeneous Earth*. New York: Springer-Verlag.

Scales, J. A., and R. Snieder. (1998). What is noise? *Geophysics* 63(4), 1122–1124.

Schmidt, R. O. (1986). Multiple emitter location and signal parameter estimation. *IEEE Trans Antenn Propag* 34(3), 276–280.

Schnabel, P. B., J. Lysmer, and H. B. Seed. (1972). *SHAKE: A Program for Earthquake Response Analysis of Horizontally Layered Sites*. Report No. EERC-72/12. Berkeley: University of California.

Scholte, J. G. (1947). The range of existence of Rayleigh and Stoneley waves. *Geophys J Int* 5, 120–126.

Schwab, F., and L. Knopoff. (1970). Surface-wave dispersion computations. *Bull Seismol Soc Am* 60(2), 321–344.

Schwab, F., and L. Knopoff. (1971). Surface waves on multilayered anelastic media. *Bull Seismol Soc Am* 61(4), 893–912.

Schwab, F., and L. Knopoff. (1972). *Fast Surface Wave and Free Mode Computations*. New York: Academic Press.

Sen, M., and P. L. Stoffa. (1995). *Global Optimization Methods in Geophysical Inversion*. Amsterdam: Elsevier.

SESAME. (2004). *Guidelines for the Implementation of the H/V Spectral Ratio Technique on Ambient Vibrations Measurements, Processing and Interpretation*. SESAME European research project. WP12—Deliverable D23.12.

Shapiro, N. M., and M. Campillo. (2004). Emergence of broadband Rayleigh waves from correlations of the ambient seismic noise. *Geophys Res Lett* 31, L07614-1–L07614-4.

Shen, W., M. H. Ritzwoller, V. Schulte-Pelkum, and F.-C. Lin. (2012). Joint inversion of surface wave dispersion and receiver functions: A Bayesian Monte-Carlo approach. *Geophys J Int* 192(2), 807–836.

Sheriff, R. E., and Geldart, L. P. (1995). *Exploration Seismology*. Cambridge: Cambridge University Press.

Shibuya, S., T. Mitachi, F. Fukuda, and T. Degoshi. (1995). Strain rate effects on shear modulus and damping of normally consolidated clay. *Geotech Test J* 18(3), 365–375.

Shirazi, H., I. Abdallah, and S. Nazarian. (2009). Developing artificial neural network models to automate spectral analysis of surface wave method in pavements. *J Mater Civil Eng* 21(12), 722–729.

Snieder, R. (2004). Extracting the Green's function from the correlation of coda waves: A derivation based on stationary phase. *Phys Rev E* 69, 046610-1–046610-8.

Socco, L. V., and C. Strobbia. (2004). Surface Wave Methods for near-surface characterisation, a tutorial: Near Surface Geophysics, 2, 165–185.

Socco, L. V., and D. Boiero. (2008). Improved Monte Carlo inversion of surface wave data. *Geophys Prospect* 56(3), 357–371.

Socco, L. V., D. Boiero, C. Comina, S. Foti, and R. Wisén. (2008). Seismic characterization of an Alpine site. *Near Surf Geophys* 6(8), 255–267.

Socco, L. V., D. Boiero, S. Foti, and R. Wisén. (2009). Laterally constrained inversion of ground roll from seismic reflection records. *Geophysics* 74, G35–G45.

Socco, L. V., S. Foti, and D. Boiero. (2010). Surface wave analysis for building near surface velocity models: Established approaches and new perspectives. *Geophysics* 75, A83–A102.

Stokes, G. G. (1880). *Mathematical and Physical Papers* (J. Larmor, ed.). Reprinted from the original journals and transactions, with additional notes by the author. 5 vol. Cambridge: Cambridge University Press.

Stokoe, K. H., II, S. Nazarian, G. J. Rix, I. Sanchez-Salinero, J. Sheu, and Y. Mok. (1988). In situ seismic testing of hard-to-sample soils by surface wave method. In: Earthquake Engineering and Soil Dynamics II: Recent Advances in Ground-Motion Evaluation, ASCE, Park City, June 27–30; pp. 264–277.

Stokoe, K. H., S. G. Wright, J. Bay, and J. M. Roesset. (1994). Characterization of geotechnical sites by SASW method. In: *Geophysical Characterization of Sites* (R. D. Woods, ed.). ISSMFE Technical Committee #10. Oxford: IBH, pp. 15–25.

Stoll, R. D. (1974). *Acoustic Waves in Saturated Sediments.* New York: Plenum Press.

Stoll, R. D., and E. Bautista. (1994). New tool for studying seafloor geotechnical and geoacoustic properties. *J Acoust Soc Am* 96, 2937–2944.

Strang, G. (1988). *Linear Algebra and its Applications.* San Diego: Harcourt Brace Jovanovich.

Strobbia, C. (2005). Love wave analysis for the dynamic characterisation of sites. *Bollettino Geofisic Teor Appl* 46(2/3), 135–152.

Strobbia, C. (2010). *First Break* 28, 85–91, 2010.

Strobbia, C., and G. Cassiani. (2011). Refraction microtremors: Data analysis and diagnostics of key hypotheses. *Geophysics* 76(3), MA11–MA20.

Strobbia, C., and S. Foti. (2006). Multi-offset phase analysis of surface wave data (MOPA). *J Appl Geophys* 59(4), 300–313.

Strobbia, C., A. Laake, P. L. Vermeer, and A. Glushchenko. (2011). Surface waves: Use them then lose them. Surface wave analysis, inversion and attenuation in land reflection seismic surveying. *Near Surf Geophys* 9, 503–514.

Strobbia, C., P. L. Vermeer, A. Laake, A. Glushchenko, and S. Re. (2010). Surface waves: Processing, inversion and attenuation. *First Break* 28(8), 85–91.

Strobbia, C. L. (2003). Surface wave methods for near-surface site characterization. PhD Dissertation, Politecnico di Torino, Italy.

Szelwis, R., and A. Behle. (1987). Shallow shear-wave velocity estimation from multimodal Rayleigh waves. In: *Shear-Wave Exploration* (S. Danbom and S. N. Domenico, eds.). Society of Exploration Geophysicists, Tulsa, OK, pp. 214–226.

Takeuchi, H., and M. Saito. (1972). *Seismic Surface Waves.* New York: Academic Press.

Tarantola, A. (2005). *Inverse Problem Theory and Methods for Model Parameter Estimation.* Philadelphia, PA: Society for Industrial and Applied Mathematics.

Thomson, W. T. (1950). Transmission of elastic waves through a stratified solid medium. *J Appl Phys* 21, 89–93.

Tikhonov, A. N., and V. Y. Arsenin. (1977). *Solutions of III-Posed Problems*. Washington, DC: Winston & Sons.

Tokimatsu, K. (1995). Geotechnical site characterization using surface waves. In: Proceedings of 1st International Conference on Earthquake Geotechnical Engineering, Vol. 3, Tokyo, Japan, November 14–16; pp. 1333–1368.

Tokimatsu, K., S. Tamura, and H. Kojima. (1992). Effects of multiple-modes on Rayleigh-wave dispersion characteristics. *J Geotech Eng* 118(10), 1529–1543.

Tran, K., and D. Hiltunen. (2012). Two-dimensional inversion of full waveforms using simulated annealing. *J Geotech Geoenviron Eng* 138(9), 1075–1090.

Tran, K. T., and D. R. Hiltunen. (2011). *An Assessment of Surface Wave Techniques at the Texas A&M National Geotechnical Experimentation Site*. Geo-Risk 2011. ASCE, Reston, Virginia. pp. 859–866.

Tryggvason, A., and N. Linde. (2006). Local earthquake (LE) tomography with joint inversion for P- and S-wave velocities using structural constraints. *Geophys Res Lett* 33, L07303.

Tschoegl, N. W. (1989). *The Phenomenological Theory of Linear Viscoelastic Behavior: An Introduction*. Berlin: Springer-Verlag.

Tselentis, G. A., and G. Delis. (1998). Rapid assessment of S-wave profiles from the inversion of multichannel surface wave dispersion data. *Ann Geofisic* 41(1), 1–15.

Tuomi, K. E., and D. R. Hiltunen. (1996). Reliability of the SASW method for determination of the shear modulus of soils. In: Proceedings of Uncertainty in Geologic Environment: From Theory to Practice, American Society of Civil Engineers, Shacklford, Reston, Virginia; pp. 1125–1237.

van der Veen, M., R. Spitzer, A. G. Green, and P. Wild. (2001). Design and application of a towed landstreamer for cost-effective 2D and pseudo-3D shallow seismic data acquisition. *Geophysics* 66, 482–500.

Vangkilde-Pedersen, T., J. F. Dahl, and J. Ringgaard. (2006). Five years of experience with landstreamer vibroseis and comparison with conventional seismic data acquisition. In: Proceedings of the Symposium on the Application of Geophysics to Engineering and Environmental Problems (SAGEEP'06), Seattle, WA, USA, 2–6 April 2006, pp. 1086–1093.

Vignoli, G., and G. Cassiani. (2009). Identification of lateral discontinuities via multioffset phase analysis of surface wave data. *Geophys Prospect* 58, 389–413.

Vignoli, G., C. Strobbia, G. Cassiani, and P. Vermeer. (2011). Statistical multioffset phase analysis for surface-wave processing in laterally varying media. *Geophysics* 76(2), U1–U11.

Visintin, A. (1994). *Differential Models of Hysteresis*. Berlin: Springer-Verlag.

von Terzaghi, K. (1936). The shearing resistance of saturated soils and the angle between the planes of shear. In: First International Conference on Soil Mechanics, Vol. 1, Harvard University, pp. 54–56.

Vrettos, C. (1991). Time-harmonic Boussinesq problem for a continuously non-homogeneous soil. *Earthquake Eng Struct Dynam* 20, 961–977.

Vucetic, M. (1994). Cyclic threshold shear strains in soils. *J Geotech Eng* 120(12), 2208–2228.

Wapenaar, K. (2004). Retrieving the elastodynamic Green's function of an arbitrary inhomogeneous medium by cross correlation. *Phys Rev Lett* 93(25), 254301.

Wapenaar, K., D. Draganov, R. Snieder, X. Campman, and A. Verdel. (2010a). Tutorial on seismic interferometry: Part 1: Basic principles and applications. *Geophysics* 75(5), 75A195–75A209.

Wapenaar, K., and J. Fokkema. (2006). Green's function representations for seismic interferometry. *Geophysics* 71(4), SI33–SI46.

Wapenaar, K., E. Slob, R. Snieder, and A. Curtis. (2010b). Tutorial on seismic interferometry: Part 2: Underlying theory and new advances. *Geophysics* 75(5), 75A211–75A227.

Weaver, R. L., and O. I. Lobkis. (2001). Ultrasonics without a source: Thermal fluctuation correlations at MHz frequencies. *Phys Rev Lett* 87, 134301-1–134301-4.

Weaver, R. L., and O. I. Lobkis. (2002). On the emergence of the Green's function in the correlations of a diffuse field: Pulse-echo using thermal phonons. *Ultrasonics* 40, 435–439.

White, J. E. (1983). *Underground Sound. Application of Seismic Waves.* Elsevier Science, Amsterdam, The Netherlands.

Whitham, G. B. (1999). *Linear and Nonlinear Waves.* New York: Wiley.

Wilmanski, K. (2005). Elastic modelling of surface waves in single and multicomponent systems. In: *Surface Waves in Geomechanics: Direct and Inverse Modelling for Soil and Rocks* (C. G. Lai and K. Wilmanski, eds.). New York: Springer-Verlag, pp. 203–276.

Winsborrow, G., D. G. Huws, and E. Muyzert. (2003). Acquisition and inversion of Love wave data to measure the lateral variability of geoacoustic properties of marine sediments. *J Appl Geophys* 54, 71–84.

Wisén, R., and A. V. Christiansen. (2005). Laterally and mutually constrained inversion of surface wave seismic data and resistivity data. *J Environ Eng Geophys* 10(3), 251–262.

Wyllie, M. R. J., A. R. Gregory, and L. W. Gardner. (1956). Elastic wave velocities in heterogeneous and porous media. *Geophysics* 21, 41–70.

Xia, J. G., Y. X. Xu, R. D. Miller, and C. Zeng. (2010). A trade-off solution between model resolution and covariance in surface-wave inversion. *Pure Appl Geophys* 167(12), 1537–1547.

Xia, J. H., R. D. Miller, C. B. Park, J. A. Hunter, J. B. Harris, and J. Ivanov. (2002). Comparing shear-wave velocity profiles inverted from multichannel surface wave with borehole measurements. *Soil Dynam Earthquake Eng* 22(3), 181–190.

Xu, J. C. Q., S. D. Butt, and P. J. C. (2008). Seismic Rayleigh wave method for localizing and imaging subsurface cavities in extensively exploited districts, Proc. of 21st Symposium on the Application of Geophysics to Engineering and Environmental Problems (SAGEEP 2008), Philadelphia, Pennsylvania, April 6–10; pp. 662–678.

Yamanaka, H., and H. Ishida. (1996). Application of genetic algorithms to an inversion of surface-wave dispersion data. *Bull Seismol Soc Am* 86(2), 436–444.

Yilmaz, O. (1987). *Seismic Data Processing.* Tulsa: Society of Exploration Geophysicists.

Yoon, S., and G. Rix. (2009). Near-field effects on array-based surface wave methods with active sources. *J Geotech Geoenviron Eng* 135(3), 399–406.

Zywicki, D. J. (1999). Advanced signal processing methods applied to engineering analysis of seismic surface waves. PhD Dissertation, Georgia Institute of Technology, Atlanta, USA.

Zywicki, D. J., and G. J. Rix. (2005). Mitigation of near-field effects for seismic surface wave velocity estimation with cylindrical beamformers. *J Geotech Geoenviron Eng* 131, 970–977.

Index

Printed in the United States
by Baker & Taylor Publisher Services